Lecture Notes in Artificial Intelligence   13067

Subseries of Lecture Notes in Computer Science

Series Editors

Randy Goebel
  University of Alberta, Edmonton, Canada
Yuzuru Tanaka
  Hokkaido University, Sapporo, Japan
Wolfgang Wahlster
  DFKI and Saarland University, Saarbrücken, Germany

Founding Editor

Jörg Siekmann
  DFKI and Saarland University, Saarbrücken, Germany

More information about this subseries at http://www.springer.com/series/1244

Ildar Batyrshin · Alexander Gelbukh ·
Grigori Sidorov (Eds.)

Advances in Computational Intelligence

20th Mexican International Conference
on Artificial Intelligence, MICAI 2021
Mexico City, Mexico, October 25–30, 2021
Proceedings, Part I

Springer

Editors
Ildar Batyrshin ⓘ
Instituto Politécnico Nacional
Centro de Investigación en Computación
Mexico City, Mexico

Alexander Gelbukh ⓘ
Instituto Politécnico Nacional
Centro de Investigación en Computación
Mexico City, Mexico

Grigori Sidorov ⓘ
Instituto Politécnico Nacional
Centro de Investigación en Computación
Mexico City, Mexico

ISSN 0302-9743   ISSN 1611-3349 (electronic)
Lecture Notes in Artificial Intelligence
ISBN 978-3-030-89816-8   ISBN 978-3-030-89817-5 (eBook)
https://doi.org/10.1007/978-3-030-89817-5

LNCS Sublibrary: SL7 – Artificial Intelligence

This Springer imprint is published by the registered company Springer Nature Switzerland AG
The registered company address is: Gewerbestrasse 11, 6330 Cham, Switzerland

Preface

The Mexican International Conference on Artificial Intelligence (MICAI) is a yearly international conference series that has been organized by the Mexican Society for Artificial Intelligence (SMIA) since 2000. MICAI is a major international artificial intelligence (AI) forum and the main event in the academic life of the country's growing AI community.

MICAI conferences publish high-quality papers in all areas of AI and its applications. The proceedings of the previous MICAI events have been published by Springer in its Lecture Notes in Artificial Intelligence (LNAI) series, vol. 1793, 2313, 2972, 3789, 4293, 4827, 5317, 5845, 6437, 6438, 7094, 7095, 7629, 7630, 8265, 8266, 8856, 8857, 9413, 9414, 10061, 10062, 10632, 10633, 11288, 11289, 11835, 12468 and 12469. Since its foundation in 2000, the conference has been growing in popularity and improving in quality.

The proceedings of MICAI 2021 are published in two volumes. The first volume, Advances in Computational Intelligence, contains 30 papers structured into three sections:

- Machine and Deep Learning
- Image Processing and Pattern Recognition
- Evolutionary and Metaheuristic Algorithms

The second volume, Advances in Soft Computing, contains 28 papers structured into two sections:

- Natural Language Processing
- Intelligent Applications and Robotics

The two-volume set will be of interest for researchers in all fields of artificial intelligence, students specializing in related topics, and for the public in general interested in recent developments in AI.

The conference received for evaluation 129 submissions from authors in 22 countries: Algeria, Argentina, Bangladesh, Belgium, Brazil, Canada, Colombia, Costa Rica, Croatia, Cuba, the Czech Republic, India, Japan, Kazakhstan, Mexico, the Netherlands, Peru, Portugal, Russia, Spain, Sweden, and the USA. From these submissions, 58 papers were selected for publication in these two volumes after a peer-review process carried out by the international Program Committee. The acceptance rate was 45%.

The international Program Committee consisted of 188 experts from 16 countries: Brazil, Colombia, France, Iran, Ireland, Japan, Kazakhstan, Malaysia, Mexico, Pakistan, Philippines, Portugal, Russia, Spain, the UK, and the USA.

MICAI 2021 was honored by the presence of renowned experts who gave excellent keynote lectures:

- Fabio A. González O., Universidad Nacional de Colombia, Colombia
- Eyke Hüllermeier, Paderborn University, Germany

- Piero P. Bonissone, Piero P. Bosissone Analytics, USA
- Marta R. Costa-Jussà, Universitat Politècnica de Catalunya, Spain
- Hugo Jair Escalante, National Institute of Astrophysics, Optics and Electronics, Mexico
- María Vanina Martínez, Consejo Nacional de Investigaciones Científicas y Técnicas, Argentina

Five workshops were held jointly with the conference:

- SC-AIS 2021: International Workshop on Soft Computing and Advances in Intelligent Systems
- WILE 2021: 14th Workshop on Intelligent Learning Environments
- HIS 2021: 14th Workshop of Hybrid Intelligent Systems
- CIAPP 2021: 3rd Workshop on New Trends in Computational Intelligence and Applications
- WIDSSI 2021: 7th International Workshop on Intelligent Decision Support Systems for Industry

The authors of the following papers received the Best Paper Awards based on the paper's overall quality, significance, and originality of the reported results:

- First place: "Multi-objective Release Plan Rescheduling in Agile Software Development," by Abel García Nájera, Saúl Zapotecas Martínez, Jesús Guillermo Falcón Cardona, and Humberto Cervantes, Mexico
- Second place: "Deep Learning Approach for Aspect-Based Sentiment Analysis of Restaurants Reviews in Spanish," by Bella-Citlali Martínez-Seis, Obdulia Pichardo-Lagunas, Sabino Miranda-Jiménez, Israel-Josafat Perez-Cazares, and Jorge-Armando Rodriguez-González, Mexico
- Second place: "Question Answering for Visual Navigation in Human-centered Environments," by Daniil Kirilenko, Alexey Kovalev, Evgeny Osipov, and Aleksandr Panov, Russia
- Third place: "Sign Language Translation using Multi Context Transformer," by M Badri Narayanan, Mahesh Bharadwaj K, Nithin G R, Dhiganth Rao Padamnoor, and Vineeth Vijayaraghavan, India
- Third place: "Comparing Machine Learning based Segmentation Models on Jet Fire Radiation Zones," by Carmina Pérez, Adriana Palacios, Gilberto Ochoa-Ruiz, Christian Mata, Miguel Gonzalez-Mendoza, and Luis Eduardo Falcón-Morales, Mexico/Spain

We want to thank all the people involved in the organization of this conference: the authors of the papers published in these two volumes – it is their research work that gives value to the proceedings – and the organizers for their work. We thank the reviewers for their great effort spent on reviewing the submissions, the Track Chairs for their hard work, and the Program and Organizing Committee members.

We are deeply grateful to the Center for Computing Research at the Instituto Politécnico Nacional (Mexico) for their warm hospitality to MICAI 2021. We would like to express our gratitude to the General Director, Arturo Reyes Sandoval, the Secretary of Research and Postgraduate Studies, Heberto Balmori Ramírez, the

Director of Research, Laura Arreola Mendoza, the Director of Postgraduate Studies, Luis Gil Cisneros, and the Director of the Center for Computing Research, Marco Antonio Moreno Ibarra.

The entire submission, reviewing, and selection process, as well as preparation of the proceedings, was supported by the EasyChair system (www.easychair.org). Last but not least, we are grateful to Springer for their patience and help in the preparation of these volumes.

October 2021   Ildar Batyrshin
Alexander Gelbukh
Grigori Sidorov

Conference Organization

MICAI 2021 was organized by the Mexican Society for Artificial Intelligence (SMIA, Sociedad Mexicana de Inteligencia Artificial) in collaboration with the Center for Computing Research, Instituto Politécnico Nacional.

The MICAI series website is www.MICAI.org. The website of the Mexican Society for Artificial Intelligence, SMIA, is www.SMIA.mx. Contact options and additional information can be found on these websites.

Conference Committee

General Chair

Félix A. Castro Espinoza   Universidad Autónoma del Estado de Hidalgo, Mexico

Program Chairs

Ildar Batyrshin	CIC-IPN, Mexico
Alexander Gelbukh	CIC-IPN, Mexico
Grigori Sidorov	CIC-IPN, Mexico

Workshop Chair

Hiram Ponce   Universidad Panamericana, Mexico

Tutorials Chair

Roberto Antonio Vázquez   Universidad La Salle, Mexico
Espinoza de los
Monteros

Doctoral Consortium Chairs

Miguel Gonzalez-Mendoza	Tecnológico de Monterrey, Mexico
Juan Martínez-Miranda	CICESE Research Center, Mexico

Keynote Talks Chair

Noé Alejandro   Centro Nacional de Investigación y Desarrollo
Castro-Sánchez   Tecnológico, Mexico

Publication Chair

Hiram Ponce   Universidad Panamericana, Mexico

Financial Chairs

Oscar Herrera-Alcántara	Universidad Autónoma Metropolitana, Mexico
Lourdes Martínez-Villaseñor	Universidad Panamericana, Mexico

Grant Chair

Félix A. Castro Espinoza	Universidad Autónoma del Estado de Hidalgo, Mexico

Local Organizing Committee

Marco Antonio Moreno Ibarra	CIC-IPN, Mexico
Eusebio Ricárdez Vazquez	CIC-IPN, Mexico
Elvia Cruz Morales	CIC-IPN, Mexico
Alejandra Ramos Porras	CIC-IPN, Mexico
Mauricio Sebastian Martín Gascón	CIC-IPN, Mexico
Jorge Benjamín Martell Ponce de León	CIC-IPN, Mexico
Cristian Maldonado-Sifuentes	CIC-IPN, Mexico
César Jesús Núñez-Prado	CIC-IPN, Mexico

Track Chairs

Natural Language Processing

Grigori Sidorov	CIC-IPN, Mexico

Machine Learning

Alexander Gelbukh	CIC-IPN, Mexico
Navonil Majumder	CIC-IPN, Mexico

Deep Learning

Hiram Ponce	Universidad Panamericana, Mexico

Evolutionary and Metaheuristic Algorithms

Roberto Antonio Vázquez Espinoza de los Monteros	Universidad La Salle, Mexico
Oscar Herrera-Alcántara	Universidad Autónoma Metropolitana, Mexico

Soft Computing

Miguel Gonzalez-Mendoza	Tecnológico de Monterrey, Mexico

Image Processing and Pattern Recognition

Lourdes
Martínez-Villaseñor
Universidad Panamericana, Mexico

Robotics

Gilberto Ochoa-Ruiz Tecnológico de Monterrey, Mexico

Intelligent Applications and Social Network Analysis

Iris Iddaly Méndez-Gurrola Universidad Autónoma de Ciudad Juárez, Mexico

Other Artificial Intelligence Approaches

Nestor Velasco Bermeo University College Dublin, Ireland
Gustavo Arroyo-Figueroa Instituto Nacional de Electricidad y Energías Limpias,
Mexico

Program Committee

Iskander Akhmetov IICT, Kazakhstan
José David Alanís Urquieta Carrera de Tecnologías de la Información y
Comunicación Universidad Tecnológica de Puebla,
Mexico
Giner Alor-Hernández Instituto Tecnologico de Orizaba, Mexico
Joanna Alvarado-Uribe Instituto Tecnologico y de Estudios Superiores de
Monterrey, Mexico
Maaz Amjad CIC IPN, Mexico
Jason Efraín Angel Gil Instituto Politécnico Nacional, Mexico
Ignacio Arroyo-Fernández Universidad Tecnológica de la Mixteca, Mexico
Gustavo Arroyo-Figueroa Instituto Nacional de Electricidad y Energías Limpias,
Mexico
Edgar Avalos-Gauna Universidad Panamericana, Mexico
Fausto Antonio Balderas Instituto Tecnológico de Ciudad Madero, Mexico
Jaramillo
Alejandro Israel Barranco Instituto Tecnológico de Celaya, Mexico
Gutiérrez
Ramon Barraza Universidad Autonoma de Ciudad Juarez, Mexico
Ari Yair Barrera-Animas Tecnológico de Monterrey, Mexico
Ildar Batyrshin Instituto Politecnico Nacional, Mexico
Gemma Bel-Enguix UNAM, Mexico
Sara Besharati Instituto Politécnico Nacional, Mexico
Rajesh Roshan Biswal Tecnologico de Monterrey, Mexico
Vadim Borisov National Research University "Moscow Power
Engineering Institute", Smolensk, Russia
Monica Borunda Instituto Nacional de Electricidad y Energías Limpias,
Mexico

Alexander Bozhenyuk	Southern Federal University, Russia
Ramon F. Brena	Tecnologico de Monterrey, Mexico
Davide Buscaldi	LIPN, Université Paris 13, France
Sabur Butt	IPN Computing Research Center, Mexico
Hiram Calvo	Nara Institute of Science and Technology, Japan
Ruben Carino-Escobar	Instituto Nacional de Rehabilitación, Mexico
Felix Castro Espinoza	CITIS-UAEH, Mexico
Noé Alejandro Castro-Sánchez	Centro Nacional de Investigación y Desarrollo Tecnológico, Mexico
Hector Ceballos	Tecnologico de Monterrey, Mexico
Jaime Cerda Jacobo	Universidad Michoacana de San Nicolás de Hidalgo, Mexico
Ofelia Cervantes	Universidad de las Américas Puebla, Mexico
Haruna Chiroma	Federal College of Education Technical Gombe, Malaysia
Elisabetta Crescio	Instituto Tecnológico y de Estudios Superiores de Monterrey, Mexico
Laura Cruz	Instituto Tecnologico de Ciudad Madero, Mexico
Nareli Cruz Cortés	CIC-IPN, Mexico
Andre de Carvalho	University of São Paulo, Brazil
Jorge De La Calleja	Universidad Politécnica de Puebla, Mexico
Omar Arturo Domiguez Ramírez	UAEH, Mexico
Andrés Espinal	Universidad de Guanajuato, Mexico
Daniel Yacob Espinoza González	CIC-IPN, Mexico
Oscar Alejandro Esquivel Flores	UNAM, Mexico
Barbaro Ferro	BestS2S, Mexico
Karina Figueroa	Universidad Michoacana de San Nicolás de Hidalgo, Mexico
Denis Filatov	Sceptica Scientific Ltd, UK
Dora-Luz Flores	Universidad Autónoma de Baja California, Mexico
Juan Jose Flores	Universidad Michoacana, Mexico
Anilú Franco Árcega	UAEH, Mexico
Sofia N. Galicia-Haro	UNAM, Mexico
Vicente Garcia	Universidad Autónoma de Ciudad Juárez, Mexico
Leonardo Garrido	Tecnológico de Monterrey., Mexico
Alexander Gelbukh	Instituto Politécnico Nacional, Mexico
Claudia Gomez	Instituto Tecnolgico de Ciudad Madero, Mexico
Eduardo Gómez-Ramírez	Universidad La Salle, Mexico
Pedro Pablo Gonzalez	Universidad Autonoma Metropolitana, Mexico
Luis-Carlos González-Gurrola	Universidad Autonoma de Chihuahua, Mexico
Miguel Gonzalez-Mendoza	Tecnologico de Monterrey, Mexico
Gabriel Gonzalez-Serna	TecNM/CENIDET, Mexico

Fernando Gudiño	UNAM, Mexico
Rafael Guzman Cabrera	Universidad de Guanajuato, Mexico
Jorge Hermosillo	UAEM, Mexico
Yasmin Hernandez	Centro Nacional de Investigación y Desarrollo Tecnológico, Mexico
José Alberto Hernández	Universidad Autnoma del Estado de Morelos, Mexico
Betania Hernandez-Ocaña	Universidad Juárez Autónoma de Tabasco, Mexico
Oscar Herrera	UAM Azcapotzalco, Mexico
Laura Hervert-Escobar	Tecnologico de Monterrey, Mexico
Seyed Habib Hosseini Saravani	Instituto Politécnico Nacional, Mexico
Joel Ilao	De La Salle University, Philippines
Jorge Jaimes	Universidad Autónoma Metropolitana, Mexico
Olga Kolesnikova	Instituto Politécnico Nacional, Mexico
Nailya Kubysheva	Kazan Federal University, Russia
Angel Kuri-Morales	ITAM, Mexico
Carlos Lara-Alvarez	Centro de Investigación en Matemáticas, Mexico
José Antonio León-Borges	Universidad de Quintana Roo, Mexico
Victor Lomas-Barrie	UNAM, Mexico
Omar López Ortega	UAEH, Mexico
Gerardo Loreto	Instituto Tecnologico Superior de Uruapan, Mexico
Mykola Lukashchuk	Instituto Politécnico Nacional, Mexico
Yazmin Maldonado	Instituto Tecnológico de Tijuana, Mexico
Christian Efraín Maldonado Sifuentes	Tecnológico de Estudios Superiores de Cuautitlán Izcalli, Mexico
Jerusa Marchi	Federal University of Santa Catarina, Brazil
Aldo Márquez Grajales	Universidad Veracruzana, Mexico
Carolina Martín del Campo Rodríguez	CIC-IPN, Mexico
Lourdes Martínez	Universidad Panamericana, Mexico
Bella Citlali Martinez Seis	CINVESTAV-IPN, Mexico
Jose Martinez-Carranza	Instituto Nacional de Astrofísica, Óptica y Electrónica, Mexico
Juan Martínez-Miranda	Centro de Investigación Científica y de Educación Superior de Ensenada, Mexico
Iris Iddaly Méndez-Gurrola	Universidad Autónoma de Ciudad Juárez, Mexico
Efrén Mezura-Montes	University of Veracruz, Mexico
Sabino Miranda-Jiménez	INFOTEC, Mexico
Daniela Moctezuma	CONACYT - CentroGEO, Mexico
Saturnino Job Morales Escobar	Centro Universitario UAEM Valle de México, Mexico
Guillermo Morales-Luna	CINVESTAV-IPN, Mexico
Alicia Morales-Reyes	Instituto Nacional de Astrofisica, Optica y Electronica, Mexico
Masaki Murata	Tottori University, Japan
Antonio Neme	Universidad Nacional Autonoma de Mexico, Mexico

César Núñez	Centro de Investigación en Computación, Mexico
Gilberto Ochoa-Ruiz	ITESM, Guadalajara, Mexico
C. Alberto Ochoa-Zezatti	Universidad Autónoma de Ciudad Juárez, Mexico
Juan Carlos Olivares Rojas	Tecnológico Nacional de México/Instituto Tecnológico de Morelia, Mexico
José Luis Oliveira	University of Aveiro, Portugal
José Carlos Ortiz-Bayliss	Tecnológico de Monterrey, Mexico
Ismael Osuna-Galán	Universidad Politécnica de Chiapas, Mexico
Rushabh Patel	Children's Hospital of Philadelphia, USA
Karinaruby Perez Daniel	Universidad Panamericana, Mexico
Garibaldi Pineda García	University of Sussex, UK
Hiram Ponce	Universidad Panamericana, Mexico
Belem Priego-Sanchez	UAM Azcapotzalco, Mexico
Vicenc Puig	Universitat Politècnica de Catalunya, Mexico
José Federico Ramírez Cruz	Tecnológico Nacional de México/Instituto Tecnológico de Apizaco, Mexico
Tania Aglaé Ramírez Del Real	Universidad Politécnica de Aguascalientes, Mexico
Juan Ramirez-Quintana	Instituto Tecnológico de Chihuahua, Mexico
Jorge Reyes	Universidad Autónoma de Yucatán, Mexico
José A. Reyes-Ortiz	Universidad Autónoma Metropolitana, Mexico
Elva Lilia Reynoso Jardón	UACJ, Mexico
Gilberto Rivera	Universidad Autónoma de Ciudad Juárez, Mexico
Noel Enrique Rodriguez Maya	Instituto Tecnológico de Zitácuaro, Mexico
Katya Rodriguez-Vazquez	IIMAS-UNAM, Mexico
Ansel Y. Rodríquez González	Centro de Investigación Científica y de Educación Superior de Ensenada, Mexico
Alejandro Rosales	Tecnologico de Monterrey, Mexico
Alberto Rosales	National Politechnic Institute of Mexico, Mexico
Horacio Rostro Gonzalez	Universidad de Guanajuato, Mexico
Antonio Sanchez	Texas Christian University, USA
Angel Sanchez	University of Veracruz, Mexico
Luis Humberto Sánchez Medel	ITO, Mexico
Eddy Sánchez-Delacruz	Instituto Tecnológico Superior de Misantla, Mexico
Alejandro Santiago	Universidad Politécnica de Altamira, Mexico
Arsenii Shulikov	Instituto Politécnico Nacional, Mexico
Grigori Sidorov	CIC-IPN, Mexico
Rafaela Silva	UAM, Mexico
Efrain Solares	Autonomous University of Sinaloa, Mexico
Valery Solovyev	Kazan University, Russia
Juan Humberto Sossa Azuela	CIC-IPN, Mexico

Israel Tabarez	ITESM, Mexico
Antonio Jesús Tamayo Herrera	Universidad de Antioquia, Colombia
Eric S. Tellez	CONACyT - INFOTEC, Mexico
David Tinoco Varela	UNAM, Mexico
Nasim Tohidi	K. N. Toosi University of Technology, Iran
Aurora Torres	Universidad Autonoma de Aguascalientes, Mexico
Diego Uribe	Instituto Tecnologico de la Laguna, Mexico
José E. Valdés	CIC-IPN, Mexico
Jose Valdez	Instituto Politecnico Nacional, Mexico
Genoveva Vargas Solar	CNRS/LAFMIA, France
Nestor Velasco Bermeo	University College Dublin, Ireland
Juan Villegas-Cortez	UAM Azcapotzalco, Mexico
Yenny Villuendas-Rey	CIDETEC-IPN, Mexico
Saúl Zapotecas Martínez	UAM Cuajimalpa, Mexico
Alisa Zhila	NTENT, USA

Additional Reviewers

Ewin Cordoba
Raúl Dalí Cruz Morales
José Yaír Guzmán-Gaspar
Mukhtat Hamza
Nidiyare Hevia-Montiel
Andrey Labunets

Jesús-Adolfo Mejía-de-Dios
Erick Esteven Montelongo González
Alexandr Pak
Jose Fabian Paniagua Reyes
David Tinoco Varela
Gustavo Adolfo Vargas Hakim

Contents – Part I

Best Paper Award, First Place

Contents – Part II

Intelligent Applications and Robotics

Best Paper Award, Third Place

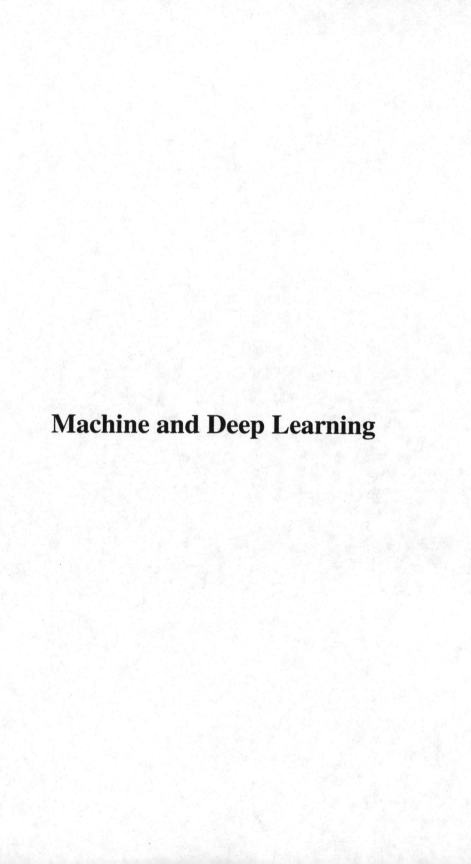

Machine and Deep Learning

Identifying Optimal Clusters in Purchase Transaction Data

L. Cleofas-Sanchez[1] , A. Pineda-Briseño[2]([ENV]) , and J. S. Sanchez[3]

[1] Escuela Superior de Ingenieria Mecanica y Electrica, Zacatenco,
Instituto Politecnico Nacional, CDMX, Mexico
[2] Tecnologico Nacional de Mexico/Instituto Tecnologico de Matamoros
H. Matamoros, Tamps, Mexico
anabel.pb@matamoros.tecnm.mx
[3] Department Computer Languages and Systems, Institute of New Imaging
Technologies, Universitat Jaume I, Castello de la Plana, Spain
sanchez@uji.es

Abstract. Clustering in transaction databases can find potentially useful patterns to gain some insight into the structure of the data, which can help for effective decision-making. However, one of the critical tasks in clustering is to identify the appropriate number of clusters, which will determine the performance of any process further applied to the transaction database. This paper presents a methodology to discover the optimal structure of purchase transaction data using the Davies-Bouldin and Calinski-Harabasz validity indices to obtain the number of clusters and formed them with the farthest-first traversals algorithm. The quality of the structures previously formed is evaluated with data complexity measures such as F1, F2, F3, N1 and IR. In this work, we use the support vector machine and multi-layer perceptron classification algorithms, to determine recognition ability in classification problems of more than two classes, and in the context of separability and imbalance of classes present in the groups previously obtained. The experimental results exhibit the viability of the proposed methodology for decision-making.

Keywords: Cluster validity indices · Farthest-first traversals ·
Transaction data · Complexity measures of classification problems

1 Introduction

A key issue in pattern recognition and data mining problems is to automatically summarize huge amounts of data into simpler and more understandable categories. One of the most common and well-studied ways of data categorization consists of splitting the data objects into disparate groups or clusters. Transaction clustering is now an important application of clustering that has received increasing attention due to the rapid development and popularity of e-commerce, along with its relevance in marketing and customer relationship management. In essence, transaction clustering is meant to divide a set of data items into some proper groups in such a way that items in the same group are as similar as possible.

© Springer Nature Switzerland AG 2021
I. Batyrshin et al. (Eds.): MICAI 2021, LNAI 13067, pp. 3–22, 2021.
https://doi.org/10.1007/978-3-030-89817-5_1

In recent years, a number of studies have discussed the usage of clustering techniques to discover useful patterns in transaction databases. In the paper by Han et al. [10], items are first partitioned based on association rules into clusters of items, and each transaction is then assigned to the cluster with the highest percentage of intersection with the respective transaction. Chen et al. [4] presented a graph-based clustering algorithm in large transaction databases. The self-organizing map was modified to cluster transaction data by incorporating a normalized dot product norm based dissimilarity measure and a modified weight adaptation function [11]. Xu et al. [26] described a clustering algorithm for transaction data, which first attempts to group customers with similar behavior and then transactions. With a new measurement called the small-large ratio, Yun et al. [27] designed an algorithm for clustering market-basket data items. Huang and Song [13] proposed a modification of the K-means algorithm for the analysis of e-commerce transactions. Tsai and Chiu [22] developed a clustering algorithm and a clustering quality function for market segmentation of customers with similar purchasing behaviors. Xiao and Dunham [25] proposed an interactive clustering algorithm for transaction data that allows the user to adjust the number of clusters during the process. Wu and Chou [24] developed a soft clustering method that uses the latent Dirichlet allocation model to create customer segments with the purpose of classifying on-line customers based on their purchasing data across categories.

Although many proposals for item clustering in transaction databases have been introduced, the problem of determining the optimal number of clusters still remains as an open problem. This is a very important issue in the context of clustering purchase transaction data because knowing the optimal structure of data can be critical for accurate decision-making. In this paper, we present a methodology to find the optimal number of clusters in transaction data sets using two well-suited cluster validity indices, whose solutions are then used by the farthest-first traversals algorithm to generate the correct partitions. The quality of the clusters is determined by data complexity measures. Finally, we measure the recognition capability with support vector machine and a multi-layer perceptron neural network, in the context of clusters separability, as well as in the context of class imbalance in problems with more than two classes.

Hereafter, the paper is organized as follows. Section 2 gives a brief description of clustering and presents a classification of the clustering algorithms. Section 3 introduces the clustering validity indices that will be used in the experiments. Section 4 briefly shows the data complexity measures used in this work. Section 5 presents the data sets and the experimental methodology. Section 6 discusses the results. Finally, Sect. 7 summarizes the main conclusions of this work and outlines possible avenues for further research.

2 Clustering Taxonomies

Clustering (or unsupervised learning) is defined as a problem that involves the partitioning of a set of data points into subsets in such a manner that points with similar features fall inside the same cluster and those with different features end up into others clusters [15,16,19]. This simple rule guides the design of most clustering algorithms and also the evaluation of the clustering solution. Clustering analysis is a useful tool with application in many different fields such as engineering, finance, medicine, biology, computer vision, marketing, document retrieval, web mining, among others [14,15,19].

In the literature, many taxonomies to classify the different clustering algorithms have been proposed [6,7,14,16]. This work adopts the categorization framework developed by Nisha and Kaur [16], which groups the various clustering algorithms into two distinct categories: hierarchical methods and partitioning methods, as shown in Fig. 1.

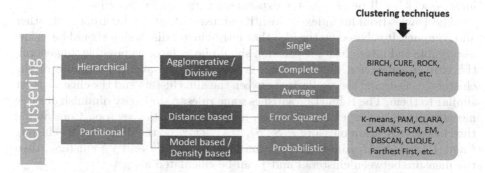

Fig. 1. Taxonomy of the clustering algorithms proposed by Nisha and Kaur [16]

The hierarchical techniques consider a hierarchy strategy to build several groups from data sets. In turn, the hierarchical algorithms are divided into agglomerative (bottom-up) or divisive (top-down). An agglomerative clustering starts with one data point for each cluster and recursively merges two or more of the most similar clusters in terms of proximity or closeness. A divisive clustering starts with the data set as a unique cluster and recursively splits the most appropriate cluster. The process continues until a stopping criterion is reached (frequently, the requested number of clusters, K). A tree or dendrogram is used to show the structure of the clusters. Some of the most representative state-of-the-art algorithms of this category are BIRCH, CURE, ROCK, ScaleKM and Chameleon.

On the other hand, the partitioning method initially treats all data points as a single cluster, which is divided into several additional non-overlapping subsets. K-means, PAM, mean-shift, CLARA, CLARANS, Expectation-Maximization and farthest-first traversals are some of the well-known algorithms of this category. In this paper, we will take the farthest-first traversals method proposed by Hochbaum and Shmoys [12], which is similar to the K-means algorithm. The idea is to pick any data point to start with, then choose the points furthest from it, then the object furthest from the first two, and so on until K points are obtained. These data points are taken as cluster centers and each remaining point is assigned to the closest center.

3   Cluster Validity Indices

The two fundamental questions that should be addressed in any clustering scenario are: (i) how many groups are actually present in the data, and (ii) how good is the clustering result. In other words, one has to determine the true number of groups and also the validity of the groups formed [17]. This is often analyzed by using some cluster validity index, which aims to find out the optimal cluster (in terms of some goodness measure) with very little knowledge about the data.

The many different cluster validity indices proposed can be classified into internal and external indices [2,9]: the former are usually based on information intrinsic to the data, while the latter are based on prior knowledge about the data. In this section, the Davies-Bouldin and Calinski-Harabasz cluster validity indices, which will be used in the experiments, are briefly described.

The Davies-Bouldin index [5] identifies clusters that are far from each other and compact. It is based on the idea that intra-cluster dispersion should be as low as possible and inter-cluster separation should be as large as possible. In general, the cluster separation measures aim is to obtain the average measure of the clusters, considering the similarity between the aim clusters and the clusters most similar to them. The R function satisfies some rules to similarity diminish of three measures such as dispersion, distance, and characteristics vectors. Considering the cluster separation measure $R(S_i, S_j, M_{ij})$, let S_i be the dispersion of clusters i and j., and let $M_{i,j}$ be the separation between the ith and j^{th} clusters., then the measure between clusters i and j can be calculated as,

$$R_{i,j} = \frac{S_i + S_j}{M_{i,j}} \tag{1}$$

Then, the Davies-Bouldin index is defined as follows:

$$DB = \frac{1}{K} \sum_{i=1}^{K} R_i \tag{2}$$

where $R_i = \max_{i \neq j} R_{i,j}$, and K is the number of clusters.

The Calinski-Harabasz index [3] evaluates the goodness of the clustering solution based on the between-cluster sum of squares and the within-cluster sum of squares. Well-separated and compact clusters should maximize this ratio. Let N the number of data points and $C = \{C_1, \ldots, C_K\}$ the set of clusters, this index is defined as follows:

$$CH = \frac{SS_B}{SS_W} \cdot \frac{N - K}{K - 1} \tag{3}$$

where SS_B and SS_W are the overall between-cluster and the within-cluster variances respectively defined as,

$$SS_B = \sum_{i-1}^{K} |C_i| \cdot ||\mu_i - \mu||^2 \tag{4}$$

$$SS_W = \sum_{i=1}^{K} \sum_{x_j \in C_i} ||x_j - \mu_i||^2 \tag{5}$$

where μ_i the the mean of i-th cluster, μ if the overall mean of the data set, and $||x_j - \mu_i||$ is the L^2 norm between the two vectors.

4   Data Complexity Measures

The analysis of the classifier's performance has received much attention from the scientific community. Although empirical studies have shown a little bit the behavior of classifiers, the patterns recognition ability strongly depends on the quality of the data sets, which may be observe with the help of data complexity measures F1, F2, F3, N1, and IR [18,20,21].

Among the measures that can determine the separability of the classes, we may find the Fishers' Discriminant Ratio (F1), which may obtain the degree of overlap between two classes due to the quality of the characteristic [18,21].

$$F1 = \frac{(m_1 - m_2)^2}{\sigma_1^2 + \sigma_2^2} \tag{6}$$

where m_1 and m_2 represent the means, and the variances of two classes are represented by σ_1^2 and σ_2^2. It is important to mention that not all the features help in the separation of classes.

The Volume of Overlap Region (F2) helps us to understand whether two classes may have/may not have a clean decision border. The process to obtain the measure F2 is to determine for each feature (f_k) the level of overlap that exists. The volume of the overlap region is represented by the multiplication of the length of the overlap region normalized of all features [20,21].

$$F2 = \prod_k \frac{minmax_k - maxmin_k}{maxmax_k - minmin_k} \tag{7}$$

where $k = 1, ...d$ for a d-dimensional problem of classification.

In high-dimension classification problems, it is very important to know which features contribute more to the separation of the classes. In this context, with the help of a measure called Maximum Feature Efficiency (F3), we may determine the efficiency of the features to observe which of them contributes more to obtain a clean decision border. This measure determines the number the points that may be separated by the quality of each feature [21].

The works [21] and [20] describe another complexity measure that represents the fraction of points on the boundary of the opposite class (N1). Its process consists of building a Minimum Spanning Tree (MST).

Another complexity in the data sets is the imbalance rate (IR) of the classes, which is presented when one class or more classes are less represented in the number patterns (minority class) than other classes (majority class). In this case, the classifiers tend to learn the most represented class, and the patterns' recognition may be incorrectly obtained [8].

5 Data Sets and Experimental Methodology

Experiments over four data sets taken from the KEEL Database Repository [1] were carried out. These are simulated data sets modeling a number of transactions where each sample represents the purchase of an item in a given transaction. The number of purchased items in a transaction was randomly generated in a uniform distribution over the range of 1–19, the purchased items in each transaction were then selected from the 64 items in an exponential distribution with the rate parameter set at 16, and their quantities were assigned from an exponential distribution with the rate parameter set at 5. Table 1 reports the number of attributes and the number of unlabeled examples for the data sets.

Table 1. Transactions data sets

	Attributes	Examples		Attributes	Examples
Trans10k	3	120427	Trans30k	3	284284
Trans50k	3	475649	Trans70k	3	665470

The experiments aim to discover the optimal structure (number of clusters) of the purchase transaction data sets in the context clusters quality considering complexity measures. To this end, the experimental methodology consisted of the following steps:

1. The Gaussian mixture and K-means (using the Euclidean distance) algorithms were used to generate different partitions (with K ranging from 1 to 50) of each original training set. The k-means used in this work forms groups where the patterns of the same group are very similar and disjoint from other groups.
2. The partitions obtained in the previous step were evaluated by the Davies-Bouldin (DB) and Calinski-Harabasz (CH) validity indices in order to identify the best structure for each training data.
3. The farthest-first traversals algorithm was used to construct the groups suggested by the two cluster validity indices.
4. The structures' quality of the clusters previously obtained, were evaluated with complexity measures called Fisher's Discriminant Ratio (F1), Volume of Overlap (F2), Feature Efficiency (F3), Mixture Identifiability (N1), and Imbalance rate (IR).
5. The recognition capacity in the context of structures' complexity of the clusters were performed by the support vector machine (SVM) with a linear kernel and a multi-layer perceptron (MLP).

Both the clustering algorithms in Step 1 and the validity indices were implemented in Matlab, whereas farthest-first traversals algorithm and the classifiers were taken from the WEKA environment [23] (a big collection of statistical and machine learning algorithms for preprocessing, classification and regression in data mining problems) with the default parameter values. We adopted the commonly-used five-fold cross-validation method for the experiments, where we randomly divided each original data set into five parts of equal (or approximately equal) size. For each fold, the examples from four blocks were used as the training data and the examples from the remaining part were employed as an independent test set.

6   Results and Discussions

For each data set, Table 2 reports the number of groups that the Davies-Bouldin and Calinski-Harabasz validity indices estimated as the best partitions after applying the Gaussian mixture and K-means clustering methods. As can be seen, the Calinski-Harabasz index shows more stable results than the Davies-Bouldin method, that is, the former determined the same optimal number of groups independently of the clustering algorithm used in Trans10k, Trans30k and Trans50k databases.

Table 2. Number of groups given by the validity indices

	Trans10k		Trans30k		Trans50k		Trans70k	
	DB	CH	DB	CH	DB	CH	DB	CH
Gaussian mixture	4	2	7	2	12	2	10	5
K-means	2	2	2	2	5	2	2	2

Table 3. Imbalance rate for each cluster

Clusters	Trans10k	Trans30k	Trans50k	Trans70k
2	5.7	4.35	1.27	**10.33**
4	8.55	–	–	–
7	–	6.17	–	–
5	–	–	2.99	**23**
10	–	–	–	**28.13**
12	–	–	7.59	–

Next, the farthest-first traversals algorithm was applied to split the original training sets into the number of groups suggested by the cluster validity indices. For instance, in the case of the Trans10k database, the clustering algorithm was used to construct two different data sets: one with two groups and another one with four groups. The groups quality was evaluated with complexity measures F1, F2, F3, N1, and IR. Afterwards, the patterns' recognition capacity, in the context of data sets' complexity, was performed with the SVM and MLP algorithms.

Table 3 reports the level of imbalance in each cluster previously formed. Although all of them have a certain imbalance level, the Trans70k shows the highest level of imbalance in their clusters. On the other hand, the Trans50k presents the lowest level of imbalance in their clusters 2 and cluster 5. This imbalance indicates that the classifier may learn more about the majority class than the minority class, which may affect the interpretation of the patterns' recognition. In Table 3 we can observe in the first column the number of the clusters, the first row indicate each data set, and the results in the bold present the highest imbalance.

The Tables (4, 5, 6,7, 8, 9, 10, 11, 12 and 13) exhibit the results of the complexity measures of each cluster of the data sets, which determine the separability of classes. The measures can be classified into two groups by definition: F1, F2, and F3 are based on the data topology, and N1 is based in the Nearest-neighbor where a Minimum Spanning Tree is built. F1 exhibits the overlap ratio that exists between the classes, where if values are small, the overlap will be high. F2 shows the Volume of the Overlap Region, where if values are small, the overlap will be small. F3 presents the Feature Efficiency, where if the values are small, the overlap will be high. N1 exhibits the fraction of points on the boundary that belong to the opposite class. To do this experiments we considered pair classes of each cluster, obtaining the combination of all of them. The results are shown as half-filled results matrices because the rest of them are reciprocal. The number of the clusters is indicated in the Tables with the letter "c" plus the corresponding number of clusters.

Although there is some overlap level in the Trans10k clusters (Tables 4 and 5). The results exhibit that while the cluster number increase (cluster 4) the separation between classes is more possible in terms of the measure F1, F2, and F3. Whereas N1 shows that the cluster 2 has more points of the opposite class than in cluster 4.

Table 4. Class separability, Trans10k_c2

F1	F2	F3	N1
1.09	1.00	9.24E-04	3.54E-03

Table 5. Class separability, Trans10k_c4

F1	c0	c1	c2	c3	F2	c0	c1	c2	c3
c0	–	2.36	3.40	3.99	c0	–	0.10	0.20	0.10
c1	–	–	7.89	8.13	c1	–	–	0.0	3.38E-03
c2	–	–	–	3.74	c2	–	–	–	0.10

F3	c0	c1	c2	c3	N1	c0	c1	c2	c3
c0	–	0.67	0.45	0.53	c0	–	1.45E-03	2.65E-03	2.11E-03
c1	–	–	0.95	0.94	c1	–	–	2.67E-04	4.71E-03
c2	–	–	–	0.59	c2	–	–	–	1.31E-03

The Tables 6, 7 and 8 exhibit the results of complexity measures of each cluster of the Trans50k. Although the results show a certain level of overlap. They show that while the number clusters are greater than 2, the borders between classes are cleaner (in the context of F1 measure), except in Trans50k–12c where the values of the measures in the classes c0–c9 exhibit more overlap. In the context of the F2 and F3 measures, the separation of classes is more presented when the cluster number is greater than 2. The N1 measure shows more points of the opposite class in cluster 2 than in cluster 5, except in the values of measures in the classes c0–c4. However, cluster 12 has more points of the opposite class than in cluster 2. This situation depends on the distribution of the data sets in each cluster.

Table 6. Class separability, Trans50k_c2

F1	F2	F3	N1
1.70	0.99	7.88E-06	1.73E-03

The Tables 9 and 10 exhibit the results of the complexity measures of each cluster of Trans30k. In the context of the measures F1, F2, and F3, if the clusters' number increase, the classes overlap will decrease. The measure N1 exhibits that cluster 2 has more points of the opposite class than cluster 7, except in the values the classes c1–c4 and c2–c4 of the latter cluster.

The Tables 11, 12 and 13 show the values of the complexity measures of each cluster of Trans70k. In the context of the F1, F2, and F3 measures exhibit more overlap in cluster 2 than in cluster 5 (it excepts in the results c0–c3) and cluster 10 (it excepts in the results c1–c8 and c3–c9). The measure N1 exhibits in class 2 more points of the opposite class than in classes 5 and 10.

The confusion matrices of each cluster of the imbalanced data sets are shown in appendix: 17, 18, 19, 20, 21, 22, 23, 24, 25 and 26. Each confusion matrix

Table 7. Class separability, Trans50k_c5

F1						F2				
	c0	c1	c2	c3	c4		c0	c1	c2	c3	c4
c0	–	8.09	6.46	4.59	2.54	c0	–	0.01	0.22	0.08	0.19
c1	–	–	7.89	6.62	7.83	c1	–	–	0.0	2.89E-03	-0.10
c2	–	–	–	6.21	6.40	c2	–	–	–	-0.10	0.02
c3	–	–	–	–	5.20	c3	–	–	–	–	3.45E-03
F3						N1				
	c0	c1	c2	c3	c4		c0	c1	c2	c3	c4
c0	–	0.87	0.60	0.59	0.33	c0	–	1.18E-05	1.12E-03	8.76E-04	1.87E-03
c1	–	–	0.95	0.93	0.92	c1	–	–	2.67E-04	1.27E-03	1.29E-05
c2	–	–	–	0.74	0.69	c2	–	–	–	6.37E-04	1.23E-04
c3	–	–	–	–	0.79	c3	–	–	–	–	2.17E-04

represents the correct and incorrect results of five partitions for each cluster. The main diagonal results of the matrices represent the correct classification, and the rest of them exhibit incorrect classification. Finally, the classes are represented by the letter "c" plus their corresponding number. The percentage of classification per class ("Pc") are exhibited in the Tables of the appendix, and in the Tables 14 and 15.

In Trans30k, the imbalance in cluster 2 is not too high to harm the performance of the classifiers because the number of correct patterns is greater than the incorrect patterns number in the MLP and SVM performance (Table 17). The Table 18 shows the results of the Trans30k in the cluster 7, where the imbalance is greater than in the cluster 2, and the performance of the classifiers decreases in some classes (MLP and SVM).

The Tables 19, 20 and 21 show the confusion matrix results of each cluster of Trans50k. Wherein cluster 2 exhibits a suitable classification of classes in the context of the low imbalance (IR 1.27), in clusters 5 and 12, the number of correct patterns slightly decreases when the imbalance increases (IR 2.99 and 7.59), this situation is presented with the MLP, but with SVM only in some cases.

The Tables 22, 23 and 24 exhibit the results of the confusion matrices of each cluster of Trans70k. Where a suitable recognition by MLP and SVM is presented in cluster 2 when the imbalance is 10.23. However, in cluster 5, each class slightly decreases its performance when the imbalance increase (IR 23), except in the results of SVM. Also, in cluster 10, the imbalance is more represented than in the cluster 5 and 2, in this sense, the results show that the recognition of the minority class is more affected than the recognition of the majority class.

The Tables 25 and 26 show the results confusion matrices of each cluster of Trans10k, where we observe that with a classes' imbalance of 5.7 in cluster 2, the recognition of each class is not very good, except in the class0 of the results

Table 8. Class separability, Trans50k_c12

F1

	c0	c1	c2	c3	c4	c5	c6	c7	c8	c9	c10	c11
c0	–	1.51E+01	3.36E+01	1.44E+01	1.22E+01	1.72E+01	4.72E+00	5.44E+00	4.73E+00	5.18E-05	2.89E+00	6.74E+00
c1	–	–	2.33E+01	1.12E+01	1.20E+01	4.50E+01	1.23E+00	1.89E+01	6.90E+00	5.79E+00	4.05E+01	1.48E+01
c2	–	–	–	1.16E+01	2.48E+01	2.34E+01	5.65E+00	3.15E+00	1.96E+01	3.28E+00	7.36E+00	3.28E+01
c3	–	–	–	–	1.33E+01	1.87E+01	3.34E+00	8.54E+00	6.57E+00	1.70E+01	1.75E+01	1.41E+01
c4	–	–	–	–	–	4.57E+01	3.70E+00	2.35E+01	7.89E+00	1.47E+01	4.11E+01	3.70E+00
c5	–	–	–	–	–	–	8.92E+00	3.53E+01	3.55E+01	1.42E+01	3.08E+00	5.95E+00
c6	–	–	–	–	–	–	–	5.82E+00	1.94E+00	2.11E+00	8.43E+00	4.62E+00
c7	–	–	–	–	–	–	–	–	1.41E+01	3.19E+01	5.41E+00	1.22E+01
c8	–	–	–	–	–	–	–	–	–	1.16E+01	3.24E+01	4.43E+00
c9	–	–	–	–	–	–	–	–	–	–	4.57E+00	1.83E+01
c10	–	–	–	–	–	–	–	–	–	–	–	5.56E+00

F2

	c0	c1	c2	c3	c4	c5	c6	c7	c8	c9	c10	c11
c0	–	0.00E+00	-3.05E-02	-3.77E-02	0.00E+00	2.34E-03	5.92E-03	2.34E-03	3.55E-02	-8.72E-04	2.70E-02	4.84E-03
c1	–	–	-5.84E-02	3.73E-05	-1.71E-01	-7.18E-04	8.35E-02	-1.70E-03	1.93E-03	0.00E+00	-1.50E-02	-3.90E-02
c2	–	–	–	-1.28E-02	1.86E-02	-8.49E-03	3.66E-02	2.53E-02	-1.98E-04	7.06E-02	3.31E-03	-2.21E-02
c3	–	–	–	–	-1.17E-04	-6.93E-06	5.57E-02	0.00E+00	4.46E-02	1.36E-03	-1.97E-03	-5.93E-04
c4	–	–	–	–	–	-1.86E-01	1.23E-02	-5.15E-04	6.69E-03	0.00E+00	-7.33E-02	1.65E-01
c5	–	–	–	–	–	–	0.00E+00	1.03E-02	-6.13E-04	1.14E-03	1.93E-01	-3.41E-02
c6	–	–	–	–	–	–	–	1.15E-02	6.26E-02	5.27E-02	0.00E+00	3.27E-02
c7	–	–	–	–	–	–	–	–	-1.26E-02	1.84E-02	8.12E-03	9.39E-05
c8	–	–	–	–	–	–	–	–	–	5.19E-05	-6.67E-02	1.16E-02
c9	–	–	–	–	–	–	–	–	–	–	1.08E-02	-5.86E-02
c10	–	–	–	–	–	–	–	–	–	–	–	3.10E-02

F3

	c0	c1	c2	c3	c4	c5	c6	c7	c8	c9	c10	c11
c0	–	9.90E-01	4.78E-01	9.99E-01	1.00E+00	1.00E+00	7.43E-01	7.61E-01	8.53E-01	1.00E+00	5.57E-01	9.24E-01
c1	–	–	3.01E-01	9.98E-01	9.73E-01	9.74E-01	2.37E-01	1.00E+00	8.81E+01	9.43E-01	6.92E-01	9.22E-01
c2	–	–	–	3.63-01	4.54E-01	6.16E-01	9.02E-01	5.77E-01	9.98E-01	4.14E-01	9.68E-01	7.07E-01
c3	–	–	–	–	1.00E+00	1.00E+00	5.56E-01	9.58E-01	8.20E-01	1.00E+00	6.67E-01	9.40E-01
c4	–	–	–	–	–	8.99E-01	7.07E-01	1.00E+00	9.28E-01	9.62E-01	6.80E-01	5.45E-01
c5	–	–	–	–	–	–	9.77E-01	1.00E+00	9.88E-01	9.89E-01	4.93E-01	8.74E-01
c6	–	–	–	–	–	–	–	8.49E-01	2.95E-01	4.51E-01	9.68E-01	7.39E-01
c7	–	–	–	–	–	–	–	–	3.97E-01	1.00E+00	8.15E-01	1.00E+00
c8	–	–	–	–	–	–	–	–	–	9.78E-01	3.35E-01	7.61E-01
c9	–	–	–	–	–	–	–	–	–	–	7.44E-01	6.98E-01
c10	–	–	–	–	–	–	–	–	–	–	–	8.25E-01

N1

	c0	c1	c2	c3	c4	c5	c6	c7	c8	c9	c10	c11
c0	–	3.37E-05	6.46E-05	2.68-05	1.73E-03	7.31E-05	5.64E-04	1.76E-03	2.92E-04	5.18E-05	1.40E-03	1.95E-03
c1	–	–	4.73E-05	3.72E-04	8.43E-05	5.17E-05	1.23E-03	4.40E-05	1.86E-05	2.00E-04	3.68E-05	2.00E-05
c2	–	–	–	3.47-05	4.67E-05	1.94E-04	3.34E-04	4.11E-03	2.53E-05	3.44E-03	1.30E-02	5.09E-05
c3	–	–	–	–	2.31E-05	3.70E-05	1.33E-03	6.58E-05	9.95E-04	3.07E-05	3.33E-05	2.41E-05
c4	–	–	–	–	–	5.10E-05	5.92E-04	4.35E-05	4.64E-04	3.97E-05	4.41E-05	1.30E-03
c5	–	–	–	–	–	–	2.24E-05	1.49E-04	2.65E-04	2.23E-04	4.68E-03	1.04E-03
c6	–	–	–	–	–	–	–	6.15E-04	9.12E-04	1.42E-03	4.20E-05	2.05E-03
c7	–	–	–	–	–	–	–	–	2.44E-05	8.11E-05	2.86E-03	4.71E-05
c8	–	–	–	–	–	–	–	–	–	2.31E-05	2.45E-05	5.46E-04
c9	–	–	–	–	–	–	–	–	–	–	1.37E-03	4.26E-05
c10	–	–	–	–	–	–	–	–	–	–	–	5.74E-04

Table 9. Class separability, Trans30k_c2

F1	F2	F3	N1
0.79	0.99	2.55E-04	1.94E-03

Table 10. Class separability, Trans30k_c7

F1	c0	c1	c2	c3	c4	c5	c6	F2	c0	c1	c2	c3	c4	c5	c6
c0	–	7.84	8.13	9.76	7.02	5.65	5.03	c0	–	2.2E-04	6.1E-02	1.4E-03	6.9E-02	0.03	5.2E-02
c1	–	–	8.71	9.08	8.60	9.97	5.30	c1	–	–	1.6E-03	0	2.3E-03	-2.7E-02	2.2E-02
c2	–	–	–	1.16	5.84	5.60	1.01	c2	–	–	–	0	9.3E-02	0.05	8.1E-04
c3	–	–	–	–	1.07	4.49	4.02	c3	–	–	–	–	0	0.04	0.04
c4	–	–	–	–	–	6.67	5.77	c4	–	–	–	–	–	5.7E-04	5.2E-02
c5	–	–	–	–	–	–	1.16	c5	–	–	–	–	–	–	-3.9E-02

F3	c0	c1	c2	c3	c4	c5	c6	N1	c0	c1	c2	c3	c4	c5	c6
c0	–	9.4E-01	8.6E-01	9.7E-01	8.6E-01	0.79	7.5E-01	c0	–	3.0E-05	7.3E-04	2.0E-04	5.3E-04	1.7E-03	1.8E-03
c1	–	–	9.7E-01	9.6E-01	9.6E-01	9.2E-01	8.3E-01	c1	–	–	4.8E-04	8.4E-05	3.4E-03	3.9E-05	6.2E-04
c2	–	–	–	9.7E-01	6.8E-01	0.78	9.6E-01	c2	–	–	–	3.9E-05	2.4E-03	9.7E-04	2.9E-05
c3	–	–	–	–	9.9E-01	0.72	0.66	c3	–	–	–	–	4.1E-05	1.9E-03	1.9E-03
c4	–	–	–	–	–	8.6E-01	8.2E-01	c4	–	–	–	–	–	3.8E-05	7.0E-04
c5	–	–	–	–	–	–	4.2E-01	c5	–	–	–	–	–	–	2E-05

Table 11. Class separability, Trans70k_c2

F1	F2	F3	N1
2.4	0.79	0.44	7.81

Table 12. Class separability, Trans70k_c5

F1	c0	c1	c2	c3	c4	F2	c0	c1	c2	c3	c4
c0	–	3.24	6.29	1.83	4.30	c0	–	0.05	0.03	0.08	0.05
c1	–	–	10.76	8.31	10.98	c1	–	–	-1.08E-04	0.01	-0.16
c2	–	–	–	11.75	8.69	c2	–	–	–	-0.10	1.40E-03
c3	–	–	–	–	11.96	c3	–	–	–	–	-2.60E-03

F3	c0	c1	c2	c3	c4	N1	c0	c1	c2	c3	c4
c0	–	0.55	0.85	0.48	0.72	c0	–	3.80E-04	2.26E-04	5.12E-04	3.68E-04
c1	–	–	0.99	0.93	0.99	c1	–	–	2.00E-05	6.02E-04	1.72E-05
c2	–	–	–	0.95	0.98	c2	–	–	–	7.73E-06	4.18E-04
c3	–	–	–	–	0.9	c3	–	–	–	–	2.74E-05

SVM. However, cluster 4 obtains the best performance with SVM and MLP, although the classes' imbalance increases.

Table 16 provides the classification accuracy averaged across the five runs for each transaction data set. The values for the structure with the highest performance for each classifier are highlighted in bold. The results reveal that the highest classification accuracy rates were achieved with the training sets formed by two clusters, except in the case of the Trans10k database where the optimal structure corresponded to the set with four groups. However, it is also worth remarking that differences in accuracy were not meaningful in most cases,

Table 13. Class separability, Trans70_c10

F1

	c0	c1	c2	c3	c4	c5	c6	c7	c8	c9
c0	–	10.92	15.69	4.94	9.15	3.70	4.15	3.52	4.44	3.3
c1		–	29.99	26.44	17.63	10.9	14.39	18.32	1.72	30.98
c2			–	11.57	35.68	16.78	18.34	29.25	3.73	2.68
c3				–	31.19	15.01	5.64	25.76	3.29	21.73
c4					–	12.43	16.26	9.15	4.09	3.99
c5						–	8.94	12.94	4.45	20.19
c6							–	14.68	5.14	23.46
c7								–	4.47	3.46
c8									–	6.27

F2

	c0	c1	c2	c3	c4	c5	c6	c7	c8	c9
c0	–	0	-8.8E-03	7.1E-03	0	0.05	0.07	0.04	0.09	0.08
c1		–	-2.5E-03	6.1E-03	-0.2	0	-8.7E-05	0.02	0.09	-0.02
c2			–	-0.11	0.03	1.8E-04	0.03	-7.1E-03	7.5E-03	0.11
c3				–	-3.6E-03	3.1E-04	0	0	0.08	-0.05
c4					–	0	1.6E-04	-0.10	0.01	0.05
c5						–	3.4E-11	-0.11	0.03	-6.4E-03
c6							–	-1.8E-03	0.01	-7.6E-03
c7								–	1.7E-03	-7.1E-03
c8									–	0.08

N1

	c0	c1	c2	c3	c4	c5	c6	c7	c8	c9
c0	–	9.9E-06	9.9E-06	6.1E-05	9.8E-06	0.02	4.9E-04	8.9E-04	4.3E-04	2.4E-04
c1		–	1.4E-04	7.8E-05	1.1E-04	3.5E-04	3.1E-05	3.5E-05	1.7E-06	3.8E-05
c2			–	8.0E-05	1.2E-04	2.6E-05	3.0E-05	3.5E-05	8.1E-06	1.6E-06
c3				–	7.0E-05	2.3E-05	2.2E-04	3.0E-05	2.3E-06	3.2E-05
c4					–	2.5E-05	2.8E-05	0.38	1.4E-06	1.4E-06
c5						–	6.2E-04	1.7E-05	8.5E-04	1.7E-05
c6							–	1.8E-05	2.2E-04	1.9E-05
c7								–	1.7E-04	3.7E-04
c8									–	2.2E-06

F3

	c0	c1	c2	c3	c4	c5	c6	c7	c8	c9
c0	–	0.99	0.48	0.91	0.99	0.7	0.71	0.69	0.74	0.74
c1		–	1	1	0.99	0.98	1	0.99	0.39	0.66
c2			–	0.99	1	1	0.77	1	0.75	0.59
c3				–	1	1	0.95	1	0.59	0.36
c4					–	0.98	1	0.99	0.75	0.73
c5						–	1	0.87	0.66	0.62
c6							–	1	0.79	0.73
c7								–	0.77	0.69
c8									–	0.86

Table 14. The percentage of classification per class., MLP

MLP (PC)	C0	C1	C2	C3	C4	C5	C6	C7	C8	C9	C10	C11
Trans30k2c	99.98	99.71										
Trans50k2c	99.94	99.98										
Trans70k2c	99.98	99.70										
Trans10k2c	75.37	6.29										
Trans10k4c	97.56	92.99	96.16	95.61								
Trans50k5c	97.04	96.44	93.89	95.01	98.93							
Trans70k5c	96.50	92.91	93.59	96.32	92.75							
Trans30k7c	96.81	83.37	90.00	97.23	83.65	97.84	98.47					
Trans70k10c	98.04	75.75	75.62	88.84	85.88	96.62	97.67	95.70	87.06	93.97		
Trans50k12c	97.18	99.78	97.26	99.06	98.93	99.95	98.90	98.55	99.59	99.16	96.86	98.27

Table 15. The percentage of classification per class., SVM

SVM (PC)	C0	C1	C2	C3	C4	C5	C6	C7	C8	C9	C10	C11
Trans30k2c	99.95	99.98										
Trans50k2c	99.96	99.99										
Trans70k2c	99.97	99.98										
Trans10k2c	89.13	64.59										
Trans10k4c	99.84	99.07	99.77	99.11								
Trans50k5c	99.90	99.70	99.05	99.64	99.80							
Trans70k5c	99.86	99.93	99.88	99.96	99.95							
Trans30k7c	99.33	98.65	99.53	99.20	98.33	99.44	99.39					
Trans70k10c	99.83	96.25	98.22	98.94	98.37	99.77	99.87	99.42	98.59	98.59		
Trans50k12c	93.70	94.25	84.12	96.43	94.63	85.04	89.18	92.35	94.32	89.40	83.89	90.85

Table 16. Classification performance

	Trans10k		Trans50k		
	2 groups	4 groups	2 groups	5 groups	12 groups
MLP	39.04	**96.72**	**99.96**	96.54	92.33
SVM	87.25	**99.71**	**99.98**	99.70	98.90
	Trans30k		Trans70k		
	2 groups	7 groups	2 groups	5 groups	10 groups
MLP	**99.93**	96.31	**99.95**	96.01	94.67
SVM	**99.96**	99.32	**99.97**	99.88	99.39

especially with the SVM classifier. Also, with the accuracy averaged does not possible to observe the contribution of each class, which is important when the classes' imbalance and classes' overlap are presented in the data sets. In this situation, we may assume a suitable classification accuracy, but in the context of the complexity measures of data sets, this it is not possible.

7   Conclusions

In this paper, we have presented some experiments to find a set of candidate optimal number of clusters in transaction data sets using the Davies-Bouldin and Calinski-Harabasz cluster validity indices and the farthest-first traversals algorithm, whose solutions are then evaluated through complexity measures such as F1, F2, F3, N1 and IR, after that the recognition of the data sets is performed by the SVM with a linear Kernel and an MLP neural network. This issue is of special interest in the context of clustering purchase transaction data because knowing the optimal structure of data can lead to more accurate decisions.

The experimental results have shown the validity of this methodology when applied to four purchase transaction databases. It was observed in the experiments that if the number of clusters increases in each data set, the overlap between the classes will decrease. Besides, the results show that the accuracy of each class influences over the accuracy averaged of each classifier. In the context of class imbalance, if the classes' imbalance increases, the majority class will contribute more to the performance of the classifiers, this situation happens in the majority of the experiments.

Future investigations will be addressed to extend the methodology here proposed with other cluster validity indices and clustering algorithms. Besides, adaptation of this methodology to large-scale and multimodal databases is another potential issue that deserves further research.

Acknowledgment. This work was partially supported by the E.S.I.M.E., Zacatenco, Instituto Politecnico Nacional, the TecNM/Instituto Tecnologico de Matamoros, and the Universitat Jaume I under grant UJI-B2018-49.

A   Appendix

Table 17. Confusion matrix of Trans30k_2c

MLP			SVM		
	c0	c1		c0	c1
c0	46195	30	c0	46224	1
c1	8	10622	c1	20	10610
Pc	99.98	99.71	Pc	99.95	99.98

Table 18. Confusion matrix of Trans30k_7c

MLP								SVM						
	c0	c1	c2	c3	c4	c5	c6		c0	c1	c2	c3	c4	c5	c6
c0	14503	0	35	12	31	26	27	c0	14528	0	5	0	27	1	75
c1	0	1726	3	202	24	14	79	c1	0	1682	0	0	2	0	7
c2	142	143	2604	0	65	164	0	c2	3	1	3106	0	1	9	0
c3	15	13	0	9788	0	20	8	c3	20	0	0	9783	0	40	3
c4	78	44	134	0	1053	0	99	c4	1	7	3	0	2269	0	0
c5	205	0	116	52	0	10343	0	c5	16	0	7	58	0	8969	0
c6	36	142	0	11	85	1	13735	c6	58	15	0	21	9	0	13970
Pc	96.81	83.37	90.00	97.23	83.65	97.84	98.47	Pc	99.33	98.65	99.53	99.20	98.33	99.44	99.39

Table 19. Confusion matrix of Trans50k_2c

MLP			SVM		
	c0	c1		c0	c1
c0	53303	9	c0	53310	2
c1	31	41785	c1	20	41796
Pc	99.94	99.98	Pc	99.96	99.99

Table 20. Confusion matrix of Trans50k_5c

MLP					SVM				
	c0	c1	c2	c3	c4	c0	c1	c2	c3	c4
c0	23133	0	208	384	127	c0	23778	0	21	42	12
c1	0	17576	191	628	0	c1	0	18348	42	5	0
c2	256	307	7367	202	38	c2	8.20	52.40	8068.80	35	6
c3	350	336	80	23579	52	c3	5	3	13	24356	21
c4	98	5	0	22	20182	c4	10	0	0	6	20292
Pc	97.04	96.44	93.89	95.01	98.93	Pc	99.90	99.70	99.05	99.64	99.80

Table 21. Confusion matrix of Trans50k_12c

MLP

	c0	c1	c2	c3	c4	c5	c6	c7	c8	c9	c10	c11
c0	5961	0	0	0	0	0	30	1	0	0	4	6
c1	0	8792	0	4	0	0	31	0	0	4	0	0
c2	0	0	1659	0	0	0	0	15	0	11	49	0
c3	0	2	0	12636	0	0	0	0	29	0	0	0
c4	0	0	0	0	8912	0	17	0	12	0	0	26
c5	0	0	0	0	0	804	0	0	0	0	3	28
c6	30	17	43	25	50	0	21220	16	6	0	0	50
c7	35	0	0	0	0	0	14	2469	0	0	5	0
c8	99	0	0	90	2	0	18	0	17763	0	0	23
c9	0	0	3	0	0	0	34	0	0	3593	9	0
c10	6	0	0	0	0	0	0	3	0	13	2343	4
c11	1	0	0	0	42	0	89	0	26	0	3	7926
Pc	97.18	99.78	97.26	99.06	98.93	99.95	98.90	98.55	99.59	99.16	96.86	98.27

SVM

	c0	c1	c2	c3	c4	c5	c6	c7	c8	c9	c10	c11
c0	5415	0	0	4	0	0	102	48	275	0	62	96
c1	0	8313	0	0	44	0	470	0	0	4	0	0
c2	0	0	1445	16	0	0	100	31	0	95	46	0
c3	6	96	0	11667	0	0	494	16	385	0	0	0
c4	0	17	0	0	8617	0	105	0	110	0	0	118
c5	0	0	0	0	0	494	0	0	0	0	49	292
c6	62	389	108	203	179	0	19923	38	157	207	26	165
c7	68	0	72	104	0	0	87	2034	0	17	134	4
c8	69	0	0	104	62	0	262	0	17479	0	0	20
c9	0	4	89	0	0	0	518	3	0	2986	39	0
c10	41	0	1	0	0	26	2	31	0	28	2207	31
c11	116	0	0	0	203	59	275	0	123	0	65	7246
Pc	93.70	94.25	84.12	96.43	94.63	85.04	89.18	92.35	94.32	89.40	83.89	90.85

Table 22. Confusion matrix of Trans70k_2c

MLP			SVM		
	c0	c1		c0	c1
c0	121310	35	c0	121344	2
c1	25	11722	c1	38	11709
Pc	99.98	99.70	Pc	99.97	99.98

Table 23. Confusion matrix of Trans70k_5c

MLP						SVM				
	c0	c1	c2	c3	c4	c0	c1	c2	c3	c4
c0	101647	58	91	183	385	102355	2	0	4	3
c1	1031	6401	0	161	17	38	7572	0	0	0
c2	522	21	4237	9	83	50	0	4823	0	0
c3	1697	287	45	9265	1	38	3	0	11256	0
c4	434	120	153	0	6236	19	0	5	0.0	6920
Pc	96.50	92.91	93.59	96.32	92.75	99.86	99.93	99.88	99.96	99.95

Table 24. Confusion matrix of Trans70k_10c

MLP										
	c0	c1	c2	c3	c4	c5	c6	c7	c8	c9
c0	48012	0	0	63	0	30	187	219	116	111
c1	0	1223	0	0	2	148	0	0	472	1
c2	0	0	1363	10	0	0	0	0	123	235
c3	0	2	13	4369	0	0	0	0	168	0
c4	0	1	0	0	1884	0	0	15	582	79
c5	252	26	0	0	0	16841	146	4	147	0
c6	248	0	0	8	0	72	14711	0	21	0
c7	144	0	0	0	0	11	0	12162	58	31
c8	172	360	229	466	44	327	18	24	15723	162
c9	140	1	197	0	262	0	0	282	647	9709
Pc	98.04	75.75	75.62	88.84	85.88	96.62	97.67	95.70	87.06	93.97

SVM										
	c0	c1	c2	c3	c4	c5	c6	c7	c8	c9
c0	48506	0	0	33	0	35	18	16	0	130
c1	0	1829	0	0	0	0	0	0	17	0
c2	0	0	1714	0	0	0	0	0	12	6
c3	0	0	0	4538	0	0	0	0	15	0
c4	0	0	0	0	2514	0	0	0	41	8
c5	5	68	0	0	0	17293	1	0	49	0
c6	10	0	0	1	0	3	15028	0	18	0
c7	14	0	0	0	0	0	0	12394	0	0
c8	52	2	29	13	32	0	0	25	17359	14
c9	0	0	2	0	9	0	0	30	94	11104.40
Pc	99.83	96.25	98.22	98.94	98.37	99.77	99.87	99.42	98.59	98.59

Table 25. Confusion matrix of Trans10k_2c

MLP			SVM		
	c0	c1		c0	c1
c0	8606	11868	c0	19822	653
c1	2813	796	c1	2418	1191
Pc	75.37	6.29	Pc	89.13	64.59

Table 26. Confusion matrix of Trans10k_4c

MLP				SVM			
	c0	c1	c2	c3	c0	c1	c2	c3
c0	13436	54	186	32	c0	13676	2	14	16
c1	113	1414	55	19	c1	2	1600	0	0
c2	71	23	6695	28	c2	14	7	6795	1
c3	152	28	25	1748	c3	5	4	1	1943
Pc	97.56	92.99	96.16	95.61	Pc	99.84	99.07	99.77	99.11

References

1. Alcalá-Fdez, J., et al.: KEEL data-mining software tool: data set repository, integration of algorithms and experimental analysis framework. J. Multiple-Valued Logic Soft Comput. **17**(2), 255–287 (2011)
2. Arbelaitz, O., Gurrutxaga, I., Muguerza, J., Pérez, J.M., Perona, I.: An extensive comparative study of cluster validity indices. Pattern Recogn. **46**(1), 243–256 (2013)
3. Calinski, T., Harabasz, J.: A dendrite method for cluster analysis. Commun. Stat. **3**(1), 1–27 (1974)
4. Chen, N., Chen, A., Zhou, L., Lu, L.: A graph-based clustering algorithm in large transaction databases. Intell. Data Anal. **5**(4), 327–338 (2004)
5. Davies, D.L., Bouldin, D.W.: A cluster separation measure. IEEE Trans. Pattern Anal. Mach. Intell. **1**(2), 224–227 (1979)
6. Fahad, A., et al.: A survey of clustering algorithms for big data: taxonomy and empirical analysis. IEEE Trans. Emerg. Top. Comput. **2**(3), 267–279 (2014)
7. Fraley, C., Raftery, A.E.: How many clusters? Which clustering method? Answers via model-based cluster analysis. Comput. J. **41**(8), 578–588 (1998)
8. Garcia, V., Mollineda, R., Sánchez, J.: On the KNN performance in a challenging scenario of imbalance and overlapping. Pattern Anal. Appl. **11**, 269–280 (2007)
9. Halkidi, M., Batistakis, Y., Vazirgiannis, M.: On clustering validation techniques. J. Intell. Inf. Syst. **17**(2–3), 107–145 (2001)
10. Han, E.H., Karypis, G., Kumar, V., Mobasher, B.: Hypergraph based clustering in high-dimensional data sets: a summary of results. IEEE Bulletin Tech. Committee Data Eng. **21**, 01–08 (1998)
11. He, Z., Xu, X., Deng, S.: TCSOM: clustering transactions using self-organizing map. Neural Process. Lett. **22**(3), 249–262 (2005)

12. Hochbaum, D.S., Shmoys, D.B.: A best possible heuristic for the k-center problem. Math. Oper. Res. **10**(2), 180–184 (1985)
13. Huang, X., Song, Z.: Clustering analysis on e-commerce transaction based on K-means clustering. J. Netw. **9**(2), 443–450 (2014)
14. Jain, A.K., Murty, M.N., Flynn, P.J.: Data clustering: a review. ACM Comput. Surv. **31**(3), 264–323 (1999)
15. Kokate, U., Deshpande, A., Mahalle, P., Patil, P.: Data stream clustering techniques, applications, and models: comparative analysis and discussion. Big Data Cogn. Comput. **2**(4), 32 (2018)
16. Kaur, P.J.: A survey of clustering techniques and algorithms. In: Proceedings of the 2nd International Conference on Computing for Sustainable Global Development, pp. 304–307. New Delhi (2015)
17. Pakhira, M.K., Bandyopadhyay, S., Maulik, U.: Validity index for crisp and fuzzy clusters. Pattern Recogn. **37**(3), 487–501 (2004)
18. Sánchez, J.S., Mollineda, R.A., Sotoca, J.M.: An analysis of how training data complexity affects the nearest neighbor classifiers. Pattern Anal. Appl. **10**(3), 189–201 (2007)
19. Saxena, M.P.A., et al.: A review of clustering techniques and developments. Neurocomputing **267**, 664–681 (2017)
20. Sotoca, J., Mollineda, R.A., Sánchez, J.: A meta-learning framework for pattern classification by means of data complexity measures. Inteligencia Artif. Revista Iberoamericana de Inteligencia Artif. **29**, 31–38 (2006)
21. Tin, K., Mitra, B.: Complexity measures of supervised classification problems. IEEE Trans. Pattern Anal. Mach. Intell. **24**(3), 289–300 (2002)
22. Tsai, C.Y., Chiu, C.C.: A purchase-based market segmentation methodology. Exp. Syst. Appl. **27**(2), 265–276 (2004)
23. Witten, I., Frank, E., Hall, M.: Data Mining: Practical Machine Learning Tools and Techniques. Morgan Kaufmann, Burlington (2011)
24. Wu, R.S., Chou, P.H.: Customer segmentation of multiple category data in e-commerce using a soft-clustering approach. Electron. Commer. Res. Appl. **10**(3), 331–341 (2011)
25. Xiao, Y., Dunham, M.H.: Interactive clustering for transaction data. In: Proceedings of the 3rd International Conference on Data Warehousing and Knowledge Discovery, pp. 121–130. Munich (2001)
26. Xu, J., Xiong, H., Sung, S.Y., Kumar, V.: A new clustering algorithm for transaction data via caucus. In: Proceedings of the 7th Pacific-Asia Conference on Advances in Knowledge Discovery and Data Mining, pp. 551–562. Seoul (2003)
27. Yun, C.H., Chuang, K.T., Chen, M.S.: An efficient clustering algorithm for market basket data based on small large ratios. In: Proceedings of the 25th Annual International Computer Software and Applications Conference. pp. 505–510. Chicago (2001)

Artificial Organic Networks Approach Applied to the Index Tracking Problem

Enrique González N.[✉][iD] and Luis A. Trejo[✉][iD]

School of Engineering and Sciences, Tecnologico de Monterrey, Carretera al Lago de Guadalupe Km. 3.5, Atizapán, Edo. de México C.P, 52926 State of Mexico, Mexico
A00457801@itesm.mx, ltrejo@tec.mx

Abstract. The present work aims to adapt the Artificial Organic Networks (AON), a nature-inspired, supervised, metaheuristic, machine learning class, for computational finance purposes, applied as an efficient stock market index forecasting model. Thus, the proposed model aims to forecast a stock market index, with the aid of other economic indicators, employing a historic dataset of at least eleven years for all the variables. To accomplish this, a target function is proposed: a multiple non-linear regressive model. The relevance of computational finance is discussed, pointing out that is an area that has developed significantly in the last decades with different applications, some of these are: rich portfolio optimization, index-tracking, credit risk, stock investment, among others. Specifically, the Index Tracking Problem (ITP) concerns the prediction of stock market prices, being this a complex problem of the kind NP-hard. In this work, is discussed the undertaken innovative approach to implement the AON method, its main properties, as well as its implementation using the topology defined as Artificial Hydrocarbon Network (AHN), to tackle the ITP. Finally, we present the results of using a hybrid method based on K-means and the AHN configuration; within the result, the relative error obtained with this hybrid method was 0.0057.

Keywords: Machine learning · Nature-inspired · Metaheuristic · Stock market index · Forecast

1 Introduction

One of the most important concerns for finance theory is the management of money to reach an expected return under conditions of risk or uncertainty. The management of uncertainty and risk factors in every topic of finance has led researchers to the development of different models and techniques. Their goal is to optimize the resources using them effectively and efficiently, according to the corresponding objective or aim that it is pursued. Computer science has contributed to the finance area by solving complex situations, obtaining proficient

Supported by ITESM-CEM and CONACyT.

I. Batyrshin et al. (Eds.): MICAI 2021, LNAI 13067, pp. 23–43, 2021.
https://doi.org/10.1007/978-3-030-89817-5_2

solutions. As a consequence, computational finance was born as a new discipline to use advances in computer technology to solve economic and financial problems. Some research in this topic can be found in [2,4,8,9,29]. Alternate studies have been presented in literature applying different Machine Learning methods for finance purposes, including the task of forecasting a stock market index or variants of this problem [11,12,29,30]. The prediction of a stock market or index tracking problem is well-known in finance; however, it has received less attention in contrast to other problems like portfolio optimization and selection problems, among others [28–30]. Furthermore, since Artificial Organic Networks (AON) is a novel machine learning model framework, an algorithm from this computational class has not been implemented before with the specific approach stated in this work [14–24]. As explained by Zheng et al. [32], the stock markets are the place where shares are traded, this includes issuing, buying, and selling shares; these occur within other aspects, due to the allured potential of significant profit that may be generated from these markets. As explained by Elliot and Timmermann [5], asset allocation involves a real-time forecast of stock returns, and improved stock return forecast help to enhance investment performance. Thus, the ability to forecast returns has important implications for tests of market efficiency and to produce more realistic asset pricing models that enlighten better the data. Additionally, Elliot and Timmermann state that "stock returns inherently contain a sizable unpredictable component so that the best forecasting models can only explain a relatively small part of stock returns". Furthermore, Miranda [9] explains the existence of models that in the aim of capturing all the entanglements in real-world systems behavior, cannot be solved analytically using applied mathematical techniques (closed-form solutions).

Index Tracking Problem (ITP) is a buy-and-hold trading strategy of assets that considers the value of another asset: the underlying. The underlying is an arbitrary quantity that represents the risk factor to an investor. The trading strategies use the aid of an index tracker, which seeks to passively mimic the behavior of the securities found in the capital markets, like a stock market index [2,6,7,11,12,28–30]. The objective of the index tracker is to obtain the same returns as the one attained by a previously selected stock market index, such as the Standard & Poor 500 return index (the S&P 500, is a stock market index that tracks the stock of 500 large-cap companies from the U.S.A.). Hence, the achievement of this objective requires seeking a functional equation capable of describing the behavior and forecast of the selected index, for the sake of defining the trading strategy of the assets. Essentially, ITP aims to reproduce the performance of a specific stock market index, by developing models that should be able to predict its future value and determine the best moments to buy and/or sell in a capital market. ITP or stock market price prediction, as reviewed by different authors [2,5,7,28,30,32], has been an ever-going challenge for economists and machine learning researchers. The problem has been described as a dynamic, non-linear, and complex chaotic process (system), but -still- an important financial problem; as a time series phenomenon, one of the greatest issues of ITP is the amount of non-linear dependencies with other macroeconomic variables (MEVs) across time.

Here, it is proposed the approach of using the AON framework method classified as a naturally-inspired supervised metaheuristic machine learning class [14–24], for developing an efficient algorithm based on this model capable of performing a short-term market price trend forecast, identifying the buy and sell conditions within the predicted period of the stock market index. To identify a function for the forecasting model, historical data would be used to approximate a polynomial function using least-squares regression: a multiple non-linear regressive model [1,3,25,26]. It has not been reported before the development or the use of an algorithm based on the AON framework for a dynamic model like the finance task stated; nevertheless, prior successful results have been reported using algorithms based on the AON class to tackle NP-hard problems (nondeterministic polynomial-time hardness: in simple terms NP-hard problems cannot be solved in polynomial time). The results include modeling complex multivariate systems for other disciplines (similar to the stock market index rate prediction) [14–24].

1.1 Objectives and Limitations

The main purpose of this research is to develop an efficient algorithm based on the Artificial Organic Networks machine learning class, capable of performing a short-term market price trend forecast, identifying the buy and sell conditions, contemplating technical and fundamental analysis. The outcome of the model will consider the rates of the stock market index. Like any other MEV, a stock market index is affected by many factors, and their variety has grown due to globalization, increasing the difficulty of the analysis [11,12]. The complexity of the model will be narrowed by two constraints: i) the prediction of the index rate will be done using the historical data of at least three variables (the historic prices of the concerning index and a minimum of two additional MEV), ii) the MEVs would be chosen depending on their correlation rate to the index.

2 The Proposed Approach

The main objective is to develop an efficient algorithm based on the AON machine learning class, so it can be employed on ITP to forecast a stock market index. As defined by Ponce et al. [14–24], the AON framework method, is a supervised metaheuristic machine learning class bio-inspired on organic chemistry (compounds of carbon, hydrogen, oxygen, nitrogen, sulfur, and halogens).

2.1 AON Properties

An AON is a set of graphs built based on heuristic rules to form organic compounds. Each graph represents a molecule with atoms as vertices and chemical bonds as edges. These molecules interact via the chemical balance to form a mixture of compounds. An AON is a combination of compounds that have four components: atomic units, molecular units, compounds, and mixtures. The molecules can be seen as packages of information, the bonds between the structures of the model describe the complex relationship in the information. Any topological structure inspired by organic compounds that respect the next statements is an AON [14–24].

1. Ground-state and electronegativity preserve stability in organic compounds.
2. The octet rule in molecules assures they are well-formed.
3. Structures of organic compounds promote easily-spanning to generate new stable compounds, simple or complex.

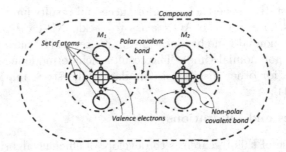

Fig. 1. Example of the structure of an Artificial Organic Network.

In Fig. 1 the structure of an AON is illustrated. The simplest unit is the *atom*, which can produce connections between each other through *bonds*. These links are created using the *valence electrons*. When two or more atoms are associated via a collection of bonds, they form *molecules* (M_1 and M_2 in the figure), and when two or more *molecules* are connected, they form *compounds*. Covalent bonds are classified by the related atoms: a) if two different atoms are linked, the bond is known as a *non-polar covalent bond*, b) if two similar atoms are linked, the bond is then a *polar covalent bond*. Molecules are also classified into two types: *stable molecules* and *unstable molecules*; if an atom inside a molecule has free valence electrons (no link with another atom), then it is called an unstable molecule [14–24].

2.2　Artificial Hydrocarbon Networks Algorithm

Organic compounds are divided into functional groups; the AON method has been implemented -up to now- based on hydrocarbons, specifically the alkanes (CH group, chains of combined atoms of carbon and hydrogen). This means that the rest of the organic compounds (oxygen, nitrogen, sulfur, and halogens) have not been used to produce other kinds of AON. The main reason for this is because hydrocarbons are the simplest and most stable organic components by strong CH bonds [14–24]. The mathematical formalization of the AON algorithm has been based on the organizations of alkanes. This particular arrangement is named Artificial Hydrocarbon Networks (AHN). The AHN algorithm can be referred to as a specific configuration or *topology* to produce an AON, no other topologies have yet been defined. The objective of the mathematical method is to constrain an AON to a simple and stable state supported by hydrocarbons, to allow them to model unknown engineering schemes and to partially understand unknown information inside systems. This is done considering the inputs and the outputs of the unknown system or process being modeled [14–24].

Diverse fields of applications of the AHN algorithm have been applied with good results since Ponce proposed this technique [14]. Some of the reported uses have covered the next areas: function approximation and modeling, robust human activity recognition systems, signal processing in denoising audio and face recognition, intelligent control systems for robotics and mechatronics, bio/medical applications, amongst others [14–24].

3   Implementation Considerations

When talking about financial markets, it should be considered that they have many fluctuations since they are susceptible to several events like political, social, or economic conditions, among others. This means that the financial markets have significant variability (volatility) due to a random (probabilistic) nature, where time is an important factor that must be considered; this stochastic volatility can be understood as the heteroscedasticity concept also used in econometrics [26]. Heteroscedasticity also addresses time-varying variance in the time series [32]. In these circumstances, approximating a target function that fits the shape or general trend employing a polynomial interpolation is inappropriate and may yield unsatisfactory results when used for modeling a time series; instead, stochastic -random- simulations are implemented [3,10,25,26].

3.1   System Identification

Considering that every system generates a response (output) from a previous stimulation (input), a simulation of a system depends on the accuracy of the dynamic model [1]. The proposed system will be set up by an agent that receives external data gotten from the economic environment; the agent will be organized in two phases: i) a training phase, and ii) a forecasting phase (see Fig. 2). First, the training phase will calculate the target function f; then, the forecasting phase will compute the prediction of the index rate, and it will identify the buy and sell conditions within the prediction, using the obtained function and the new algorithm based on the AON class.

To calculate the target function f, the training (or testing) phase performs two operations. One operation concerns the estimation of the error between the true value and the value approximated, this is the quality metric. On the second hand, the training phase extracts the parameters needed for the forecasting phase; the extraction of the parameters considers the selection of the most suited MEVs for the model. After the parameters are computed, the information can be used

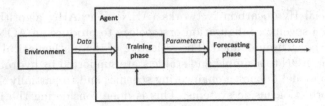

Fig. 2. Diagram of the complete system.

to approximate f, to predict a stock market index; as defined in the objectives, a short-term market price trend forecast and the identification of the buy and sell conditions has been stated.

3.2 Target Function Mathematical Formulation

Zheng et al. [32] review the most common categories found for time series analysis, mainly divided into two: 1) univariate, and 2) multivariate models. Within the univariate models the most commonly used are: a) the autoregressive (AR) models, b) the moving average (MA) models, and c) the autoregressive moving average (ARMA) models that are a combination of the last two. One important assumption of these univariate time series models is that they encompass all the useful data, including the effects of underlying explanatory variables. In counterpart, the multivariate models help to understand how the economic factors correlate to the stock markets. As Zheng remarks, all the previous models assume stationary time series, which means that their mean, variance, covariance remain unchanged over time; in other words, the variance and other statistical properties keep identically and independently distributed.

Pursuing the interest of employing an effective and efficient economic forecasting method, a model that combines technical and fundamental analysis will be defined; so model uncertainty (volatility) can be considered, as well as turning points, macroeconomic factors, within other elements. Therefore, as formerly declared, the model will use the prior information of three variables: the historic prices of the index, and a minimum of two MEVs (chosen depending on how much they correlate to the index being analyzed). To reduce parameter instability due to business cycle fluctuations, then a historic dataset of at least eleven years would be established for all the variables, to assure that no less than one short economic cycle is used for the analysis and prediction of the market index.

To identify a system according to the objectives, the training phase will approximate the target function f doing a regression to the stock market index rate, using least-squares regression (LSR); specifically, a multiple non-linear regressive (MNLR) model is proposed [1,3,26]. Therefore, the model will consider the relationship between a dependent variable y: the market index, and at least two independent variables x_1 and x_2: two correlated MEVs. From the econometric point of view, the volatility or heteroscedasticity of the index rate due to the incidence of external variables is being pondered. Along with the advantages of using LSR, it should also be remarked that, as a numerical method (analytical model) that minimizes the error, the true value of the dependent variable is calculated (approximated), being with it a very efficient method [1,3,26]. Intending to establish a target function $f(x_1, x_2)$ through the MNLR model, it has to be considered the relationship between the variables that have been defined: y dependent of x_1 and x_2 (see Fig. 3):

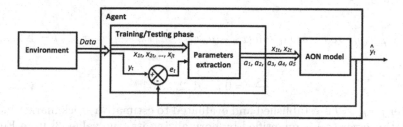

Fig. 3. Diagram of the complete proposed system.

$$y_t = f(x_{1t}, x_{2t}) \quad t = 1, 2, ..., n \tag{1}$$

And,

$$y_t = a_0 + a_1 x_{1t} + a_2 x_{2t} + a_3 x_{1t}^2 + a_4 x_{2t}^2 + a_5 x_{1t} x_{2t} \quad t = 1, 2, ..., n \tag{2}$$

Also,

$$y_t = \hat{y}_t + e_t \quad t = 1, 2, ..., n \tag{3}$$

And,

$$\hat{y}_t = \hat{a}_0 + \hat{a}_1 x_{1t} + \hat{a}_2 x_{2t} + \hat{a}_3 x_{1t}^2 + \hat{a}_4 x_{2t}^2 + \hat{a}_5 x_{1t} x_{2t} \quad t = 1, 2, ..., n \tag{4}$$

Where y_t is the actual value (true value or observed data), $\hat{y}_t$ is the value obtained from the model, and e_t is the true error between both. Rearranging Eq. 3 and 4:

$$y_t = \hat{a}_0 + \hat{a}_1 x_{1t} + \hat{a}_2 x_{2t} + \hat{a}_3 x_{1t}^2 + \hat{a}_4 x_{2t}^2 + \hat{a}_5 x_{1t} x_{2t} + e_t \quad t = 1, 2, ..., n \tag{5}$$

Considering that the true error (residual, random disturbance, or stochastic term) is the discrepancy or difference between the value observed y_t, and the approximated value by the model $\hat{y}_t$, then:

$$\sum_{t=1}^{n} e_t = \sum_{t=1}^{n} (y_t - \hat{y}_t) \quad t = 1, 2, ..., n \tag{6}$$

When f is estimated using LSR, the objective is to minimize the sum of the squares of the residuals (Sr) between y_t and $\hat{y}_t$; in other words, Sr represents the square of the discrepancy between the data and a single estimate of the measure:

$$Sr = \sum_{t=1}^{n} e_t^2 = \sum_{t=1}^{n}(y_t - \hat{y}_t)^2 \quad t = 1, 2, ..., n \tag{7}$$

To determine the values of $\hat{a}_0, \hat{a}_1, ..., \hat{a}_5$, expressions 4 and 7 are rearranged obtaining Eq. 8, then expression 8 is differentiated with respect to each of the coefficients. Each derivative is equal to zero to have a minimum Sr. One of the most important advantages of this criteria is that it yields a unique trend (polynomial expression) for a given set of data [3, 26].

$$Sr = \sum_{t=1}^{n}(y_t - \hat{a}_0 - \hat{a}_1 x_{1t} - \hat{a}_2 x_{2t} - \hat{a}_3 x_{1t}^2 - \hat{a}_4 x_{2t}^2 - \hat{a}_5 x_{1t} x_{2t})^2 \quad t = 1, 2, ..., n \tag{8}$$

Where,

$$\frac{\partial Sr}{\partial \hat{a}_i} = 0 \quad i = 1, 2, ..., 5 \tag{9}$$

After $f(x_1, x_2)$ is established and evaluated to estimate a stock market index, the relative error (ε_t) is quantified by normalizing its true value [3] from Eq. 6:

True fractional relative error = true error/true value

Or,

$$\sum_{t=1}^{n} \varepsilon_t = \sum_{t=1}^{n} |1 - \frac{\hat{y}_t}{y_t}| \quad t = 1, 2, ..., n \tag{10}$$

3.3   Financial Analysis and Strategy

As remarked by Hou et al. [7] and Zheng [32], among the financial tools for forecasting stock market index returns, the two of the most relevant approaches are: a) Technical Analysis (TA), and b) Fundamental Analysis (FA). The *Technical Analysis* focuses mainly on historical price trends (opening, closing stock prices, and volume) to study stock prices. In counterpart, *Fundamental Analysis* mainly considers endogenous and exogenous financial variables (fundamental factors) of a company [6, 7, 32].

An important implication are *business cycles*, which are a concept introduced by Samuelson [32], in his multiplier accelerator model; they are economic fluctuations that consist of the periodic but irregular downward and upward deviations of the production. The business cycles are measured from the fluctuations of the Gross Domestic Product (GDP). To overcome the business cycles difficulties, TA and FA financial techniques for stock market analysis will be used. TA helps to identify the time to buy and sell moments; in contrast, FA studies and explains the considerations of the stock market movements, these include macroeconomic factors.

Technical Analysis. In his text, Focardi [6] depicts the existence of different kinds of investors, regarding the preference for managing their investment portfolio. Thus, two main types of management are exposed: a) active portfolio management, b) passive portfolio management. *Active portfolio management* considers tracking the performance of a portfolio; it mainly consists of regularly buying, holding, and selling the assets to outperform the overall market, in this case, a particular market index. In contrast, *Passive portfolio management's* main objective is to replicate as much as possible the investment returns of an index. Passive portfolio management includes the buy-and-hold strategy and indexing strategy. Due to their specific goals, active portfolio strategies intrinsically require taking greater market risks.

It was expounded in the problem statement that ITP can be described as a buy-and-hold strategy, that uses the aid of an index tracker to passively mimic the behavior of the assets [6,29]; based on this definition, and because the stated objective pursues the identification of the time to buy and sell stock market assets, rather than exhaustively analyze the risk or pursue an expected return, the rest of this work will remain attached to the buy-and-hold trading strategy. The Sharpe ratio will be used, which considers the returns and associated risks, as it is the measure of risk-adjusted return of a financial portfolio. It is the average return earned over the risk-free rate per unit of volatility (total risk) [5,25]:

$$Sharpe\ ratio = \frac{r - r_f}{\sigma} \tag{11}$$

Where r is the expected return, r_f is the risk-free rate, σ is the standard deviation or total risk. It is important to notice that since a financial portfolio is not being explicitly constructed or established, an expected return has not been formally defined; consequently, for the value r, the annual return of the index will be used instead.

The Buy-and-Hold Strategy. The *Buy-and-Hold* trading strategy, or *double crossover method*, is, in general, a trend-following technique [10]: certain stocks for a portfolio are bought based on some criterion like their historical average price, and are mainly held until the end of an investment horizon. The strategy is implemented by computing two simple moving averages for the historic data of the stock market index; each moving average considers different time periods: a) a smaller period or fast moving average (FMA), and b) a longer period or slow moving average (SMA). To identify the buy and sell moments of an asset, both moving averages frames are plotted in the same chart, and whenever the fast MA crosses the slow MA from below a buy signal is triggered, and the sell signal is triggered in the other way.

Fundamental Analysis. The MEVs for the forecast model will be chosen depending on their correlation to the selected stock market index. The correlation coefficient (CC) will be estimated via the Pearson r correlation coefficient since Pearson CC is one the most used to measure the grade of the relationship between parametric variables (e.g. the relationship between two stocks) [25,26]. In other circumstances where non-parametric information is not excluded, like qualitative data, the Spearman correlation coefficient would be more suited as explained by Salvatore [26].

4   Preliminary Results

Different preliminary backtest experiments were performed on Python with the help of SciPy [31], Scikit-learn [13], and statsmodels [27] libraries; within the advantages of using these libraries is the ability to exploit vector and parallel operations. These experiments aimed to explore and evaluate the viability and effectiveness of implementing the initial considerations: a) the presumption of approximating a target function f using the proposed MNLR mathematical expression, b) analyze the correlation with the other MEVs to apply FA, c) carry out the buy-and-hold strategy and compute the Sharpe ratio for the FA, and d) implement a hybrid K-Means with AHN algorithm to test the AON approach on ITP tasks. In this section, the tests considered most relevant will be described; in this regard, along with the objectives of experiments 1 and 2 are measuring the performance of the MNLR model following the criteria of using it as a viable target function while considering FA features. Likewise, experiment 2 presents the comparison of the MNLR model with other ML techniques to benchmark its performance, while the purpose of experiment 3 is to cover TA details. Finally, as a first approach to assess the viability of using AON for the stated objectives of this work, experiment 4 covers the details of implementing the K-Means with the AHN algorithm.

It is important to remark that the following experiments have not been performed (yet) with the consideration of splitting the dataset into training and testing subsets; this procedure would be done in future experiments. Likewise, it can be noticed that the experiments also include an approximation using a least-squares polynomial regression (LSP) based only on historic data of the IPC Mexico index, this was for baseline purposes. Naturally, the LSP function does not consider heteroscedasticity due to the influence of other economic factors; as already mentioned, it is just employed for reference purposes. Likewise, the MNLR model can be explained as multiple inputs, single output system (MISO system); and that the LSP model can be understood as a single input, single output system (SISO system). The experiments were done using existing data for Mexico: the daily reported (labor days) IPC Mexico stock market index, the quarterly reported GDP, the daily (labor days) reported MXN-USD foreign exchange rate (FX), the monthly reported consumer price index (CPI) or the inflation rate, the monthly risk-free rate (RFR), the monthly unemployment rate (UR). The IPC data was retrieved from Yahoo Finance, the FX was acquired from the USA's Federal Reserve Board, the rest of the variables were obtained from the OECD; all the data used comprehends the period from 1/6/2006 to 30/7/2020.

4.1   Experiment 1: Establishing a Regression

For the first experiment, the MNLR model for the IPC Mexico was established using the five exogenous MEVs that have been defined: FX, GDP, CPI, RFR, and UR. The data was preprocessed in three steps. As first step, the data of the five exogenous variables were treated as "continuous signals" instead of discrete information, so for each of these inputs an independent approximation was done using a LSP based on their historic data. For the second step the data was scaled, for this step, the dataset was standardized removing the mean. For the third step a principal component analysis (PCA) was performed considering three components. Figure 4 shows the data of the five MEVs after they have been smoothed by the LSP regression.

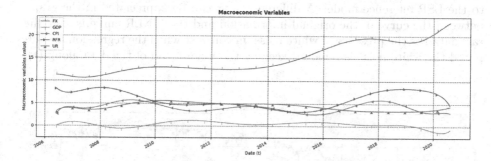

Fig. 4. Historical data for the FX, GDP, CPI, RFR, and UR values after their independent LSP regression. All values are rates %, except the FX price.

After the LSP for each variable was estimated, the correlation coefficient was calculated for each MEV with respect to the IPC Mexico; in Table 1 can be observed that the FX with a CC of 0.69, and the UR with a CC of −0.31, are the most correlated variables (from the five that have been established).

Table 1. CC of the five smoothed exogenous variables with respect to the IPC Mexico.

Correlation coefficient					
	Index	FX	GDP	CPI	RFR	UR
Index	1.00	0.69	0.20	−0.16	−0.25	−0.31

Following the regression and standardization done for each independent variable, the PCA was performed; in Table 2 is shown the explained variance ratio per component, the three components together explain in 97% the variation present in the original five MEVs.

Once the dataset was preprocessed, the MNLR model was obtained employing the three principal components (PC); next, the obtained target function was evaluated with the data of PC1, PC2, and PC3, to obtain an estimation from the regression, using the same interval from which it was calculated. Figure 5 shows the graph of three curves: a) the curve in blue depicts the original data with respect to the IPC Mexico, b) the curve in red shows the estimated MNLR model using the three preprocessing steps, c) the smoothed green curved corresponds

Table 2. Explained variance ratio for the PCA using smoothed variables.

Explained variance ratio		
PC 1	PC 2	PC 3
72.03%	22.03%	3.71%

to the LSP reference model. A difference that can be appreciated in the graph between the curve of the original information and the MNLR curve is that the raw data has the presence of white noise, in contrast when the regression is computed the white noise is removed keeping the main trend of the data along (t).

Fig. 5. Curves for the IPC Mexico using the original data (blue line), a MNLR model using PCA and smoothed variables (red line), and the LSP model (green line). (Color figure online)

Subsequently $f(x_1, x_2, x_3)$ was evaluated to estimate the IPC Mexico, the relative error ε_t between y_t and $\hat{y}_t$ was computed, finding satisfactory results. Table 3 shows the mean, the standard deviation (SD), the mean absolute deviation (MAD), the maximum (Max) value, the minimum (Min) value, and the range of the relative error obtained in this first experiment; Fig. 6 shows the relative error of the MNLR model.

Table 3. Mean, SD, MAD, Max, Min, and Range of the relative error from the MNLR model.

Relative error (MNLR)					
Mean	SD	MAD	Max	Min	Range
0.009314	0.00618	0.00523	0.03517	0.00001	0.03516

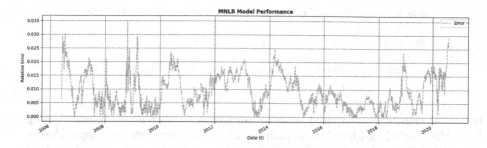

Fig. 6. Relative error for the estimated IPC Mexico using the MNLR model.

At this point, two more cases were additionally analyzed for experiment one. In the first instance, the experiment was also done using only one PC to estimate the MNLR model; on the second hand, the experiment was done once more using two PCs to estimate the MNLR model. In Table 4 the results of the relative error are compared for the three scenarios explored in experiment one; it can be observed that the case where three components are used performs better than the other two scenarios.

Table 4. Mean, SD, MAD, Max, Min, and Range of the relative error for scenarios 1–3 in experiment one.

Relative error (MNLR)						
	Mean	SD	MAD	Max	Min	Range
1 PC	0.011882	0.011862	0.008316	0.071436	0.000018	0.071418
2 PC	0.010192	0.008702	0.006567	0.055073	0.000005	0.055068
3 PC	**0.009314**	**0.00618**	**0.00523**	**0.03517**	**0.00001**	**0.03516**

In Eq. 2 it was defined the mathematical expression of the MNLR target function, considering the relationship between the variables y dependent of x_1 and x_2; nonetheless, with the experiments it was observed that the performance of the model improved significantly considering and extra independent variable x_3, increasing the model from six to eleven coefficients. In Table 5 are shown the coefficients of the target function after training the MNLR; it can be observed that the intercept $\hat{a}_0$ (independent constant) is the biggest term in magnitude and that the rest of the terms remain in a range of $|\hat{a}_i| < 0$, where $i = 1, \ldots 10$.

4.2 Experiment 2: Comparing MNLR Performance Vs. Other ML Techniques.

In experiment two the proficiency of the MNLR model implemented is compared, using the same data set preprocessed in three steps but the further methods only

Table 5. Coefficients of the target function after training the MNLR.

Coefficients of the MNLR										
$\hat{a}_0$	$\hat{a}_1$	$\hat{a}_2$	$\hat{a}_3$	$\hat{a}_4$	$\hat{a}_5$	$\hat{a}_6$	$\hat{a}_7$	$\hat{a}_8$	$\hat{a}_9$	$\hat{a}_{10}$
8.378	0.051	0.048	−0.123	−0.003	0.027	−0.084	−0.010	−0.003	−0.078	0.005

use PC1; thus, it is compared to an Autoregressive Integrated Moving Average (ARIMA) model and five ML techniques: a) Neural Network Multilayer Perceptron (NN MLP), b) Decision Trees (DT), c) Support Vector Machines (SVM), d) Random Forest (RF), and e) Gaussian Process Regression (GPR). The ARIMA model was implemented using the statsmodels library and the five ML techniques were implemented using the Scikit-learn libraries suggested for regression. The parameters of the ML methods were set by heuristics, following some suggested default parameters and examples included in the Scikit-learn library documentation; in this regard, the NN MLP method was trained using the default number of layers and neurons per layer, with a maximum of 500 iterations. The DT method used a maximum depth of five levels. The RF method used three estimator forests. The GPR method used a kernel function. Since the MNLR model did not change, the CCs of Table 1 and the explained variance ratios of Table 2 remain the same. Figure 7 shows a graphic comparing the outcomes of all regression techniques, including the LSP model used as a reference.

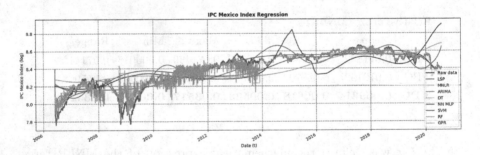

Fig. 7. Regression result comparison of all implemented techniques.

The relative error of each technique was calculated; the mean, the SD, the MAD, the Max value, the Min value, and the range were compared accordingly. The results are reported in Table 6, it can be appreciated that RF outperformed the rest of the models considering the relative error.

Also, the performance of each model was compared mutually computing the Residual Sum of Squares (RSS), the Sum of Squares Regression (SSR), the Total Sum of Squares (TSS), the coefficient of determination R-square, and the execution time. The results are reported in Table 7, it can be noticed that the ARIMA and DT techniques performed better than the other methods, surpassing as well the proposed MNLR model.

Table 6. Mean, SD, MAD, Max, Min, and Range of the relative error of the implemented techniques.

Relative error						
	Mean	SD	MAD	Max	Min	Range
LSP	0.006840	0.006877	0.004579	0.051185	0.000001	0.051184
MNLR	0.009314	0.00618	0.00523	0.03517	0.00001	0.03516
ARIMA	0.000819	0.001415	0.000609	0.066847	0.0	0.066847
NN MLP	0.016464	0.013214	0.009889	0.069876	0.000001	0.069875
DT	0.00812	0.011061	0.007305	0.074277	0.0	0.0742773
SVM	0.011452	0.012347	0.00829	0.074927	0.000003	0.074925
RF	**0.004509**	**0.006716**	**0.004689**	**0.057476**	**0.0**	**0.057476**
GPR	0.014347	0.012451	0.009365	0.069442	0.000001	0.057476
Average	0.008983	0.008783	0.006245	0.0624	0.000002	0.060903

As remarked, the MNLR model was surpassed by other methods; nevertheless, the next observations can be derived from the results:

1. The MNLR executes better than the average, considering both the relative error and the model performance.
2. The mean, the SD, and MAD of the relative error in the MNLR model are smaller than 1×10^{-2}.
3. The MNLR model behaves better than three ML techniques (NN MLP, SVP, and GPR).
4. For the training, the MNLR performs more than $25\times$ times faster than the ARIMA model.
5. The difference between the relative error of the MNLR model and the other two techniques that had better results (ARIMA and RT) is smaller than 1×10^{-2}.
6. The soft outcome of the MNLR model permits to easily read the Index and to identify their different trends across time.

Based on these important results, mainly the 6th statement, we can claim that the proposed mathematical MNLR model is adequate to perform the training phase, to implement the agent described at the beginning of this chapter (see Fig. 3).

4.3   Experiment Three: Buy-and-Hold Strategy

The aim of this experiment seeks the implementation of the Buy-and-Hold strategy and to compute the Sharpe ratio for the TA. As explained previously, the approach uses two simple moving averages for the historic data of the index with different time periods: a slow MA and a fast MA. There are many common combinations used [10]: five-day and 20-day averages, 12 and 24, 10 and 30, 10

Table 7. RSS, SSR, TSS, R-square and the execution time of the implemented techniques.

Model performance					
	RSS	SSR	TSS	R-square	Time (sec)
LSP	22.091212	118.121539	140.212751	0.842445	0.005724
MNLR	30.92239	131.673941	162.596331	0.809821	0.158883
ARIMA	**0.617502**	**139.606867**	**140.224369**	**0.995596**	2.725227
NN MLP	108.484266	95.753617	204.237883	0.468834	1.213106
DT	43.791905	96.420846	140.212751	0.687675	**0.004155**
SVM	66.029838	59.163475	125.193313	0.472577	0.235732
RF	15.425571	113.652944	129.078515	0.880495	0.06571
GPR	85.813126	54.436535	129.078515	0.880495	11.926022
Average	46.646976	101.10372	146.354303	0.754742	2.022385

and 50-day, and so forth; for the development of the current experiments the pair 10 and 30-day average was chosen. Figure 8 shows the IPC Mexico with the SMA and FMA frames; from this figure is possible to see the crosses between these two frames.

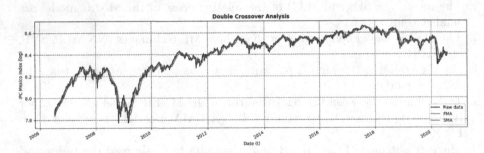

Fig. 8. The IPC Mexico with the 10-day FMA and 30-day SMA frames.

As stated before, to calculate the Sharpe ratio the value of the annual return of the index was used instead of an expected return for r. Therefore, the task required the computation of the annual return, the volatility of the index, and an expected risk-free rate. Table 8 shows these values and the Sharpe ratio computed for the IPC Mexico comprehending the period from 1/6/2006 to 30/7/2020.

4.4   Experiment 4: A Hybrid K-Means with AHN Algorithm

AON as a framework [14,15] defines components, interactions and a heuristic rule inspired on chemical organic compounds. Hence, AON has a structure and

Table 8. Financial analysis for the IPC Mexico based on annual return, volatility, and Sharpe ratio.

IPC Mexico financial analysis				
Daily return %	Annual return %	Volatility %	Sharpe ratio	Daily P. lose +5%
0.026203	6.825133	0.01231	0.71856	0.002217

behavior that states its two main characteristics: modeling nonlinear systems and partial interpretation of unknown information. AON as a learning method needs a two-step implantation: a training process to build their structure and estimate all the parameter values inspired on organic chemistry rules, and an inference process that consists of using the obtained structure to find an output considering a certain input value. The AON algorithms are defined in three types: a) chemically inspired algorithms with defined heuristic rules, based on functional groups and molecular structures of chemical organic compounds, b) artificial basis algorithms that define specific functional groups and molecular structures independently to chemical organic compounds, and c) hybrid algorithms that define structures based on a mixture of chemically inspired and artificial basis algorithms.

AON requires to use a topological structure for its implementation, since AHN is its first based and only formal configuration defined up to now, this topology was used in the present experiment to test the AON approach for ITP purposes. AHN model was originally defined as a chemically inspired algorithm that performs an optimization of a cost-energy function in two levels. First, it uses LSR to define the structure of the molecules; second, it uses gradient descent to optimize the position of the molecules in the futures space. For the current experiment a hybrid K-Means with AHN model was implemented; this kind of structure can be considered in the third kind of AON as a hybrid inspirational algorithm. In this hybrid approach the structure of the molecules is still defined via LSR, but instead of optimizing the position of the molecules using gradient descent, the futures space is segmented (clustered) using K-Means; so each time an iteration occurs, the data is segmented as many times as the same number of molecules that are going to be computed, and the structure of each molecule is computed based the corresponding segment. Although the experiment is performed based on a hybrid model, it still complies with the characteristics defined for an AON: structural and behavioral properties, encapsulation, inheritance, organization, mixing properties, stability and robustness. It also follows the two main inspirational characteristics of the AHN algorithm: they define the simplest set of organic compounds and they are the most stable compounds. Figure 9 shows the graph of two curves: a) the curve in blue depicts the IPC Mexico considering the original data, b) in red it is shown the curved estimated with the AHN algorithm. The experiment was done using two stop conditions, whichever happens first between: a) building a unique compound with a maximum number of 100 molecules b) reaching a relative error smaller than 0.005; in this case, the first condition was reached first. Table 9 shows the mean, the SD, the MAD, the Max value, the Min value, and the range of the relative error obtained.

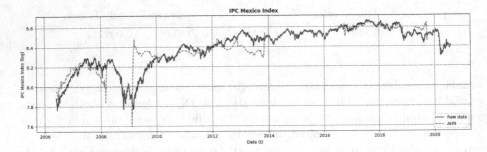

Fig. 9. Curves for the IPC Mexico using the original data (blue line), and the AHN estimation (red line). (Color figure online)

Table 9. Mean, SD, MAD, Max, Min, and Range of the relative error in experiment six.

Relative error (KMeans-AHN)					
Mean	SD	MAD	Max	Min	Range
0.005743	0.008967	0.005793	0.075664	1.395976e−07	0.075664

5   Conclusions and Future Work

In this paper, relevant notions have been reviewed to establish a mathematical model according to the objective of discussing several considerations, like replicating as much as possible the system's dynamics to define a precise model. From a financial perspective, the volatility due to external factors has been discussed; this variability can be acknowledged as the heteroscedasticity due to exogenous components, considering the analysis of at least five MEVs to choose a minimum of two to forecast an index. In this work, we proposed a model that consists of an agent with two phases: a) a training phase, and b) a forecasting phase. The training phase calculates the target function $f(x_1, x_2)$ through the MNLR model; then, the forecasting phase uses the obtained function and the new algorithm of the AON method, to do a short-term market price trend forecast, identifying the buy and sell conditions. To build the model for the training phase, two main considerations have been taken: i) the historic data of at least one short economic cycle (eleven years) is used, and ii) the model will consist of a MNLR expression; to overcome the business cycles difficulties, TA and FA financial techniques for stock market analysis were considered.

A series of backtest experiments were done to study the feasibility of implementing the initial considerations: a) the presumption of approximating a target function f using the proposed MNLR mathematical expression, b) analyze the correlation with the other MEVs to apply FA, c) carry out the buy-and-hold strategy and compute the Sharpe ratio for the FA, and d) implement a hybrid K-Means with AHN algorithm to test the AON approach on ITP tasks. Within the results of experiment two, it was remarked the effectiveness of the proposed MNLR model to perform the training phase task; also, the Buy-and-Hold trading strategy was implemented successfully, and a hybrid K-Means with AHN algorithm was applied to the ITP with satisfactory results.

Finally, for the immediate future work, the next tasks consider: i) further experiments would be done including the procedure of splitting the dataset into training and testing subsets to measure the performance of the forecasting model on unseen data, ii) a Long Short-Term Memory (LSTM) Network will be implemented to compare its performance to the present results since LSTM is a commonly used method for the current task, iii) to continue with the design of a new algorithm (topology) based on the AON machine learning class, capable of performing the stated objectives.

References

1. Billings, S.A.: Nonlinear System Identification, NARMAX Methods in the Time, Frequency, and Spatio-Temporal Domains, 1st edn. Wiley, Chichester (2013)
2. Chacón, H., et al.: Improving financial time series prediction accuracy using ensemble empirical mode decomposition and recurrent neural networks. IEEE Access **8**, 117133–117145 (2020)
3. Chapra, S.C.: Numerical Methods for Engineers, 4th edn. McGraw-Hill, Singapore (2003)
4. Dunis, C.L., et al.: Artificial Intelligence in Financial Markets, Cutting-Edge Applications for Risk Management, Portfolio Optimization and Economics. Palgrave Macmillan, London (2016)
5. Elliott, G., et al.: Handbook of Economic Forecasting, 1st edn. Elsevier Ltd., Amsterdam (2013)
6. Focardi, S.M., et al.: The Mathematics of Financial Modeling and Investment Management. Wiley, Hoboken (2004)
7. Hou, X., et al.: An enriched time-series forecasting framework for long-short portfolio strategy. IEEE Access **8**, 31992–32002 (2020)
8. Hu, Y.J., et al.: Deep reinforcement learning for optimizing finance portfolio management. In: 2019 Amity International Conference on Artificial Intelligence (AICAI), Dubai, United Arab Emirates. IEEE (2019)
9. Miranda, M.J., et al.: Applied Computational Economics and Finance. The MIT Press, Cambridge (2002)
10. Murphy, J.J.: Technical Analysis Financial Markets. New York Institute of Finance (1999)
11. Ordóñez, J.M.: Predicción del comportamiento de los mercados bursáitiles usando redes neuronales. Technical report, Universidad de Sevilla, Sevilla, España (2017)
12. Ortiz, F., et al.: Pronóstico de los índices accionarios dax y s&p 500 con redes neuronales diferenciales. Contaduría y administración **58**, 203–225 (2013)

13. Pedregosa, F., et al.: Scikit-learn: machine learning in Python. J. Mach. Learn. Res. **12**, 2825–2830 (2011)
14. Ponce, H.: A new supervised learning algorithm inspired on chemical organic compounds. Ph.D. thesis, Instituto Tecnológico y de Estudios Superiores de Monterrey, Mexico, December 2013
15. Ponce, H., et al.: Artificial Organic Networks: Artificial Intelligence Based on Carbon Networks, 1st edn. Springer, Heidelberg (2014). https://doi.org/10.1007/978-3-319-02472-1
16. Ponce, H., Miralles-Pechúan, L., de Lourdes Martínez-Villaseñor, M.: Artificial hydrocarbon networks for online sales prediction. In: Lagunas, O.P., Alcántara, O.H., Figueroa, G.A. (eds.) MICAI 2015. LNCS (LNAI), vol. 9414, pp. 498–508. Springer, Cham (2015). https://doi.org/10.1007/978-3-319-27101-9_38
17. Ponce, H., et al.: A novel wearable sensor-based human activity recognition approach using artificial hydrocarbon networks. Sensors **16**(7), 1033 (2016)
18. Ponce, H., et al.: Interpretability of artificial hydrocarbon networks for breast cancer classification. In: 2017 International Joint Conference on Neural Networks (IJCNN). IEEE (2017)
19. Ponce, H., Acevedo, M.: Design and equilibrium control of a force-balanced one-leg mechanism. In: Batyrshin, I., Martínez-Villaseñor, M.L., Ponce Espinosa, H.E. (eds.) MICAI 2018. LNCS (LNAI), vol. 11289, pp. 276–290. Springer, Cham (2018). https://doi.org/10.1007/978-3-030-04497-8_23
20. Ponce, H., et al.: Doubly fed induction generator (DFIG) wind turbine controlled by artificial organic networks. Soft. Comput. **22**, 2867–2879 (2018). https://doi.org/10.1007/s00500-017-2537-3
21. Ponce, H., Martínez-Villaseñor, M.L.: Versatility of artificial hydrocarbon networks for supervised learning. In: Castro, F., Miranda-Jiménez, S., González-Mendoza, M. (eds.) MICAI 2017. LNCS (LNAI), vol. 10632, pp. 3–16. Springer, Cham (2018). https://doi.org/10.1007/978-3-030-02837-4_1
22. Ponce, H., Acevedo, M., Morales-Olvera, E., Martínez-Villaseñor, L., Díaz-Ramos, G., Mayorga-Acosta, C.: Modeling and control balance design for a new bio-inspired four-legged robot. In: Martínez-Villaseñor, L., Batyrshin, I., Marín-Hernández, A. (eds.) MICAI 2019. LNCS (LNAI), vol. 11835, pp. 728–739. Springer, Cham (2019). https://doi.org/10.1007/978-3-030-33749-0_58
23. Ponce, H., González-Mora, G., Morales-Olvera, E., Souza, P.: Development of fast and reliable nature-inspired computing for supervised learning in high-dimensional data. In: Rout, M., Rout, J.K., Das, H. (eds.) Nature Inspired Computing for Data Science. SCI, vol. 871, pp. 109–138. Springer, Cham (2020). https://doi.org/10.1007/978-3-030-33820-6_5
24. Ponce, H., et al.: Stochastic parallel extreme artificial hydrocarbon networks: an implementation for fast and robust supervised machine learning in high-dimensional data. Eng. Appl. Artif. Intell. **89**, 103427 (2020)
25. Ruttiens, A.: Mathematics of Financial Markets, Financial Instruments and Derivatives Modeling, Valuation and Risks Issues, 1st edn. Wiley, Chichester (2013)
26. Salvatore, D., et al.: Statistics and Econometrics, 2nd edn. McGraw-Hill, New York (2002)
27. Seabold, S., et al.: Statsmodels: econometric and statistical modeling with Python. In: 9th Python in Science Conference (2010)
28. Sheta, A.F., et al.: Evolving stock market prediction models using multi-gene symbolic regression genetic programming. Artif. Intell. Mach. Learn. (AIML) **15**(1), 11–20 (2015)

29. Soler-Dominguez, A., et al.: A survey on financial applications of metaheuristics. ACM Comput. Surv. **50**, 1–23 (2017)
30. Stoean, C., et al.: Deep architectures for long-term stock price prediction with a heuristic-based strategy for trading simulations. PLoS ONE **14**, e0223593 (2019)
31. Virtanen, P., et al.: SciPy 1.0: fundamental algorithms for scientific computing in Python. Nat. Methods **17**, 261–272 (2020)
32. Zheng, X., et al.: Stock Market Modeling and Forecasting: A System Adaptation Approach, 1st edn. Springer, London (2013). https://doi.org/10.1007/978-1-4471-5155-5

Supervised Learning Approach for Section Title Detection in PDF Scientific Articles

Gustavo Bartz Guedes[1,2]([envelope]) [iD] and Ana Estela Antunes da Silva[1] [iD]

[1] University of Campinas, Limeira SP 13484-332, Brazil
`aeasilva@unicamp.br`
[2] Federal Institute of São Paulo, São Paulo SP 01109-010, Brazil
`gubartz@ifsp.edu.br`
`https://www.unicamp.br`
`https://www.ifsp.edu.br`

Abstract. The majority of scientific articles is available in Portable Document Format (PDF). Although PDF format has the advantage of preserving layout across platforms it does not maintain the original metadata structure, making it difficult further text processing. Despite different layouts, depending on the applied template, articles have a hierarchical structure and are divided into sections, which represent topics of specific subjects, such as methodology and results. Hence, section segmentation serves as an important step for a contextualized text processing of scientific articles. Therefore, this work applies binary classification, a supervised learning task, for section title detection in PDF scientific articles. To train the classifiers, a large dataset (more than 5 millions samples from 7,302 articles) was created through an automated feature extraction approach, comprised by 17 features, where 4 were introduced in this work. Training and testing were made for ten different classifiers for which the best *F1 score* reached 0.94. Finally, we evaluated our results against CERMINE, an open-source system that extracts metadata from scientific articles, having an absolute improvement in section detection of 0.19 in *F1 score*.

Keywords: Scientific article segmentation · Section title detection · Text segmentation · Supervised learning

1 Introduction

The amount of scientific articles available electronically on the internet is constantly growing. As an example, IEEE Explorer database aggregates more than 195 journals and 1,400 conferences, with approximately 20,000 new documents added monthly [10].

Most scientific articles is in Portable Document Format (PDF). The advantage of this format is that it preserves the formatting of the document across platforms, since texts and figures are displayed according to fixed coordinates values within each page [1]. However, extracting information is a challenge, since there is little or none metadata regarding the text organization.

© Springer Nature Switzerland AG 2021
I. Batyrshin et al. (Eds.): MICAI 2021, LNAI 13067, pp. 44–54, 2021.
https://doi.org/10.1007/978-3-030-89817-5_3

Scientific articles, although differ in format depending on the applied template, have a hierarchical structure divided into sections. Article segmentation is an important step towards further contextualized text based analysis, specially to perform text mining.

This work used a supervised machine learning approach, through a classification task, for detecting section titles in scientific articles in PDF format files. In total, we trained ten classification algorithms, where the best one reached a *F1 score* of 0.94. Finally, in order to compare our results we used the Content ExtRactor and MINEr (CERMINE) tool [12], which solves a similar problem.

This paper is organized as follows: Sect. 2 presents the related works; the proposed methodology is described in Sect. 3; results are presented in Sect. 4; and Sect. 5 concludes the paper and indicates future works.

2   Related Works

Since PDF files became the standard format to publish scientific articles, a variety of studies for metadata extraction has been conducted. Mainly, the works can be divided into two categories: the ones that use rule-based approaches and the ones which use machine learning approaches.

Rule-based approaches use a predetermined set of rules, defined by analyzing the documents and identifying patterns in advance. Although they can achieve a good performance, they are highly dependent on the specific document layouts from which the rules were derived. Also, identifying and defining the rules beforehand is a time-consuming task [12]. On the other hand, machine learning approaches offer more flexibility and adaptability considering a scenario with a heterogeneous set of document layouts. For this reason, we chose to present related works that use machine learning.

Automated metadata extraction from scientific articles was used by Kovacevic et al. [4] to support the scientific production monitoring system of the University of Novi Sad. The work extracted eight categories of metadata by analyzing the first page of articles. Four standard classification algorithms where used: Decision Tree, Naive Bayes, k-nearest Neighbours (Knn) and Support Vector Machine (SVM).

Kovacevic et al. experiments resulted in the definition of eight separated SVM models, one for each category of metadata. The result was a F1-score of 0.85 for almost all of the classifiers [4].

Tkaczyk et al. [12] created the Content ExtRactor and MINEr (CERMINE), an open-source system for extracting structured metadata from scientific articles in PDF format. Authors used machine learning, both supervised and unsupervised approaches, to segment parts of the text into zones. A zone "is a consistent fragment of the document's text, geometrically separated from surrounding fragments and not divided into paragraphs or columns". CERMINE model was trained using 2,500 documents from the GROund Truth for Open Access Publications (GROTOAP2) database, an heterogeneous dataset comprised by 208 publishers and 13,210 documents [11].

CERMINE used a SVM model that achieved an average *F1 score* of 77.5% in zone classification. According to the authors, the extraction of structured full text of sections and subsections was in an experimental phase.

The work of Budhiraja and Mago [2] explored supervised learning to detect headings in course outline documents. The dataset was generated from 500 PDF documents converted to the HyperText Markup Language (HTML). Each sample was manually labeled. Nine classifiers were trained and tested, where the best performance was reached by the Decision Tree Classifier, with a 0.97 *F1-Score*.

In general, the constraints of the related works are: (i) Kovacevic et al. [4] solution extracts metadata only from the first document page and does not detect section titles; (ii) Tkaczyk et al. [12] work is a general tool for zone classification that uses a subset of 2,500 files from the GROTOAP2 dataset; (iii) Budhiraja and Mago [2] applied supervised learning in a distinct context, that is, for heading course outline detection, thus, not for scientific section title detection. All solutions convert PDF files to HTML as an intermediate step for feature extraction.

Finally, the following features distinguish our approach from related works:

- Creation of a larger labeled dataset, considering section titles that spans multiple rows;
- A different set of training features, introducing four new ones;
- Automated feature extraction directly from PDF files, with no need for intermediate format conversion;
- Training and testing using a deep learning model.

3   Methodology

The methodology used in this work comprises three main tasks: (i) creation of a dataset via automated feature extraction from PDF files; (ii) training and testing of ten classifiers and their subsequent results comparison; and (iii) evaluate our results against CERMINE. Figure 1 shows a visual representation of the methodology.

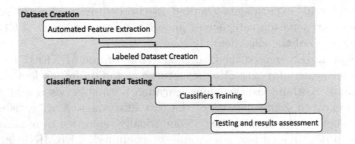

Fig. 1. Visual representation of methodology

3.1 Dataset Creation

We used GROund Truth for Open Access Publications (GROTOAP2) [11] in order to get the PDF files, since its heterogeneity provides generality for classifiers training. The dataset is a subset of PubMed Central Open Access [14] papers and provides a script to download the scientific articles packages, each package contains a PDF file with the corresponding metadata file in nxml format.

A nxml is a file in Extensible Markup Language (XML) format that uses the Journal Article Tag Suite (JATS), whose Document Type Definition (DTD) definition is available online [6]. Three tags are related to section titles:

- body: defines the begin and end of the article body text;
- sec: defines an article section;
- title: defines a section title when hierarchically positioned within a *sec* tag.

GROTOAP2 has 13,210 documents, however the following issues aroused when downloading files from PubMed Central Open Access:

- 15 packages were not present;
- 26 PDFs were missing from packages. Therefore, separately downloads were made directly from the journals websites. One PDF file was not available;
- 260 nxml files did not have section titles, that is a title tag within a body tag.

Therefore, we end up with 12,934 consistent packages for further processing.

Next, each PDF file was parsed with PyMuPDF [5], a modern Python library for PDF viewer. PyMuPDF parses each page and returns text data and metadata in a Python dictionary format.

Each PDF row was defined as a data sample. In order to automatically label then, we used a set of rules for string match using regular expressions. The match was made comparing each PDF text row with the sections contained in the corresponding nxml metadata file. The proposed rules also assured the detection of titles that spanned multiples rows.

Subsequently, from the 12,934 PDF files, we checked the total number of existing titles on nxml files against the total number of detected section titles with the proposed rules. We removed the files where the totals differ. This was necessary to ensure the correct labeling throughout each entire file. In this way, the final dataset was comprised by 7,302 PDF files of 967 journals with 5,125,562 samples, where 2,04% of samples were section titles (positive samples) and the remaining non section titles (negative samples).

Instead of using class balancing, we opted for another approach that is to adjust the imbalance through parameter settings. This option was based on the work of Weiss and Provost [15] who affirms that "results further show that as the amount of training data increases the differences in performance for different class distributions lessen (for both error rate and AUC), indicating that as more data becomes available, the choice of marginal class distribution becomes less and less important – especially in the neighborhood of the optimal distribution". Also, the use of parameters to adjust the imbalance avoids the need to artificial data augmentation. The parameters settings is mentioned in Sect. 3.2.

We divided the extracted features into *formatting* and *text-specific* categories. Figure 2 shows a PDF excerpt with an example of the corresponding *formatting* features values as provided by PyMuPDF and Fig. 3 shows the values for the corresponding *text-specific*.

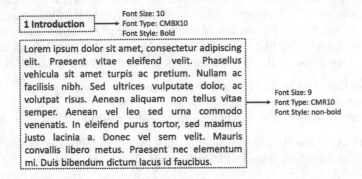

Fig. 2. PDF excerpt with formatting features

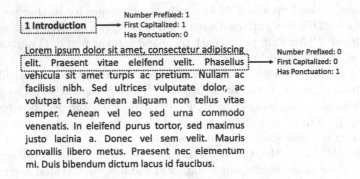

Fig. 3. PDF excerpt with text-specific features

Table 1 describes the *text-specific* features and Table 2 the *formatting* features ones. In total there are 17 features, extracted from each PDF file row, where *most_frequent_font*, *font_magnitude*, *font_variety* and *roman_number_prefixed* were introduced by this work. The remaining features are also present in the related works of Sect. 2.

3.2 Classifiers Training and Testing

Samples from the dataset were splitted in 70% for training and 30% for testing in a stratified manner. Next, both training and test datasets were standardize

Table 1. Text-specific features

Feature	Description
roman_number_prefixed	Row starts with a roman number
word_count	The number of words in the row
char_count	The number of characters in the row
first_capitalized	Row starts with a capital letter
all_words_capitalized	Row contains all words with capital letters
number_prefixed	Row starts with a cardinal number
number_count	Total of numbers characters present in row
has_punctuation	Indicates if the row contains a full stop or question mark

Table 2. Formatting features

Feature	Description
most_frequent_font	Indicates if the row contains the most frequent font in the document
font_magnitude	Sequentially assigned number in respect to the font sizes present in the document. If more than one size was present on the row, the most frequent used was assigned
font_variety	Quantity of different fonts used in the row
above_font_threshold	Indicates whether the font size used in the row is above a threshold. The threshold is defined as the most frequent font size used in the document
all_italic	Row contains all words in italic
all_bold	Row contains all words in bold
font_size	The size of the font used in the row or the mean if more than one size is present
abs_left	The absolute position of the row relative to the left edge of the document page
abs_top	The absolute position of the row relative to the top edge of the document page

using scikit-learn StandardScaler [7], that uses Eq. 1, where u is the mean of the training samples and s is the standard deviation of the training samples.

$$z = \frac{x - u}{s} \tag{1}$$

Testing on each classifier was measured in terms of Recall, Specificity, Precision, Accuracy and F1 score. We chose F1 score as the main measure, since it is the harmonic mean of Precision and Recall. In this way, in terms of measurement, we prioritize section title detection in an imbalanced dataset, where the generated models are more likely to detect false positives.

For further evaluation, specially to detect overfitting, a stratified cross-validation with 10 folds was applied for each algorithm. We chose the stratified cross-validation because it keeps the original class distribution in each partition, which is adequate for our imbalanced dataset [8].

Lastly, we ran CERMINE in the same PDF file set and compared the results with our generated models.

This work trained the classifiers present in the related works of Sect. 2, except for the Support Vector Machine (SVM), due to its limitation for large datasets (explained ahead) and the deep learning TensorFlow model. Next, we present the ten classifiers along with the training parameters used for each one.

Decision Tree is a flow-chart like tree structure. Each node represents a test on a feature value, each branch represents an outcome of the test and leaves represent classes [3]. As for the parameters, *gini index* was used as a *measure of quality for the splits*; the *minimum number of samples* to split internal nodes was set to 2; the *maximum depth* of trees was set so all leaves were pure.

Random Forest uses a set of decision trees where the final label is assigned in an ensemble manner. The same parameters of the decision tree were used with a total number of trees (*estimators*) of 100.

TensorFlow is a high-level API for creating and training deep learning models. A sequential model was defined composed by *four layers* with 64 neurons, using the *Rectified Linear Activation Function* (ReLU). The last layer is composed by one neuron and a sigmoid activation function for classification; *Adam optimization algorithm* and *Binary Crossentropy* as the loss function were set. The *batch size* was set to 2,048 to increase the chance of each one to contain positive samples (section titles), therefore, impacting the learning process at each iteration. Finally, we defined 1,000 *epochs* with an early stopping based on the loss function results in the validation set.

k-Nearest Neighbor uses similarity to assign classes, which is done by sample comparison according to a distance metric. The *number of neighbors* was set to 3; the *weights* used in prediction were set to the inverse of their distance, thus, increasing the weight of closer neighbors. Finally, Minkowski function was used as the *distance* metric.

Multi-layer Perceptron (MLP) is an artificial neural network of the feed-forward class. The following parameters were used: one *hidden layer* with 100 neurons; the *maximum number of iterations* of 700; a *batch size* of 2,048, for the same reason as TensorFlow.

Logistic Regression is a linear classifier that uses a logistic function to assess the probability when assigning a predicted class. *Penalty* was set to l2 with a *maximum number of iterations* of 100.

Gaussian Naive Bayes uses a statistical approach, based on Bayesian theorem, to assign the class. We set the *prior probabilities of the classes* as 0.9 for the negative classes and 0.1 for the positives, in order to reflect samples distribution.

Quadratic Discriminant Analysis uses a quadratic decision boundary when assigning classes and assumes a Gaussian distribution. Therefore, *prior probabilities of the classes* were also 0.9 for the negative classes and 0.1 for the positives as in Gaussian Naive Bayes.

The **Support Vector Machine** (SVM) classifier "uses multi-dimensional hyperplanes to make classification" [2]. As stated by scikit-learn documentation "the fit time scales at least quadratically with the number of samples and may be impractical beyond tens of thousands of samples." [9]. Therefore, we used the suggested Stochastic Gradient Descent and Linear Support Vector as an alternative to our large dataset.

Linear Support Vector is similar to a SVM with linear *kernel* parameter and it is also implemented with a different library, which is suited for a larger dataset. *Penalty* was set to l2 with a *maximum number of iterations* of 1,000.

Stochastic Gradient Descent optimizes an objective function in an iterative manner. *Penalty* was set to l2 with a *maximum number of iterations* of 1,000.

4   Results

This section presents and discuses the testing results after applying the methodology described in Sect. 3. Table 3 presents the test results measures, ordered from the highest *F1 score*. The *F1 mean* column refers to the stratified cross-validation results.

From the collected results, the top five classifiers, considering both *F1 score* and *F1 mean* measures were: Random Forest, Decision Tree, TensorFlow, k-Nearest Neighbor and MLP Neural Net. In general all classifiers had a high specificity, including those with a poor *F1 score*, this is expected since some classifiers are more sensible to imbalanced data.

Table 3. Classifiers testing results

Classifier	Recall	Specificity	Precision	F1 score	Accuracy	F1 Mean
Random Forest	0.929	0.999	0.964	0.946	0.998	0.948
Decision Tree	0.924	0.998	0.923	0.923	0.997	0.926
TensorFlow	0.905	0.998	0.904	0.904	0.996	0.909
k-Nearest Neighbor	0.884	0.998	0.896	0.890	0.996	0.896
MLP Neural Net	0.875	0.996	0.832	0.853	0.994	0.860
Stochastic Gradient Descent	0.511	0.995	0.693	0.588	0.985	0.580
Logistic Regression	0.495	0.995	0.682	0.574	0.985	0.572
Linear Support Vector	0.444	0.996	0.718	0.549	0.985	0.546
Quadratic Discriminant Analysis	0.869	0.955	0.286	0.430	0.953	0.427
Gaussian Naive Bayes	0.855	0.952	0.269	0.410	0.950	0.409

For further evaluation we submitted the same PDF file set (7,302 files) for CERMINE classification, we used version 1.13 [13]. The results for section title classification were a *Precision* of 0.756 and a *Recall* of 0.746, thus, a *F1 score* of 0.751. This score is in consonance with the results from Tkaczyk et al. [12], where the average *F1 score* for zone classification was 0.775 [12].

An issue observed with CERMINE is that it does not have a good performance when identifying section titles that spans multiple rows. This is caused by its automated feature extraction algorithm, when the training dataset was generated. Instead, our approach uses a set of regular expression rules. In order to compare our results in detecting section titles with row spans, we created a subset dataset keeping only the samples that span multiple rows. From the 11,908 samples, CERMINE correctly classified only 5,447, thus a total of 45.74%. In our approach we reached a total of 90.89% correctly classified.

It is important to note that CERMINE is a general solution since it works with zone classification. Also, its model was trained using a subset of 2,500 files, while we used a broader set of 7,302 files for training our models.

Another aspect we investigated was feature importance, which is the impact of each feature in class prediction. Table 4 presents the feature importance as provided by the Random Forest classifier. The top 5 features were: *font_size*, *abs_left*, *char_count*, *font_magnitude* and *abs_top*. *Font_magnitude*, is a feature that was introduced in this work, and appears as the fourth most significant in class prediction.

Table 4. Feature importance

Feature	Importance
font_size	0.2329
abs_left	0.1138
char_count	0.1097
font_magnitude	0.1097
abs_top	0.0780
all_bold	0.0667
word_count	0.0636
most_frequent_font	0.0512
above_font_threshold	0.0360
all_words_capitalized	0.0326
first_capitalized	0.0307
all_italic	0.0255
number_count	0.0226
has_ponctuation	0.0143
number_prefixed	0.0069
font_variety	0.0050
roman_number_prefixed	0.0009

5 Conclusion

This work presented the use of supervised learning approach for section title detection in scientific articles in PDF format.

Our work provides an automated feature extraction approach directly from PDF files and uses a large layout heterogeneous dataset, which was used for training ten different classifiers. We also analyzed feature importance, where the fourth most relevant feature for detecting section titles was introduced in this work. Finally, for further evaluation, we submitted our dataset to CERMINE classification for a baseline comparison. While CERMINE had a *F1 score* of 0.751, the best classifier of this work reached a *F1 score* of 0.946.

A supervised approach has more adaptability to treat different PDF file layouts. Furthermore, if new layouts become available, it is possible to complement or create a new dataset by our automated feature extraction approach and re-train the classifiers.

For future works, we will add new rules to the automated feature extraction in order to create a broader dataset. Also, we intend to explore hyperparameter values as well as other deep learning models.

Acknowledgements. This study was financed in part by the Coordenação de Aperfeiçoamento de Pessoal de Nível Superior - Brasil (CAPES) - Finance Code 001. Also, the support of the Intel DevCloud program. Finally, we thank students Felipe Favaro Müller and Pedro Artico Rodrigues for their assistance.

References

1. Adobe Systems: PDF Reference: Adobe Portable Document Format Version 1.7. Adobe Systems, 6 edn (2006)
2. Budhiraja, S., Mago, V.: A supervised learning approach for heading detection. Exp. Syst. **37**, e12520 (2020)
3. Han, J., Pei, J., Kamber, M.: Data Mining: Concepts and Techniques. The Morgan Kaufmann Series in Data Management Systems, Elsevier Science (2011), https:// books.google.com.br/books?id=pQws07tdpjoC
4. Kovacevic, A., Ivanovic, D., Milosavljević, B., Konjovic, Z., Surla, D.: Automatic extraction of metadata from scientific publications for cris systems. Prog. Electr. Libr. Inf. Syst. **45**, 376–396 (2011). https://doi.org/10.1108/00330331111182094
5. McKie, M., Liu, R.: Pymupdf - python binding for mupdf (2020). https://pypi. org/project/PyMuPDF/
6. National Library of Medicine: Journal archiving and interchange tag suite (2008). https://www.ncbi.nlm.nih.gov/pmc/tools/openftlist/
7. Pedregosa, F., et al.: Scikit-learn: machine learning in Python. J. Mach. Learn. Res. **12**, 2825–2830 (2011)
8. Santos, M.S., Soares, J.P., Abreu, P.H., Araujo, H., Santos, J.: Cross-validation for imbalanced datasets: avoiding overoptimistic and overfitting approaches [research frontier]. IEEE Comput. Intell. Mag. **13**(4), 59–76 (2018). https://doi.org/10.1109/ MCI.2018.2866730
9. Scikit-Learn: sklearn.svm.svc - scikit-learn 0.23.2 documentation (2020). https:// scikit-learn.org/stable/modules/generated/sklearn.svm.SVC.html
10. The Institute of Electrical and Electronics Engineers: IEEE Xplore. https:// ieeexplore.ieee.org/
11. Tkaczyk, D., Szostek, P., Bolikowski, L.: Grotoap2 The methodology of creating a large ground truth dataset of scientific articles. D-Lib Mag. **20** (2014). https:// doi.org/10.1045/november14-tkaczyk

12. Tkaczyk, D., Szostek, P., Fedoryszak, M., Dendek, P.J., Bolikowski, L.: Cermine: automatic extraction of structured metadata from scientific literature. Int. J. Doc. Anal. Recog. (IJDAR) **18**(4), 317–335 (2015). https://doi.org/10.1007/s10032-015-0249-8

13. Tkaczyk, D., Szostek, P., Dendek, P., Fedoryszak, M., Bolikowski, L.: Github - content extractor and miner (2017). https://github.com/CeON/CERMINE

14. U.S. National Institutes of Healths National Library of Medicine: Open access subset (2019). https://www.ncbi.nlm.nih.gov/pmc/tools/openftlist/

15. Weiss, G.M., Provost, F.: Learning when training data are costly: the effect of class distribution on tree induction. J. Artif. Int. Res. **19**(1), 315–354 (2003). Oct

Real-Time Mexican Sign Language Interpretation Using CNN and HMM

Jairo Enrique Ramírez Sánchez(✉) ⓘ, Arely Anguiano Rodríguez(✉) ⓘ, and Miguel González Mendoza(✉) ⓘ

Technological Institute of Monterrey, School of Engineering and Sciences, Atizapán de Zaragoza, Mexico
{A01750443,A01752068}@itesm.mx, mgonza@tec.mx

Abstract. Mexican Sign Language (MSL) is the primary form of communication for the deaf community in Mexico. MSL has a different grammatical structure than Spanish; furthermore, facial expression plays a determining role in complementing context-based meaning. This turns it difficult for a hearing person without prior knowledge of the language to understand what is to be transmitted, representing an important communication barrier for deaf people. In order to face this, we present the first architecture to consider facial features as indicators of grammatical tense to develop a real-time interpreter from MSL to written Spanish. Our model uses the open source MediaPipe library to extract marks from the face, body position and hands. Three 2D convolutional neural networks are used to encode individually and extract patterns, the networks converge to a multilayer perceptron for classification. Finally, a Hidden Markov Model is used to morphosyntactically predict the most probable sequence of words based on a preloaded knowledge base. From the experiments were carried out, a precision of 94.9% was obtained with $\sigma = 0.07$ for the recognition of 75 isolated words and 94.1% with $\sigma = 0.09$ for the interpretation of 20 sentences in MSL in a medical context. Being an approach based on camera inputs and observing that even with a few samples an adequate generalization can be achieved, it would be feasible to scale our architecture to other sign languages and offer possibilities of efficient communication to millions of people with hearing disability.

Keywords: Mexican sign language · Convolutional neural networks · Hidden Markov models · Real time interpreter

1 Introduction

In Mexico, there are 2.3 million people with hearing impairment (PHI) according to data from the National Institute of Statistics and Geography for Information and the National Council for the Prevention of Discrimination 2020. The official language of the deaf community in Mexico is Mexican Sign Language (MSL), highlighting that each country has a national sign system designated for its territory.

I. Batyrshin et al. (Eds.): MICAI 2021, LNAI 13067, pp. 55–68, 2021.
https://doi.org/10.1007/978-3-030-89817-5_4

In order to achieve effective and efficient communication between a PHI and a Hearing Person (HP), it is not enough to learn to identify the hand movements to designate each of the words, since manual features (MF) represent only 20% of the total communication, while non-manuals features (NMF) figure as the rest [1]. Therefore, it is also necessary to have knowledge of its grammar and non-manual features, in particular, facial features, since these are the ones that indicate the verb tense in which the action was, is being or will be carried out. The **past tense** is expressed through a semblance that the action has been consummated, in this one a relaxed expression can be observed, where the lower lip protrudes from the upper one nodding softly with the head; on the other hand, in the **present tense**, a neutral expression is projected, slightly raising the eyebrows and in accordance with the message; finally, the **future tense** is manifested through an expression of doubt or thought, where the neck is slightly rotated and the eyes are directed towards either of the two upper corners of the visual plane. Regarding MSL grammar, the verbs 'ser' and 'estar' are omitted, so only the noun and adjective are expressed or, failing that, noun and place. Likewise, it is relevant that the movement of each sign is made with the fingers in the position of the letter with which the word assigned to the sign begins, or else, highlighting sensory characteristics of the word in question. As can be seen, for a person not immersed in an environment and without previous contact with the deaf community, it is extremely difficult to reach a full understanding of what the PHI seeks to communicate. This, along with the fact that in Mexico there are only 42 certified interpreters in MSL [2], makes communication for PHI practically closed. There is vast research in American Sign Language (ASL), British Sign Language (BSL), Chinese Sign Language (CSL), among others; however, the incursion into the study of MSL is more than limited.

In our research, we will address the analysis and recognition of signs in real time using convolutional neural networks for sign classification taking into account three aspects: facial expression, hand movements and body position; the interpretation is enhanced by a Hidden Markov Model for enrichment with context and grammar.

This article is organized as follows: Sect. 2 addresses the review of previous work on real-time sign language recognition. Section 3 contains the description of our proposed model as well as the processing phases. Section 4 explains how the data acquisition, the number of participants, and the testing conditions were performed. Section 5 presents the proposed experiments and their respective results. Finally, Sect. 6 six discusses the conclusions and future work.

2   Related Work

2.1   Methods

Studies on sign language can be divided into three approaches: based on camera input [3], external sensors [4,5], and inputs by devices such as specialized gloves [6,7]. In order, the first type of approach offers the advantage of being easy and cheap to implement at the cost of requiring large amounts of data to generalize different hand sizes and skin colors. Second, those based on external sensors (commonly, Microsoft Kinect sensor) present an improvement in generalization, however, they increase costs and implementation complexity. Finally, gloves tend to be the pinnacle of generalization and are also the most difficult to implement, as well as being considered an intrusive method.

2.2 Techniques

To achieve movement classification in sign languages, several approaches have been proposed, some of them based on **Hidden Markov Models** (HMM), such as [8], who use these statistical models to predict the most probable manual gesture of American Sign Language.

Data Time Wraping (DTW) algorithms present an improvement in classification since, when comparing against all stored movements, they are able to generalize signs independently of the time duration. In [9] an accuracy of 98.57% was achieved for 20 words of Mexican sign language; however, due to the nature of DTW, as the dataset increases, the computation time grows polynomially.

Finally, **Convolutional Neural Networks** (CNN) have been shown to perform well in real-time classification of human actions in general [10]. The advantage of CNN over other methods is the wide generalization capability, the convolution layers extract the most important features by identifying patterns in the images and the Fully Connected layers perform the classification. In [11] the VGG Network model is used to classify 26 letters of the American dactylological alphabet reaching an accuracy of 98.56%. In [12] an architecture is proposed that combines an image of the signer with the extract of his body and hand position made by a MS Kinect, which achieves an accuracy of 94.2% for 25 separate words in ASL. Additionally, in [13] a solution for CSL is presented, an architecture that combines convolutional coding with processing through a recurrent neural network with attention module, which used more than 25 thousand videos created by 50 interpreters for training, reaching an accuracy of 82.7%.

2.3 Works About MSL in Mexico

In Mexico there is little work related to national sign language, focusing mainly on the static recognition of letters of the alphabet [14–19]. In [20] the identification of 15 isolated words using a MS kinect to identify body points is addressed; AdaBoost algorithm performed the temporal classification of each one of them. On the other hand, [9] presents an approach to the recognition of 20 words with the DTW algorithm obtaining an average accuracy of 98.57%, however, it becomes unfeasible to scale the dataset due to the computation time.

Finally, only [21] addresses the interpretation of 5 sentences in real time by processing manual and body position using geometric moments, these outputs are integrated by a HMM reaching a sensitivity of 86%.

In our work, we propose the first architecture for MSL that uses facial expression as a verb tense determinant based on camera inputs with the ability to morphosyntactically interpret **75 words** in real time, whose combination allows us to generate more than **50 sentences**, of which, the performance of **20 sentences in the medical context** is analyzed in this study. To achieve this, a dataset of 49 words was generated (each of the 13 verbs included was conjugated in the grammatical tenses, giving rise to three classes for each one) with an average of 35 samples for each, signed by six volunteers. Additionally, a knowledge base with 100 of the common MSL sentences was obtained, which provides the patterns on the grammatical structure itself.

3   Proposal

The processing model presented is divided into 5 stages as shown in the Fig. 1.

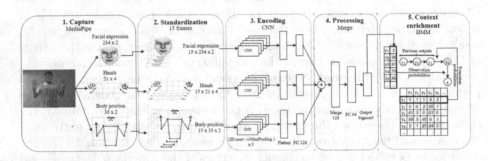

Fig. 1. Detection stages.

The capture uses the open source library *MediaPipe* presented in [22]. This library allows to build perception flows with *OpenCV* visualization tools [23]. Both allow the detection of marks on hands, body and face in real time with a wide generalization capability regardless of skin color, hand size and height of the person. Thus, the capture system blends the benefits of camera inputs (economic feasibility) and external sensors (generalization) in a non-intrusive approach.

For standardization, the coordinates of the marks are extracted to be placed in three matrices, selecting 15 frames per second. The coordinates of the facial expression with dimension 438×2, the hands 21×4 and the body position 33×2.

Subsequently, the matrices are independently encoded with a neural network with three layers of 2D convolution each one followed by MaxPooling. This performs the identification of general patterns, allowing similar signs to be encoded in a similar way. Despite the initial dimensions of the three matrices, each CNN reduces it to a Flatten vector with 128 entries in order to be proper merged in the next stage.

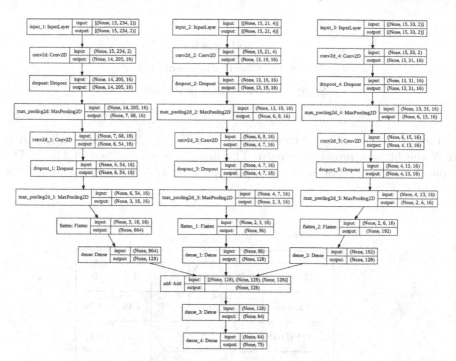

Fig. 2. Configuration for CNN and Multilayer perceptron (stages three and four).

The encodings are integrated by an addition layer. Hands coding provide meaning, body position spatial location and the face the verbal tense. In this last part of the architecture a multilayer perceptron is used to perform the classification. The last layer of the network uses a sigmoid function as activation, which assigns an servation probability vector O_i whose entry O_i^j refers to the word j in the dictionary of length N for each time step i.

Finally, a HMM is used to perform the classification based on the previous context. A transition matrix T between words of the dataset calculated with 100 common sentences in MSL extracted from [24]. With O_i and T the selection of the most probable word with a morphosyntactic meaning is performed, the set of possible states for each time step is represented by $S = (S_1, S_2, ..., S_I)$, with the value S_1 being the most probable word. Probability is calculated recursively as shown in the following equation:

$$p(O^j, S^j) = \prod_{i=1}^{I} p(O_i^j | S_i^j) p(S_i^j | S_{i-1}^j) \ \forall j \in 1, 2, ..., N \tag{1}$$

Where:

$$I = \text{Total time steps}$$
$$p(O_i^j | S_i^j) = \text{Observation probability}$$
$$p(S_i^j | S_{i-1}^j) = \text{Transition probability}$$

Finally, a manual implementation of the *Viterbi* [25] algorithm is run to track the highest probability path. The process is exemplified in Fig. 3. This implementation reduces the interpretation error by **11.7%** in contrast to isolated word identification.

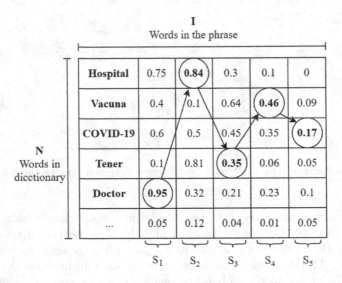

Fig. 3. Exemplification of the Viterbi algorithm applied to the determination of the most probable word sequence according to MSL grammar, in this case: *'Doctor hospital tener vacuna COVID-19'.*

4 Dataset

4.1 Description

The dataset used in the present work was generated with the help of six people using *MediaPipe* and *OpenCV* libraries. See Sect. 4.2 for participant descriptions.

The capture for the hand coordinates consists of 15 frames containing the movement of a matrix of 21 points for each hand (15, 21, 4), 33 for the body (15, 33, 2) and 438 points for the face (15, 438, 2). An average of 35 samples was made for one of the 49 signs (13 verbs and 36 words) and 15 for each facial expression for the three verb tenses used (past, present and future). In total, the classification consisted of $13 \times 3 + 36 = 75$ classes.

Table 1. Used signs

Category	Sign	Hands	Samples	Category	Sign	Hands	Samples
Basic	Hola	One	30		Hacer	Both	40
	Gracias	Both	30		Comer	One	40
	Día	Both	40		Pensar	One	30
	Horario	One	40		Creer	Both	30
People	Niño	One	35		Sentir	One	30
	Hombre	One	35		Ir	Both	40
	Mujer	One	35	Verb	Contagiar	Both	40
	Intérprete	Both	40		Pagar	Both	40
	Adulto mayor	Both	40		Trabajar	Both	40
Question	Qué	One	30		Gustar	One	40
	Quién	Both	40		Poder	One	40
	Dónde	Both	30		Tener	One	40
	Cómo estás	Both	30		Comprar	Both	35
Emergency	Accidente	One	35		Feliz	One	30
	Alergia	One	40		Bien	One	30
	Fiebre	One	35		Mal	One	30
	Vacuna	Both	35	Mood	Enfermo	Both	40
	Doctor	Both	40		Triste	One	30
	Emergencia	Both	40		Enojado	Both	30
	Infección	One	35		Nervioso	Both	40
	Medicina	One	35		Escuela	One	30
	Operación	Both	35	Places	Casa	Both	30
	COVID-19	Both	40		Hospital	Both	35
-	-	-	-		Calle	Both	30

4.2   Participants

Six volunteers with knowledge of MSL aged from 18 to 57 years participated. All participants were informed of the purpose of the investigation and gave their consent. The sex, level of knowledge, age and capture type are presented in Table 2.

4.3   Data Acquisition

To obtain the data, a code was developed in the Python programming language to perform the identification of points in real time. When the user was ready,

Table 2. Participants description

Number	Sex	Age	MSL knowledge	Capture
1	Male	57	Basic	Manual and facial
2	Female	35	Intermediate	Manual and facial
3	Male	39	Basic	Manual and facial
4	Male	19	Basic	Manual and facial
5	Male	20	Basic	Manual
6	Female	18	Intermediate	Manual

the function was run and at the end of the capture the results were stored. The participants were placed 1.2 m away from the computer with the setup shown in Fig. 4.

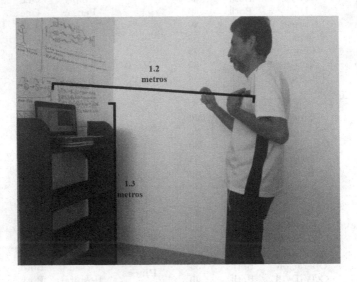

Fig. 4. Data collection setup.

4.4 Dataset Standardization

Due to the particular spatial meaning, each word has a different temporal duration in being signaled. To make the dataset uniform, n = 15 frames distributed over k frames in length were selected for each sample. Keval H. et al. show in [26] that the minimum number of frames per second to identify a human action is 8, so using almost twice as many frames ensures an acceptable level of detail.

5   Experiments and Results

5.1   Training

During the training of the proposed model, the dataset was partitioned into 75% for training (further divided into 10% for validation) and 25% for testing. The model was trained for 40 epochs on GPU supported by Google Collaboratory cloud computing software [27]. The learning curve obtained is shown in Fig. 5.

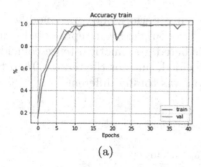 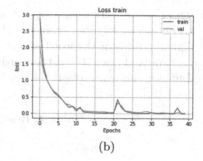

(a)   (b)

Fig. 5. Learning curves for training and validation set using dropout (0.3) as regularization. Training was considered sufficient due to the gradual asymptotic trend for cost reduction.

The F1 Score shown in Eq. 4 was used to measure the performance of the model. The results are shown in the Table 3.

$$Precision = \frac{T_p}{T_p + F_p} \tag{2}$$

$$Recall = \frac{T_p}{T_p + F_n} \tag{3}$$

$$F1Score = 2\frac{precision \cdot recall}{precision + recall} \tag{4}$$

Table 3. Partition set results

Set	F1 score
Training	0.981
Testing	0.978

5.2   Results Experiment 1: Focus on Isolated Words

Each word shown in the Table 1 was signed 10 times by each of the six volunteers, measuring performance in a binary manner, in other words, if it was predicted correctly it scored 100%, otherwise 0%.

In the case of verbs, the way of measuring performance was modified in order to obtain a weighted score based on the complexity it represented; thus, we assigned a 100% if the word and the predicted grammatical tense coincided with the expected one; 70% if the word coincided, but not the tense. 30% if the tense, but not the word. 0% if the prediction was totally different.

Table 4. Results experiment 1.

Category	Average accuracy	σ
Basic	0.987	0.03
People	0.943	0.1
Places	0.937	0.05
Mood	0.99	0.01
Present tense verb	0.91	0.09
Past tense verb	0.94	0.06
Future tense verb	0.89	0.11
Question	0.962	0.07
Emergency	0.981	0.04

An average accuracy for the nine categories of **94.9%** was obtained with $\sigma = 0.07$.

5.3   Results Experiment 2: Focus on Sentences

Participants performed the sign and facial expression of 20 sentences taking five samples. Accuracy was used as a performance metric. The results are shown in the Table 5.

An average accuracy for the 20 sentences was obtained of **94.1%** with $\sigma = \mathbf{0.09}$. In contrast, the interpretation with isolated words, in other words, without context enrichment, reached a **82.4%** with $\sigma = 0.12$.

Table 5. Results experiment 2.

Spanish sentence	MSL sentence	Average Accuracy	σ
El niño está en el hospital porque está enfermo	Niño hospital, enfermo	0.95	0.1
El adulto mayor se sintió mal y fue al hospital	Adulto mayor sentir (pasado) mal, hospital ir (pasado)	0.975	0.05
El adulto mayor se sintió bien con la vacuna	Adulto mayor sentir (pasado) bien vacuna	0.975	0.03
El niño se sintió bien y fue a la escuela	Niño sentir (pasado) bien, escuela ir (pasado)	0.925	0.15
El adulto mayor irá al hospital por la vacuna para el COVID-19	Adulto mayor ir (futuro) hospital COVID-19 vacuna	0.962	0.03
El niño enfermo puede contagiar al doctor	Niño enfermo poder (presente) contagiar (presente) doctor	0.962	0.1
El hombre contagió a la mujer en la casa	Hombre contagiar (pasado) mujer casa	0.937	0.12
El doctor tiene medicina para la operación	Doctor tener (presente) operación medicina	0.987	0.02
El niño estará feliz de ir a la escuela	Niño feliz, escuela ir (futuro)	0.977	0.02
El hombre tiene una infección	Hombre tener (presente) infección	0.8	0.14
Yo pagaré la medicina para el COVID-19	Yo pagar (futuro) COVID-19 medicina	0.83	0.11
El hombre fue a su casa porque se sintió mal	Hombre ir (pasado) casa, sentir (pasado) mal	1.0	0.0
El hombre está nervioso porque la mujer tiene COVID-19	Nervioso hombre, mujer tener (presente) COVID-19	0.825	0.12
La intérprete tuvo una emergencia y fue al doctor	Intérprete tener (pasado) emergencia, ir (pasado) doctor	0.95	0.1
El niño se fue a su casa porque se sintió mal en la escuela	Niño ir (pasado) casa, escuela niño sentir (pasado) mal	0.96	0.04
El hospital tendrá vacunas COVID-19	Hospital tener (futuro) COVID-19 vacuna	0.95	0.1
El doctor está feliz porque la mujer se podrá ir a su casa	Doctor feliz mujer poder (futuro) ir casa	1.0	0.0
¿Qué medicina tiene que comprar el hombre?	¿Medicina hombre tener comprar (presente) qué?	0.95	0.1
El niño tiene fiebre por la infección	Niño tener (presente) infección fiebre	0.862	0.11
El doctor irá a su casa porque se siente enfermo	Doctor casa ir (futuro) sentir (presente) enfermo	0.987	0.025

6 Conclusions

As has been discussed throughout this paper, real-time interpretation of sign languages is a complex process involving several factors such as body position, facial expression, hand movements and the specific context in which the above is addressed. Our study lays the groundwork for a first approach to the creation of a complete sign language to spoken language interpreter, as it is the only work in Spanish so far that considers facial expression as an indicator of grammatical tense. Moreover, in terms of feasibility, our model demonstrated that even with relatively few samples (on average, 35 for each sign and a knowledge base of 100 sentences) it is possible to obtain an accuracy of 94.9% for recognition of 75 isolated words and 94.1% for 20 test sentences in the medical context, which validates both the wide generalization capacity of the CNN coding architecture and the context-identifying HMM. This positions our work as an economically viable option - since it only uses a computer webcam -, easy to implement and fully scalable to other sign languages, especially those corresponding to countries with little or no study in the field, improving the inclusion of millions of hearing impaired people.

As future work, we intend to design, implement and include a statistical translation machine that will allow us to move from MSL grammatical structure to Spanish, which will be a significant contribution to further enhance effective communication tools.

References

1. Cruz, M.: Gramática de la Lengua de Señas Mexicana, 1st edn. Centro de Estudios Lingüísticos y Literarios, Colegio de México (2008)
2. Ordóñez, E.: Asociación de intérpretes en lengua de señas del distrito federal: Número de intérpretes de lengua de señas en México (2015)
3. Rashed, J.R.: New Method for Hand Gesture Recognition Using Wavelet Neural Network (2017)
4. Ben Jmaa, A., Mahdi, W., Ben Jemaa, Y., Ben Hamadou, A.: A new approach for hand gestures recognition based on depth map captured by RGB-D camera. Computacion y Sistemas **20**, 709–721 (2016)
5. Dong, C., Leu, M.C., Yin, Z.: American Sign Language alphabet recognition using Microsoft Kinect. In: IEEE Computer Society Conference on Computer Vision and Pattern Recognition Workshops. Volume 2015, pp. 44–52. IEEE Computer Society, October 2015
6. Fels, S.S., Hinton, G.E.: Glove-TalkII - A neural-network interface which maps gestures to parallel formant speech synthesizer controls. IEEE Trans. Neural Networks **8**, 977–984 (1997)
7. Tolba, A.S.: Arabic Glove-Talk (AGT): a communication aid for vocally impaired. Pattern Anal. Appl. **1**, 218–230 (1998)
8. Grobel, K., Assan, M.: Isolated sign language recognition using Hidden Markov Models. In: Proceedings of the IEEE International Conference on Systems, Man and Cybernetics. vol. 1, pp. 162–167. IEEE (1997)

9. García-Bautista, G., Trujillo-Romero, F., Caballero-Morales, S.O.: Mexican sign language recognition using Kinect and data time warping algorithm. In: 2017 International Conference on Electronics, Communications and Computers, CONIELE-COMP 2017, pp. 1–5. Institute of Electrical and Electronics Engineers Inc., (2017)
10. Baccouche, M., Mamalet, F., Wolf, C., Garcia, C., Baskurt, A.: Sequential deep learning for human action recognition. In: Salah, A.A., Lepri, B. (eds.) HBU 2011. LNCS, vol. 7065, pp. 29–39. Springer, Heidelberg (2011). https://doi.org/10.1007/978-3-642-25446-8_4
11. Kadhim, R.A., Khamees, M.: A real-time American sign language recognition system using convolutional neural network for real datasets. TEM J. **9**, 937–943 (2020)
12. Huang, J., Zhou, W., Li, H., Li, W.: Sign language recognition using 3D convolutional neural networks. In: Proceedings - IEEE International Conference on Multimedia and Expo, vol. 2015, pp. 1–6. IEEE Computer Society, August 2015
13. Huang, J., Zhou, W., Zhang, Q., Li, H., Li, W.: Video-based sign language recognition without temporal segmentation. In: 32nd AAAI Conference on Artificial Intelligence, AAAI 2018, pp. 2257–2264. AAAI Press (2018)
14. Carmona-Arroyo, G., Rios-Figueroa, H.V., Avendaño-Garrido, M.L.:Mexican Sign-Language static-alphabet recognition using 3D affine invariants. In: Machine Vision Inspection Systems, vol. 2, pp. 171–192. Wiley (2021)
15. Galicia, R., Carranza, O., Jimenez, E.D., Rivera, G.E.: Mexican sign language recognition using movement sensor. In: IEEE International Symposium on Industrial Electronics, vol. 2015, pp. 573–578. Institute of Electrical and Electronics Engineers Inc., (2015)
16. Luis-Pérez, F.E., Trujillo-Romero, F., Martínez-Velazco, W.: Control of a service robot using the Mexican sign language. In: Batyrshin, I., Sidorov, G. (eds.) MICAI 2011. LNCS (LNAI), vol. 7095, pp. 419–430. Springer, Heidelberg (2011). https://doi.org/10.1007/978-3-642-25330-0_37
17. Sataloff, R.T., Johns, M.M., Kost, K.M.: Reconocimiento de Imágenes del Lenguaje de Señas Mexicano, México D.F. (2012)
18. Solís, F., Martínez, D., Espinoza, O.: Automatic Mexican sign language recognition using normalized moments and artificial neural networks. Engineering **08**, 733–740 (2016)
19. Solís, F., Toxqui, C., Martínez, D.: Mexican sign language recognition using Jacobi-Fourier moments. Engineering **07**, 700–705 (2015)
20. Álvarez, N.A.: Kinect V2 como alternativa para desarrollar un traductor de ideogramas de lengua de señas mexicana (LSM) (2016)
21. Sosa-Jimenez, C.O., Rios-Figueroa, H.V., Rechy-Ramirez, E.J., Marin-Hernandez, A., Gonzalez-Cosio, A.L.S.: Real-time Mexican Sign Language recognition. In: 2017 IEEE International Autumn Meeting on Power, Electronics and Computing, ROPEC 2017, vol. 2018, pp. 1–6. Institute of Electrical and Electronics Engineers Inc. (2017)
22. Lugaresi, C., et al.: MediaPipe: a framework for building perception pipelines (2019)
23. Naveenkumar, M., Ayyasamy, V.: OpenCV for computer vision applications. In: Proceedings of National Conference on Big Data and Cloud Computing (NCBDC 2015), pp. 52–56 (2016)
24. Serafín, M., González, R.: Diccionario de Lengua de Señas Mexicana, vol. 38, México D.F. (2011)
25. Forney, G.D.: The Viterbi algorithm. Proc. IEEE **61**, 268–278 (1973)

26. Keval, H., Sasse, M.A.: To catch a thief - You need at least 8 frames per second: the impact of frame rates on user performance in a CCTV detection task. In: MM'08 - Proceedings of the 2008 ACM International Conference on Multimedia, with Co-located Symposium and Workshops, pp. 941–944 (2008)
27. Bisong, E.: Google Colaboratory, pp. 59–64. Apress, Berkeley (2019)

RiskIPN: Pavement Risk Database for Segmentation with Deep Learning

Uriel Escalona[1]([⊠]), Erik Zamora[1], and Humberto Sossa[1,2]

[1] Instituto Politécnico Nacional, CIC, Av. Juan de Dios Batiz S/N, Col. Nueva Industrial Vallejo, Gustavo A. Madero, 07738 Ciudad de México, Mexico
[2] Tecnológico de Monterrey, Campus Guadalajara. Av. Gral. Ramón Corona 2514, 45138 Zapopan, Jalisco, Mexico
{jgonzaleze2017,hsossa}@cic.ipn.mx, ezamorag@ipn.mx

Abstract. A large number of car accidents are caused by failures in the pavement. Their automatic detection is important for pavement maintenance, however, the current public datasets of images to train and test these systems contain a few hundred samples. In this paper, we introduce a new large dataset of images with more than 2000 samples that contains the five most common risks on pavement manually annotated. We analyze and describe statistically the properties of this dataset and we establish the performance of some baseline methods in order to be useful as a benchmark. We achieve up to 89.35% accuracy in the segmentation of the different types of risk on the pavement

1   Introduction

Natural wear and traffic load cause certain risks to appear on the pavement that can result in car accidents. Its early detection is one of the most important tasks in road maintenance. Formerly, this detection was carried out by manually segmenting images, but this hand-operated approach is tardy and labor-intensive [9,22,23].

Methods such as those described in [4,19,24] have focused on the rapid detection of risks on the pavement through the use of boxes. This method allows to measure the risks in an area using a low amount of computation that can be done even with the processing of a smartphone.

Automatic segmentation still remains a challenge due to texture variety, light changes and different types of noise, such as shadows, vegetation, oil and water spots, which can be easily confused with these risks.

Early approaches assumed that this kind of risks were able to be resolved by threshold value segmentation [17], or by edge detection [39]. The main disadvantage of these methods is that are strongly depended on illumination. The development of deep model improved segmentation solutions [5,6,25,29,31,32,35,36]. And new works were focused on pavement problems [3,7,11,13,18,20,21,27,28, 33,37,38] showing significant improvement compared with previous models.

© Springer Nature Switzerland AG 2021
I. Batyrshin et al. (Eds.): MICAI 2021, LNAI 13067, pp. 69–80, 2021.
https://doi.org/10.1007/978-3-030-89817-5_5

Deep learning models face the problem of data insufficiency. The models mentioned above to solve the problem of segmentation of risks in the pavement (cracks and potholes), use real images taken of the pavement, later, they are manually segmented to highlight the risk. This process is time consuming and laborious, so there are not many databases and the most used ones have few samples [8,12,26,30,33,38].

In this paper, we propose RiskIPN, a new challenging database which contain 5 risks found in pavement: holes, cracks, potholes, bumps, and drains. Describing its characteristics, and results in deep learning models for multiple segmentation.

The main contributions of this research are the following:

1. We propose a new database which contain more risks and samples than previous datasets used in the analysis of pavement.
2. We analyze the characteristics of the data such as the amount of risk per image or difference between similar risks.
3. We evaluate different models and observe the segmentation results tested in new data.

This paper is organized as follows: Sect. 2 represents a description of related public databases and RiskIPN, its description, form of obtaining and main properties. Section 3 describes the convolutional neural model that use this new database. Section 4 shows the experimental results to analyze the performance of this approach. Section 5 contains the conclusions and suggestions for future work.

2 Databases

In literature we can find databases that present some risks in the pavement, these risks are generally presented as a binary classification where the purpose is to identify the presence of the risk, such as cracks, holes and potholes. Many databases are not public, which makes it difficult to acquire and use them in models other than their original publications.

In this section we present some relevant public datasets frequently used and our proposed dataset, RiskIPN, which present 5 risk manually annotated taken from many states of Mexico with a great diversity of types of pavements.

2.1 Previous Datasets

The CFD [33] dataset is a manually segmented benchmark that has 118 RGB images of size 320×480 pixels. The images were collected with an iPhone 5 in Beijing, China. Each of the images contains cracks and noises such as oil and water stains.

AigleRN [8] dataset consists of 38 RGB images with 991×462 pixel size taken from different cracked pavements in France with intense crack texture inhomogeneity, some of the images within the dataset do not contain cracks.

Crack500 [38] is a dataset with 500 pavement pictures of size 3264×2448 collected at the Temple University campus by using a smartphone, where each image is annotated by multiple annotators. This dataset shows a high variety of pavement texture.

[12] present a database with 600 potholes taken with a ZED stereo camera with a resolution of 1080p (3840×1080) videos at 30 fps or 2.2K (4416×1242) videos at 15 fps. The baseline is 120 mm. Used a six element all-glass dual lenses and 16:9 native sensors, the video was 110° wide-angle and able to cover a scene up to 20 m.

[30] describes a collection of 55 images containing 97 potholes taken using a GoPro camera that was connected to the front windscreen of a car with the setting on 0.5 s time lapse mode and it provided footage that required no deblurring.

Recently, in [26] a database containing two risks, cracks and potholes, was presented, the database contains 2235 images taken from highways in the states of Espirito Santo, Rio Grande do Sul and Federal District, Brazil. These images were taken between 2014 and 2017. The annotation of the road consisted of demarcating the total region corresponding to the vehicle's road. The annotation of cracks and potholes consisted of the selection of the defect as a whole, maintaining its shape. To each image correspond 3 binary masks separated for each type of annotation, road, crack and pothole.

2.2   RisksIPN

RisksIPN consists of a database of 2000 images collected in Mexico, specifically from the states: Mexico City, Mexico State, Oaxaca, Cuernavaca and Morelos. The diversity of geographical areas allows a great variety of types of pavement where the following risks are present: holes, potholes, bumps, cracks and drains.

Our dataset is available at the following link for academic purpose: https://github.com/UrielEscalona/RiskIPN

These images were collected during hours of the day, in a period of one year between 2019 and 2020, under various weather conditions such as sun, light rain and high humidity. In the images you can find different types of noise that could resemble the risks sought, among which are oil stains, vegetation, inorganic garbage and waste generated by automobile use.

The images were taken through the use of different smartphones with different image resolution, under the restriction of taking the image at angles greater than 70° and less than 120° relative to the ground. At a distance between one meter and one meter and 20 cm from the ground.

A manual segmentation was carried out where the risks are identified by a color code. This segmentation produces an image of the same height and width as the original photograph of the pavement that contains 6 classes, the 5 risks and the healthy pavement. The color code for each class is presented in the Table 1.

In order to generalize the type of risks found in the pavement and their segmentation, the following rules were followed:

Table 1. Pavement risks color code

Risk	Color	Hexadecimal code	Decimal value
None	Black	000000	(0,0,0)
Crack	Red	FF0000	(1,0,0)
Hole	Green	00FF00	(0,1,0)
Pothole	Light blue	00BBBB	(0,0.73,0.73)
Bump	Yellow	AAAA00	(0.66,0.66,0)
Drain	Blue	0000FF	(0,0,1)

– Crack is a small split in the pavement with thin width and little depth.
– Hole is an opening in the pavement that has a pronounced depth.
– Bump is a protruding part of the pavement.
– Pothole is an opening in the pavement with little depth, which it is possible to see its bottom.
– Drain is a section of the pavement used for water drainage.

With these rules, certain situations were the segmentation is not clear are avoided, for example when a pothole contains water, it is difficult to measure its depth and therefore it is classified as a hole, likewise, the drains if they are uncovered or deteriorated they should be classified as a hole since these structures are generally very deep and their risk is high as seen in Fig. 1.

Another problem is the classification of cracks and potholes, where cracks that are not maintained in proper time, continue to deteriorate, increasing in depth and thickness until they become potholes, this is why it is common to find potholes surrounded by cracks, however, with the rule for classifying a crack as a risk with thin width it is possible to determine in image the difference of a crack and a pothole as seen in Fig. 2.

A manual segmentation of each image was performed by coloring the risks according to the previous table, to produce a new image in PNG format that preserves the hexadecimal values with the same image size as the photograph.

In previous datasets, having only one risk class, the amount of risk per image is very low, approximately 1.35% [8,12,33], which is identified as a high imbalance between the classes. In the case of RiskIPN, this problem is solved by containing more types of risks per class, in addition to the fact that the new risks that had not been segmented in previous datasets increase the amount of risk per image. Table 2 shows the average risk per image for this dataset. With a total mean of 17.44% of risk per image, where the pothole is the most present risk and the hole risk is only presented in very few images.

3   Segmentation Deep Model

Convolutional neural networks [16] have improved the image process in many areas of computing, their use in segmentation has been developed through very deep models [2,5]. In particular, the U-Net model [29] has been widely used to segment images related to areas such as medicine or construction [11].

Fig. 1. Images of potholes and drains that should be classified as holes.

Table 2. Properties of the RiskIPN database.

Data	Percentage
Healthy pavement	82.53%
Crack	1.61%
Pothole	7.18%
Hole	0.21%
Bump	5.60%
Drain	2.84%

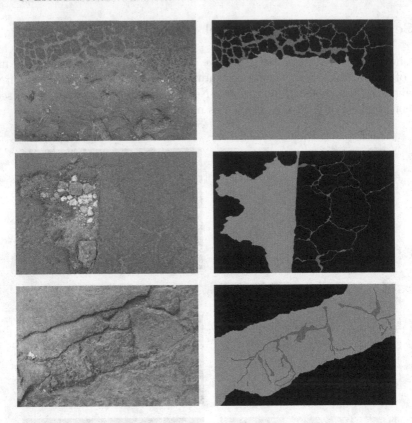

Fig. 2. Difference between a crack and a pothole.

Figure 3 shows the first proposed model based on U-Net, the encoder shows a series of convolutional layers followed by pooling layers as a mean of dimension reduction, this process is repeated three times until it is possible to reduce the dimension of the input images by up to a third, compressing the input information to a space of smaller dimension but with an increased depth.

The resulting matrix is processed through the decoder, made up of decoder layers, which are responsible for increasing the size of the matrix up to the original input size, reducing the depth of the matrix by a size proportional to the increase in length and width. This new matrix is processed through convolutional layers that can differentiate the different types of risks in the pavement from healthy pavement, assigning a different color to each of these.

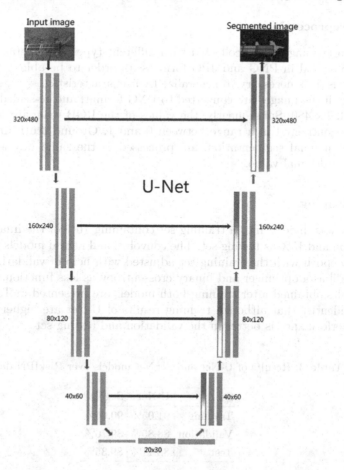

Fig. 3. Deep model for multiple segmentation on pavement.

In research such as [11,14,29,34], the interconnection between encoder and decoder layers has been proposed as a means of preserving the dimensionality of the objects within the convolution process, characteristic adopted by this model.

The second model is SegNet [5], a deep neural model used to multiple segmentation of urban environments. This model used convolutional neural networks capable of find relevant information to highlight the different classes, in this work last layers are cut to present the results as a image with size $320 \times 480 \times 3$ due to RGB properties.

4   Experiments and Results

All the experiments were carried out on a computer with an Intel I9-9900K processor, 64 Gb of RAM memory and an Nvidia 1080ti graphics card. The model was trained using Keras [10] and Tensorflow [1] to build the structure of the neural network model.

4.1 Preprocessing

The dataset images were collected with different types of cameras, at different resolutions, and in PNG and JPG formats. In order to be able to use them in the models, it is necessary to generalize their characteristics.

First all the images are converted to PNG format, and re-scaled to a resolution of 320 × 480. Subsequently, the values of the RGB scale are obtained and they are contracted in a range between 0 and 1. Ground truth images, which represent manual segmentation, are processed in the same way and risks are assigned a decimal value.

4.2 Training

RiskIPN was divided into a training set containing 70% of the images, 15% for validation and 15% as testing set. The convolutional neural models were trained over 100 epoch with the training set adjusted with help of validation set, using Adam [15] as a optimizer and binary crossentropy as loss function.

Results obtained after training both models are presented in Table 3, where we can identify that although training results of U-Net are higher, the SegNet model performance is better in the validation and testing set.

Table 3. Results of U-Net and SegNet models over RiskIPN dataset

Set	U-Net	SegNet
Training	91.05%	90.06%
Validation	88.85%	89.76%
Test	88.11%	89.35%

Figure 4 shows some segmentation in testing set. We can observe than in (a) that U-Net model segment the crack correctly and SegNet slightly identify the crack and segment some parts of healthy pavement as a drain. In (b) both models fail to segment the risk and add some false positives as bumps or drains. In (c) it is presented one of the new risk the bump, this risk is only identified with U-Net, moreover, the presence of shadows in the original image decrease the segmentation performance, and in (d) we present an image with high presence of risk, this pothole represent more than half of the pavement image, U-Net and SegNet identify the risk, nonetheless all the risk surface is not segmented.

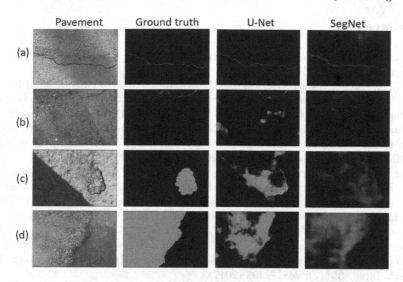

Fig. 4. U-Net and SegNet multiple segmentation results. (a) presents a centered horizontal crack, (b) shows a crack with different types of noise such as shadows and pavement stains, (c) shows a bump that is difficult to identify due to the difference in illumination and (d) presents a large pothole.

5 Conclusion

In machine learning it is essential to have data to be able to train the models, however, in specific tasks such as segmentation there are few public databases, which makes it difficult to use them in research. In this article we present the RiskIPN database, which has already studied risks in the pavement such as cracks and potholes, and add three new risks, holes, bumps and drains which are also the objective of analysis for the maintenance of a healthy pavement. As the experiments and results show, its segmentation is possible through deep learning models, achieving a percentage higher of 88%, facilitating its detection for correct maintenance.

Acknowledgments. Authors would like to acknowledge the support provided by the Instituto Politecnico Nacional under projects: SIP 20210788, SIP 20210316 and CONACYT under projects: 065 (Fronteras de la Ciencia) and 6005 (FORDECYT PRONACES) to carry out this research. Uriel Escalona thanks CONACYT for the scholarship granted towards pursuing his PhD studies. Authors acknowledge the support of social service students of the Instituto Politecnico Nacional for the creation of the database.

References

1. Abadi, M., et al.: TensorFlow: large-scale machine learning on heterogeneous distributed systems (2016)
2. Aich, S., van der Kamp, W., Stavness, I.: Semantic binary segmentation using convolutional networks without decoders. In: 2018 IEEE/CVF Conference on Computer Vision and Pattern Recognition Workshops (CVPRW). IEEE, June 2018
3. Amhaz, R., Chambon, S., Idier, J., Baltazart, V.: Automatic crack detection on two-dimensional pavement images: an algorithm based on minimal path selection. IEEE Trans. Intell. Transp. Syst. **17**(10), 2718–2729 (2016)
4. Angulo, A., Vega-Fernández, J.A., Aguilar-Lobo, L.M., Natraj, S., Ochoa-Ruiz, G.: Road damage detection acquisition system based on deep neural networks for physical asset management. In: Martínez-Villaseñor, L., Batyrshin, I., Marín-Hernández, A. (eds.) MICAI 2019. LNCS (LNAI), vol. 11835, pp. 3–14. Springer, Cham (2019). https://doi.org/10.1007/978-3-030-33749-0_1
5. Badrinarayanan, V., Kendall, A., Cipolla, R.: SegNet: a deep convolutional encoder-decoder architecture for image segmentation. IEEE Trans. Pattern Anal. Mach. Intell. **39**(12), 2481–2495 (2017)
6. Budak, U., Guo, Y., Tanyildizi, E., Sengur, A.: Cascaded deep convolutional encoder-decoder neural networks for efficient liver tumor segmentation. Med. Hypotheses **134**, 109431 (2020)
7. Cha, Y.-J., Choi, W., Büyüköztürk, O.: Deep learning-based crack damage detection using convolutional neural networks. Comput.-Aided Civ. Infrastruct. Eng. **32**(5), 361–378 (2017)
8. Chambon, S., Moliard, J.-M.: Automatic road pavement assessment with image processing: review and comparison. Int. J. Geophys. **1–20**, 2011 (2011)
9. Cheng, H.D., Chen, J.-R., Glazier, C., Hu, Y.G.: Novel approach to pavement cracking detection based on fuzzy set theory **13**, 270–280 (1999)
10. Chollet, F.: Keras (2015). https://keras.io
11. Escalona, U., Arce, F., Zamora, E., Sossa Azuela, J.H.: Fully convolutional networks for automatic pavement crack segmentation. Computacion y Sistemas **23**(2) (2019)
12. Fan, R., Ai, X., Dahnoun, N.: Road surface 3d reconstruction based on dense subpixel disparity map estimation. IEEE Trans. Image Process. **27**(6), 3025–3035 (2018)
13. Fan, Z., Wu, Y., Lu, J., Li, W.: Automatic pavement crack detection based on structured prediction with the convolutional neural network (2018)
14. Gopalakrishnan, K., Khaitan, S.K., Choudhary, A., Agrawal, A.: Deep convolutional neural networks with transfer learning for computer vision-based data-driven pavement distress detection. Constr. Build. Mater. **157**, 322–330 (2017)
15. Kingma, D.P., Ba, J.: Adam: a method for stochastic optimization. In: 3rd International Conference for Learning Representations (2014)
16. LeCun, Y., et al.: Backpropagation applied to handwritten zip code recognition. Neural Comput. **1**(4), 541–551 (1989)
17. Li, Q., Liu, X.: Novel approach to pavement image segmentation based on neighboring difference histogram method. In: 2008 Congress on Image and Signal Processing, vol. 2, pp. 792–796, May 2008
18. Liu, Y., Yao, J., Xiaohu, L., Xie, R., Li, L.: DeepCrack: a deep hierarchical feature learning architecture for crack segmentation. Neurocomputing **338**, 139–153 (2019). Apr

19. Maeda, H., Sekimoto, Y., Seto, T., Kashiyama, T., Omata, H.: Road damage detection using deep neural networks with images captured through a smartphone (2018)
20. Mazzini, D., Napoletano, P., Piccoli, F., Schettini, R.: A novel approach to data augmentation for pavement distress segmentation. Comput. Ind. **121**, 103225 (2020)
21. Mei, Q., Gül, M.: A cost effective solution for pavement crack inspection using cameras and deep neural networks. Constr. Build. Mater. **256**, 119397 (2020)
22. Nguyen, T.S., Avila, M., Begot, S.: Automatic detection and classification of defect on road pavement using anisotropy measure. In: 2009 17th European Signal Processing Conference, pp. 617–621, August 2009
23. Nguyen, T.S., Begot, S., Duculty, F., Avila, M.: Free-form anisotropy: a new method for crack detection on pavement surface images. In: 2011 18th IEEE International Conference on Image Processing, pp. 1069–1072, September 2011
24. Ochoa-Ruiz, G., et al.: An asphalt damage dataset and detection system based on RetinaNet for road conditions assessment. Appl. Sci. **10**(11), 3974 (2020)
25. Oliveira, H., Correia, P.L.: Crackit - an image processing toolbox for crack detection and characterization. In: 2014 IEEE International Conference on Image Processing (ICIP). IEEE, October 2014
26. Passos, B.T.: Cracks and potholes in road images (2020)
27. Rajagopal, M., Balasubramanian, M., Palanivel, S.: An efficient framework to detect cracks in rail tracks using neural network classifier. Computación y Sistemas **22**(3) (2018)
28. Ren, Y., et al.: Image-based concrete crack detection in tunnels using deep fully convolutional networks. Constr. Build. Mater. **234**, 117367 (2020). Feb
29. Ronneberger, O., Fischer, P., Brox, T.: U-Net: convolutional networks for biomedical image segmentation. In: Navab, N., Hornegger, J., Wells, W.M., Frangi, A.F. (eds.) MICCAI 2015. LNCS, vol. 9351, pp. 234–241. Springer, Cham (2015). https://doi.org/10.1007/978-3-319-24574-4_28
30. Nienaber, S., Booysen, M.J., Kroon, R.S.: Detecting potholes using simple image processing techniques and real-world footage (2015)
31. Shao, C., Chen, Y., Xu, F., Wang, S.: A kind of pavement crack detection method based on digital image processing. In: 2019 IEEE 4th Advanced Information Technology, Electronic and Automation Control Conference (IAEAC). IEEE, December 2019
32. Shi, J., Li, Z., Zhu, T., Wang, D., Ni, C.: Defect detection of industry wood veneer based on NAS and multi-channel mask r-CNN. Sensors **20**(16), 4398 (2020)
33. Shi, Y., Cui, L., Qi, Z., Meng, F., Chen, Z.: Automatic road crack detection using random structured forests. IEEE Trans. Intell. Transp. Syst. **17**(12), 3434–3445 (2016)
34. Simonyan, K., Zisserman, A.: Very deep convolutional networks for large-scale image recognition. Computer Vision and Pattern Recognition (2014)
35. Urbonas, A., Raudonis, V., Maskeliūnas, R., Damaševičius, R.: Automated identification of wood veneer surface defects using faster region-based convolutional neural network with data augmentation and transfer learning. Appl. Sci. **9**(22), 4898 (2019)
36. Yang, F., et al.: Feature pyramid and hierarchical boosting network for pavement crack detection. IEEE Trans. Intell. Transp. Syst. **21**(4), 1525–1535 (2020)
37. Yusof, N.A.M., et al.: Automated asphalt pavement crack detection and classification using deep convolution neural network. In: 2019 9th IEEE International Conference on Control System, Computing and Engineering (ICCSCE). IEEE, November 2019

38. Zhang, L., Yang, F., Zhang, Y.D., Zhu, Y.J.: Road crack detection using deep convolutional neural network. In: 2016 IEEE International Conference on Image Processing (ICIP). IEEE, September 2016
39. Zhao, H., Qin, G., Wang, X.: Improvement of canny algorithm based on pavement edge detection. In: 2010 3rd International Congress on Image and Signal Processing, pp. 964–967. IEEE, October 2010

A Comparative Study on Approaches to Acoustic Scene Classification Using CNNs

Ishrat Jahan Ananya[iD], Sarah Suad[iD], Shadab Hafiz Choudhury[(✉)][iD],
and Mohammad Ashrafuzzaman Khan[iD]

North South University, Dhaka, Bangladesh
{ishrat.jahan16,sarah.suad,shadab.choudhury,
mohammad.khan02}@northsouth.edu

Abstract. Acoustic scene classification is a process of characterizing and classifying the environments from sound recordings. The first step is to generate features (representations) from the recorded sound and then classify the background environments. However, different kinds of representations have dramatic effects on the accuracy of the classification. In this paper, we explored the three such representations on classification accuracy using neural networks. We investigated the spectrograms, MFCCs, and embeddings representations using different CNN networks and autoencoders. Our dataset consists of sounds from three settings of indoors and outdoors environments – thus, the dataset contains sounds from six different kinds of environments. We found that the spectrogram representation has the highest classification accuracy while MFCC has the lowest classification accuracy. We reported our findings, insights, and some guidelines to achieve better accuracy for environment classification using sounds.

Keywords: Acoustic scene classification · Signal processing · Deep learning convolutional neural network · Autoencoders

1 Introduction

Acoustic Scene Classification (ASC) is the process of understanding and classifying scenes and environments from ambient audio. It has plenty of use cases such as autonomous monitoring, environment perception in self-driving vehicles and robotics, analyzing multimedia recordings, and helping the visually and audibly challenged people better interpret the environment. However, research on a generic and scalable methodology for solving audio classification problems is somewhat spotty. The lack of a generic approach is a significant issue, as it has restricted the development and widespread application of ASC.

In ASC problems, we generally start with unstructured data in the form of audio files. Audio can be represented as 2D matrices of amplitude, energy, or another sound property against time. Convolutional Neural Networks (CNNs) are very effective at extracting features from such data formats. While CNNs are commonly used for images, they can be applied to any data represented as 2D matrix forms. To carry out ASC using CNNs,

© Springer Nature Switzerland AG 2021
I. Batyrshin et al. (Eds.): MICAI 2021, LNAI 13067, pp. 81–91, 2021.
https://doi.org/10.1007/978-3-030-89817-5_6

we first need to convert the audio files to a 2D representation of acoustic features such as volume, pitch, and sequence of sounds. There are three common ways to doing that – Spectrograms, MFCCs, and Embeddings. Researchers have used these approaches to optimize specific solutions in the field of ASC. However, they did not create any general solution that is applicable on a broader scale.

We hypothesized that a more generic approach or technique would improve scalability and reproducibility. Our goal was to figure out what kind of preprocessing and feature representation gives us the best results. This would be the first step in developing a general approach to CNN-based acoustic scene classification. We used indoor and outdoor scene classification problems to compare different feature representation techniques. The dataset we used consists of six types of scenes. We converted them to each of the three feature representations mentioned above and compared their performances using several popular CNN models.

The experimental results show that Spectrograms offer the best results, reaching up to 90% accuracy on this problem. The MFCCs were less effective, as the type of features they represent are not so distinguishable when applied to ASC. These two approaches are commonly used in audio classification. Embeddings are a less common approach, but they proved to be a very lightweight and efficient solution. At around 80% accuracy, the results are acceptable but could be improved further. They are more suitable for less powerful devices such as mobile phones.

The paper is organized in the following way: first, we discuss previous research done in classifying audio using artificial intelligence and ASC in general. Then we cover the experimental process, starting from the preparation of the data and details on the CNN models used. Finally, we evaluate and discuss the results and put forward some suggestions for future work.

2 Related Work

Analysis and classification of auditory signals with artificial intelligence have a long history. Initially, research focused on simply detecting and distinguishing acoustic events [1], such as distinct noises like claps and speech or different individuals speaking [2]. These early examples of the use of neural networks in audio classification developed from the intersection of signal processing and artificial intelligence.

More mature artificial intelligence techniques such as sophisticated convolutional neural networks have enabled further exploration of Acoustic Scene Classification through different approaches. The DCASE Challenges, initially started in 2013, offer datasets and a platform for investigating Acoustic Scene Classification [3]. DCASE 2013 highlighted the use of large datasets for acoustic scene classification in various scenes such as a bus, office, and market.

Early research yielded good outcomes with machine learning models. Good results were achieved in the DCASE 2013 challenge using algorithms such as support vector machines and decision trees [4]. However, as the sophistication and size of datasets increased, neural networks became a compelling choice. In this specific dataset, Valenti et al.'s approach using a custom CNN model resulted in higher accuracy compared to earlier work – up to 9.7% depending on the technique it is compared to [5].

The majority of recent works on acoustic scene classification have followed up on the CNN approach. Hussein et al. developed a more in-depth technique using a deep neural network with only three hidden layers that achieved up to 90% accuracy on the DCASE 2016 challenge [6].

In the DCASE 2020 challenge, several attempts were able to reach 96% test accuracy by implementing modern deep convolutional neural networks such as ResNets. While both these and the previous approach were highly accurate, they also made use of specific preprocessing techniques and model designs that could be difficult to implement on a larger scale.

Finally, an excellent overview of the development and use of deep learning in acoustic scene classification between 2013 and 2020 is given in a review paper by Abeßer [7].

3 Methodology

The first step is to convert the audio file, an uncompressed.wav file, into a numerical form. This returns a one-dimensional array of length equal to the sampling rate – in this case, 44100. This array is entirely impractical to use in any deep learning application, so we must use a different representation. Initial audio preprocessing was carried out on this data, and then it was converted to the three different feature representations.

The CNN models utilized in this experiment were a simple autoencoder, AlexNet, ResNet-18, and ResNet-50. ResNets were used because they are an extremely robust and powerful architecture for classification problems. Due to the number of layers, it would be able to extract features from inputs like Spectrograms that don't have many obvious features. While they are mainly used for images, they could also be used for other types of inputs with acceptable performance. The number of deep layers in ResNet makes it extremely suitable for generalizing, which is one of our goals. Alterations to the structure of the models were kept to a minimum in order to ensure a fair comparison. Any changes made were in order to ensure that the input could be appropriately fed into the model.

3.1 Data Organization and Collection

The dataset was collected from DCASE 2020 challenge [8, 9]. It had three classes of indoor, outdoor and transport. These classes were subdivided into nine more subclasses. The raw dataset contained 10-s audio clips in 24-bit.wav format taken from ten different cities worldwide.

For this paper, we chose to reduce it to two classes: indoor and outdoor. This left us with six subclasses, three from each. For Indoor Scenes, the subclasses were 'Metro Station', 'Shopping Mall' and 'Airport'. For Outdoor Scenes, the subclasses were 'Park', 'Pedestrian Street' and 'Public Square'.

There were 8640 data samples in total, adding up to 24 h of audio. Each subclass had 1440 data points. We ensured that the number of audio samples for each class was equal so that the model was not biased towards any particular category due to unbalanced data. The original audio files were binaural at 44.1 kHz. This is a relatively small dataset, so we split each of the 10-s audio files into two 5 s files to double the size of the dataset. Table 1 shows the size of the dataset at different stages of collection and preprocessing.

Table 1. Data organization

Type of data	No. of data points	Hours of audio
Files in dataset (10-s clips)	8,640	24
Files per subclass	1,440	4
Indoor scenes (5-s clips)	8,640	12
Outdoor scenes (5-s clips)	8,640	12
Indoor scenes with augmentation	17,280	24
Outdoor scenes with augmentation	17,280	24
Test data per class	3,456	4.8
Train data per class	13,824	19.2
Total data (with 2 classes)	34,560	48

3.2 Data Augmentation

Before converting the data into a feature representation, some data augmentation was carried out. A small amount of random noise was added, and audio tracks were randomly shifted forward and backward.

Many other typical augmentations, such as pitch-shifting and extending silences, were determined to be detrimental to this dataset. They are typically used for speech-based datasets. Here, the audio is a continuous stream of noise rather than speech at different tones and with small silences in between.

This gave us a rich dataset of 34,560 five-second audio clips, half of which was augmented.

3.3 Feature Representations

The final part of data preprocessing required taking audio samples as input and extracting features from the audio signals. We aim to find components of the audio signals that will help us differentiate them from other categories of signals. We implemented three methods of generating feature representations: producing log-mel spectrograms [10], Mel Frequency Cepstral Coefficients (MFCC) [11] and audio embeddings.

MFCCs and Spectrograms. To generate MFCCs, first, the audio signal is sliced into 20 ms wide frames. We assume that there is no change in the signal within each 20 ms frame. We applied a short-term Fourier Transform on each frame to calculate the power spectrum. This gives us the distribution of power into frequency components that make up the signal. Next, we applied the Mel filter bank to each power spectra and summed up the energy in each filter. This step actually estimates how the human ear perceives sounds at different frequencies and different volumes. The last step in the process is to take the discrete cosine transform of the logarithm of each filter bank. All these calculations were carried out using the python library Librosa [12].

Log-mel spectrograms were produced in a similar way. The audio dataset underwent a short-time Fourier transform to get spectrograms based on the Frequency and Amplitude of the signal rather than Power. The spectrograms were then scaled to the Mel scale and saved in .png format.

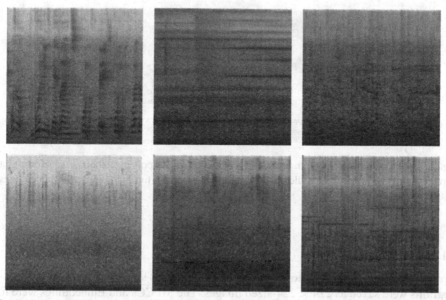

Fig. 1. Spectrograms of each class. The top row shows Indoor (Airport, Metro Station and Shopping Mall from left to right). The bottom row shows Outdoor (Park, Public Square and Pedestrian Street from left to right).

Audio Embeddings. Humans categorize or recognize things by comparing its details to previous knowledge. Many image classification algorithms use the same approach. These algorithms are trained on datasets where the input is in the form of images that have objects labelled in them. Using audio embeddings utilizes a similar approach for audio classification.

Embeddings are used to map items from a high dimensional vector space to a low dimensional vector space. In dense data such as audio, the embeddings determine similarity metrics to other sounds. Essentially, it splits the audio clip into smaller intervals. For each interval, it gives a similarity value to all the classes the original embedding model was trained on. So, while MFCCs and Spectrograms use features extracted directly from the audio signals, Embeddings simply use learned features generated by another machine learning model that has been trained to label audio files.

The embeddings we used were generated using an Audio Embedding Generator [13, 14]. The generator accepts a 16-bit PCM .wav file as input, embeds the feature labels and outputs the result as arrays of 1 s embeddings. The model was trained on Audioset, which includes 632 classes. For each second, the embeddings list the 128 classes that have the highest similarity to the sound.

Figure 2 shows an example of each type of preprocessed data, in a visual format. Note, that while the MFCCs and Embeddings are depicted visually here, they were input into the CNN as two-dimensional matrices.

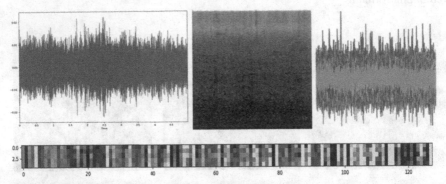

Fig. 2. Four representations of an audio file recorded at an Airport in Lisbon: a waveform plot (top left), a spectrogram (top middle), MFCCs plotted as a graph (top right) and audio embeddings plotted as a graph (bottom).

3.4 Development of CNNs

Inputting the Spectrograms into the CNNs for training was straightforward, as the spectrograms were saved as 224×224-sized images. The MFCCs and Embeddings were in the form of two-dimensional matrices and could not be input directly. They had to be reshaped before being fed into the CNNs. When necessary, the initial layer was altered to fit the dataset.

We tested four CNNs for all three feature representations: ResNet-18, ResNet-50, AlexNet, and finally, an Autoencoder. For each CNN, a limited degree of hyperparameter optimization was carried out.

The ResNet family of neural network architectures is ideal for image classification tasks. The ResNet architecture uses stacked layers of residual learning blocks using shortcuts between layers to minimize the effect of the vanishing gradient problem [15]. We used the SGD optimizer with Cross-Entropy Loss and a learning rate of 0.001. Despite being an older architecture, AlexNet [16] was also tested to see how the parameters affect the results. The results of both models are discussed in the Evaluation section.

We initially tested the autoencoder model for the Embeddings only. The structure of the embeddings is such that labels for similar sounds would be clustered together. Using a simple Autoencoder with linear layers enhanced this and exposed the largest clusters. For the sake of comparison, other autoencoder models were also tested on the other two feature representations.

4   Results and Evaluation

The models were trained and tested on a single Nvidia RTX 3070 GPU. Table 2 summarizes the results of our experimentation, and a discussion of the results follows.

Table 2. Experimental results for spectrograms

CNN architecture	Test accuracy	Training accuracy
Autoencoder	76.2%	79.3%
AlexNet	86.3%	90.7%
ResNet-18	89.7%	91.9%
ResNet-50	90.4%	93.6%
ShuffleNet	93.1%	95.2%

It's clear from Table 2 that the Spectrograms offer the best results, up to 93%. Considering that these CNNs are typically used for image classification, these results are as expected. Most models give acceptable results with the other feature representations, getting around 70–80% accuracy (Table 3).

Table 3. Confusion matrix for the best result – 93.1% with ShuffleNet

<table>
<tr><td></td><td colspan="2">predicted</td><td></td></tr>
<tr><td>n = 3456</td><td>Indoor</td><td>Outdoor</td><td></td></tr>
<tr><td>Indoor</td><td>1575</td><td>153</td><td>1728</td></tr>
<tr><td>Outdoor</td><td>88</td><td>1640</td><td>1728</td></tr>
<tr><td></td><td>1728</td><td>1728</td><td></td></tr>
</table>

Even a simple fully-connected autoencoder of 4096-2048-1024-512 parameters gives us 76.3% accuracy, implying that the spectrograms are shallow and do not have many features to extract. We used the encoder layer from the autoencoder mentioned above to improve feature extraction before feeding the parameters into the ResNets to gain a slight improvement. After getting an extremely high value from the ResNets but a lower value from AlexNet, we decided to try another state-of-the-art model with fewer parameters. We chose Shufflenet, a computationally efficient CNN architecture specifically designed for mobile devices with limited processing power [17]. While these results do not break any of the benchmarks set in previous DCASE challenges, they are all reasonably generic approaches that require minimal customization to the dataset.

This ensures that the results are applicable across different acoustic scene datasets rather than being optimized for this particular problem.

Table 4. Experimental results for MFCCs

CNN architecture	Test accuracy	Training accuracy
Autoencoder	49.9%	50.0%
AlexNet	69.8%	89.1%
ResNet-18	71.3%	86.6%
ResNet-50	72.1%	88.0%

Due to the features of the MFCCs, using an Autoencoder was utterly ineffective, as shown by the 50% accuracy on a binary classifier (Table 4). We note that ResNet models are too heavy for the MFCC, generally overfitting within a few epochs. The test accuracies recorded are before it fully overfits. Using the non-augmented half of the data set saw a slight improvement in accuracy, at the cost of even more overfitting. At this stage, regularization techniques were ineffective. Therefore, for a dataset of this size, augmentation was necessary.

MFCCs are generally used for speech classification. They can easily distinguish between high and low volumes and pitches. However, the audio in this dataset is primarily background noise at a similar energy level throughout. Therefore, it is harder to extract features using MFCCs compared to other approaches, and this feature representation has the lowest accuracy of all.

Table 5. Experimental results for embeddings

CNN model	Test accuracy	Training accuracy
Autoencoder	80.8%	82.2%
AlexNet	77.9%	96.8% (overfit)
ResNet-18	77.6%	99.7% (overfit)
ResNet-50	77.1%	99.6% (overfit)

The audio embeddings performed surprisingly well, considering the nature of the dataset. Reducing the audio dataset into a series of labels allowed the autoencoder to learn features quickly, even if there was much less data than the other approaches. The original dataset used to develop the embedding generator was focused on speech, music, and the sounds made by individual objects. Despite being a somewhat unsuitable dataset, it gave good results. A more closely related embedding generator could give results comparable to Spectrograms at a fraction of the computation power. However, no such generator for urban scenes is currently available, and developing one from scratch would be out of the scope of this paper. Finally, as seen in Table 5, heavier models like the ResNets and AlexNet led to overfitting when used with embeddings.

At this point, it should be noted that all the models that were run on individual subclasses faced significant issues—even with plenty of augmentation, classifying on six classes rather than two led to extensive overfitting and an accuracy of 70% at most. The dataset is too small to achieve good accuracy unless the classes are combined. Generally, the models trained on the Embeddings distinguished Metro Station, Shopping Mall, Park, and Public Square with high accuracy – close to 80%, while Airport and Pedestrian Street displayed lower confidence. Models trained on the Spectrograms showed relatively similar confidence per subclass.

Different objects or segments are highly visible with typical image classification problems, as they have precise contours and different colors. The Spectrograms are not so distinguishable, as seen in Fig. 1. So, a massive amount of data is needed to train the models to distinguish between different classes. The MFCCs and Embeddings do not suffer from this issue but are not suitable for typical deep networks and may require custom CNNs to improve further.

Additionally, it is clear from both our work and other research that, unlike image classification, audio classification does not always benefit from deep networks with many parameters. The number of features that can be extracted from the Spectrograms or other formats is minimal. AlexNet has 61 million parameters, while ResNet-18 has 11 million and ResNet-50 has 23 million. All three networks have very similar accuracies and are in fact prone to overfitting. By contrast, the smallest network used here, the fully-connected autoencoder with audio embeddings, only has 560 k parameters. Since datasets for ASC are relatively limited, adding more data to solve this problem is not always feasible. Therefore, higher input resolution [18] or additional preprocessing may be necessary to achieve better accuracy.

5 Conclusion

As the evaluation section shows, the task of Acoustic Scene Classification faces major hurdles when it comes to larger models. Extensive data preprocessing and augmentations are necessary to achieve high accuracies on even very limited problems. Spectrograms offered the best result for acoustic scene classification of interior and exterior urban scenes out of the three different approaches tested. It achieved over 90% accuracy, but required a lot of data to reach this accuracy. This part of the conclusion may not be novel on its own, but it provides a basis for comparison.

We suggest not using MFCCs for ASC, as they require extensively customized CNNs to get a good result.

For a lightweight approach, audio embeddings are suitable. Even though the embedding generator model was not entirely appropriate for this domain, it achieved almost 81% accuracy. Higher accuracy could easily be reached if an embedding generator model is developed that is focused on urban audio. We suggest focusing on this approach for further development of a generic approach, since it is an efficient process suitable for use in low-power mobile devices.

Future work to follow up on this paper would involve two other approaches. The first would be to increase the resolution and accuracy of our comparison of models by using a wider selection of models to see which approach to audio classification works best at different sizes of models. Secondly, there are several datasets available for ASC outside DCASE. Combining multiple datasets may enable more general conclusions to be drawn.

These avenues of future development will place the groundwork for CNNs focused on audio classification and help acoustic scene classification be used in broader contexts.

References

1. Temko, A., Nadeu, C., Macho, D., Malkin, R., Zieger, C., Omologo, M.: Acoustic event detection and classification. In: Waibel, A., Stiefelhagen, R. (eds.) Computers in the Human Interaction Loop. Human–Computer Interaction, pp. 61–73. Springer, London (2009)
2. Liu, Z., Wang, Y., Chen, T.: Audio feature extraction and analysis for scene segmentation and classification. J. VLSI Sig. Proc. Syst. Signal Image Video Technol. **20**(1), 61–79 (1998)
3. Giannoulis, D., Stowell, D., Benetos, E., Rossignol, M., Lagrange, M., Plumbley, M.D.: A database and challenge for acoustic scene classification and event detection. In: 21st European Signal Processing Conference (EUSIPCO 2013) (2013)
4. Stowell, D., Giannoulis, D., Benetos, E., Lagrange, M., Plumbley, M.D.: Detection and classification of acoustic scenes and events. IEEE Trans. Multimedia **17**(10), 1733–1746 (2015)
5. Valenti, M., Squartini, S., Diment, A., Parascandolo, G., Virtanen, T.: A convolutional neural network approach for acoustic scene classification. In: 2017 International Joint Conference on Neural Networks (IJCNN) (May 2017), pp. 1547–1554. ISSN: 2161-4407. (2017)
6. Hussain, K., Hussain, M., Khan, M.G.: An improved acoustic scene classification method using convolutional neural networks (CNNs). Am. Sci. Res. J. Eng. Technol. Sci. **44**(1), 68–76 (2018)
7. Abeßer, J.: A review of deep learning based methods for acoustic scene classification. Appl. Sci. **10**(6), 2020 (2020)
8. Heittola, T., Mesaros, A., Virtanen, T.: Acoustic scene classification in DCASE 2020 Challenge: generalization across devices and low complexity solutions. arXiv:2005.14623 [ccss] (2020)
9. Heittola, T., Mesaros, A., Virtanen, T.: TAU Urban acoustic scenes 2020 mobile, development dataset. Zenodo (2020). https://doi.org/10.5281/zenodo.3819968
10. Felipe, G.Z., Maldonado, Y., Costa, G.D., Helal, L.G.: Acoustic scene classification using spectrograms. In: 2017 36th International Conference of the Chilean Computer Science Society (SCCC), pp. 1–7 (2017)
11. Davis, S., Mermelstein, P.: Comparison of parametric representations for monosyllabic word recognition in continuously spoken sentences. IEEE Trans. Acoust. Speech Signal Process. **28**(4), 357–366 (1980)
12. McFee, B., et al.: librosa: Audio and music signal analysis in python. In: Proceedings of the 14th Python in Science Conference, pp. 18–25 (2015)
13. Gemmeke, J., et al.: Audio set: an ontology and human-labeled dataset for audio events. In: IEEE International Conference on Acoustics, Speech and Signal Processing (ICASSP) (2017)
14. Hershey, S., et al.: CNN Architectures for Large-Scale Audio Classification. arXiv:1609.09430 [cs, stat] (2017)
15. He, K., Zhang, X., Ren, S., Sun, J.: Deep Residual Learning for Image Recognition. arXiv:1512.03385 [cs] (2015)
16. Krizhevsky, A., Sutskever, I., Hinton, G.E.: ImageNet classification with deep convolutional neural networks. Commun. ACM **60**(6), 84–90 (2017)

17. Zhang, X., Zhou, X., Lin, M., Sun, J.: Shufflenet: an extremely efficient convolutional neural network for mobile devices. In: Proceedings of the IEEE Conference on Computer Vision and Pattern Recognition, pp. 6848–6856 (2018)
18. Zhang, T., Liang, J., Ding, B.: Acoustic scene classification using deep CNN with fine-resolution feature. Expert Syst. Appl. **143**, 113067 (2020). https://doi.org/10.1016/j.eswa.2019.113067

Measuring the Effect of Categorical Encoders in Machine Learning Tasks Using Synthetic Data

Eric Valdez-Valenzuela[1]([✉]), Angel Kuri-Morales[2], and Helena Gomez-Adorno[3]

[1] Posgrado en Ciencia E Ingeniería de La Computación, Universidad Nacional Autónoma de México, Ciudad de México, Mexico
[2] Instituto Tecnológico Autónomo de México, Ciudad de México, Mexico
akuri@itam.mx
[3] Instituto de Investigaciones en Matemáticas Aplicadas Y en Sistemas, Universidad Nacional Autónoma de México, Ciudad de México, Mexico
helena.gomez@iimas.unam.mx

Abstract. Most of the datasets used in Machine Learning (ML) tasks contain categorical attributes. In practice, these attributes must be numerically encoded for their use in supervised learning algorithms. Although there are several encoding techniques, the most commonly used ones do not necessarily preserve possible patterns embedded in the data when they are applied inappropriately. This potential loss of information affects the performance of ML algorithms in automated learning tasks. In this paper, a comparative study is presented to measure how the different encoding techniques affect the performance of machine learning models. We test 10 encoding methods, using 5 ML algorithms on real and synthetic data. Furthermore, we propose a novel approach that uses synthetically created datasets that allows us to know *a priori* the relationship between the independent and the dependent variables, which implies a more precise measurement of the encoding techniques' impact. We show that some ML models are affected negatively or positively depending on the encoding technique used. We also show that the proposed approach is more easily controlled and faster when performing experiments on categorical encoders.

Keywords: Supervised machine learning · Data preprocessing · Categorical encoding · Synthetic data

1 Introduction

Many machine learning (ML) algorithms only accept numerical attributes as entries to be analyzed. However, most of the real-world datasets contain both categorical and numerical variables or attributes [1] (we will use the term "variables" and "attributes" interchangeably). A categorical variable has a measurement scale consisting of a set of categories. For example, the categorical variable "choice of accommodation" might include the instances "house", "condominium", and "apartment". Likewise, a diagnostic test to detect e-mail spam might identify the instance of an incoming e-mail message as "spam" or "legitimate". Categorical variables are often referred to as qualitative,

© Springer Nature Switzerland AG 2021
I. Batyrshin et al. (Eds.): MICAI 2021, LNAI 13067, pp. 92–107, 2021.
https://doi.org/10.1007/978-3-030-89817-5_7

to distinguish them from quantitative variables, which take numerical values (such as age, income, and the number of children in a family) [2]. To make proper use of the information therein, categorical variables require a preprocessing phase for their use in most ML algorithms. This phase consists of applying an encoding technique that transforms all the instances in the categorical variable into a numerical value. This transformation must preserve the possible patterns or intrinsic information in the dataset. During this task, there is a risk of losing information that can negatively affect the performance of ML algorithms.

The research question which we address in this work is *how do categorical encoding techniques affect machine learning algorithms?* There are interesting works that have attempted to answer this question in the past. Paper [3] presents a comparative study of seven categorical encoding techniques to be used for classification in artificial neural networks on a categorical dataset. Although it achieves interesting results, that work is limited to a single ML algorithm and one dataset. A broader approach to study how categorical encoders affect neural networks can be found in the paper [4]. Another work that addresses the research question is presented in [5]. This work compares different encoding strategies for high cardinality features in a benchmark with five different machine learning algorithms using datasets from the regression, binary and multiclass classification settings. Despite being one of the most complete studies we found in the literature, it is focused on the particular case where the categorical attributes have high cardinality. In the same way, the paper [6] describes how categorical independent variables can be incorporated into regression by two coding methods: dummy and effect coding.

Here we present the development of a comparative study that measures the effect of *10* categorical encoding techniques when solving *7* real-world problems with *5* different ML algorithms. From the results of this study, it can be observed that some ML models are more affected than others depending on the technique used for encoding. We also use synthetic data as a novel approach to tackle the research question. Data created synthetically is often used to test ML techniques, for example, to analyze the performance of feature selection methods [7]. In our case, synthetic data allows us to quantify more precisely the performance of trained ML models when implementing different categorical encoding techniques.

The rest of the paper is organized as follows: in Sect. 2 we describe which encoding techniques and ML algorithms were included in the comparative study as well as the training datasets selected. In Sect. 3 we present the experimental results obtained from the real and synthetic datasets. Finally, in Sect. 4 our conclusions are presented.

2   General Methodology

The purpose of our methodology is to evaluate the effectiveness of the categorical encoding techniques (CET). To accomplish this, we propose to quantify how CETs affect ML algorithms when applied to supervised learning tasks with two types of data: natural and synthetic. Seven real-world datasets were taken from the *Kaggle* platform. In addition, *10* CETs and 5 ML algorithms were selected. For each dataset, the *5* ML algorithms were combined with the *10* CETs, resulting in *50* different model-encoder combinations (in this paper, the word *model* refers to the output of an ML algorithm applied to a certain dataset and ready to make predictions). For each ML algorithm, we recorded which

CET the best and the worst performance was obtained, as well as the difference between these two measurements. To distinguish each possible combination of model-CET, we assigned codes that synthesize the names of each model and CET. The ML algorithms used along with their code are listed in Table 1:

Table 1. ML algorithms included in the comparative study

Algorithms	Abbreviation
Logistic Regression	RL
Gaussian Naïve-bayes	GNB
Support Vector Machine	SVM
Neural Network (multilayer perceptron)	NN
XGBoost	XGB

The Scikit Learn [8] library was used for the implementation of the logistic regression, Gaussian Naïve-Bayes, and support vector machine (these models were trained with the default parameters). The model Random Forest (RF) was only used in the regression problem (*Housing California*), and it was replaced by GBN in the classification problems.

In the case of the Neural Network, a Multilayer Perceptron or MLP (we will keep the abbreviation 'NN' in tables and plots) from the Keras library was used, with TensorFlow [9] as the backend. The network's architecture consisted of two hidden layers of 13 neurons (this number was chosen by trial and error), using ReLU as the activation function. In addition, the Adam optimization algorithm was used to update the neuron weights during the training phase, taking the mean square error as the loss function.

For the implementation of the CETs, we used the Category Encoders library of Scikit-learn-contrib [10]. The selected encoding techniques are classified and abbreviated in Table 2. In this table, the column *Type* indicates if the technique uses the target (supervised) or not (unsupervised) when converting the categorical attribute into a number. Also, the column *Increased size of the dataset* means that for each instance in the categorical attribute, an extra column will be added to the training dataset.

2.1 Real-World Datasets

The datasets taken from Kaggle are described in Table 3. Six datasets were solved as a classification problem and one as a regression problem.

In Table 3, the column *cardinality of categorical attributes* indicates the total sum of instances found for each categorical attribute in the dataset. It is important to take such cardinality into account when applying those techniques which may add extra columns for the instances in a categorical variable. For example, in the case of the datasets *Employee* and *Car auction*, it is necessary to add 15,626 and 42,370 extra columns respectively when *One Hot* Encoding [11] technique is applied.

Table 2. Categorical encoding techniques

Cat. encoder technique	Abbreviation	Type	Increased size of the dataset
Ordinal	ORD	Unsupervised	No
One hot	ONE	Unsupervised	Yes
Sum	SUM	Unsupervised	Yes
Helmert	HEL	Unsupervised	Yes
Backward difference	BACK	Unsupervised	Yes
Target	TAR	Supervised	No
MEstimate	ME	Supervised	No
Leave one out	LOO	Supervised	No
CatBoost	CAT	Supervised	No
James stein	JST	Supervised	No

Table 3. Description of real-world datasets

Dataset name	Type	Size	Total attributes	Cat.attributes	Cardinality of categorical attributes	Class distribution true–false %
Adult	Class	48,843	15	9	102	23.9–76.1
Titanic	Class	891	12	3	92	38.3–61.4
Credit	Class	307,511	122	18	152	91.9–08.1
Employee	Class	32,769	10	5	15,626	94.3–05.7
Car auction	Class	72,983	34	19	42,370	12.3–87.7
Telco Churn	Class	7,043	21	5	4,303	26.5–73.5
Housing California	Reg	20,610	10	1	5	NA

The attributes of type *Date* and *IDs* were removed. In addition, null values were replaced with the mean found in the column. Likewise, a standard scaling was applied to numerical values. During the training of the models, the cross-validation technique was used with a 3-fold repetition. The performance criterion for classification problems was the accuracy; the performance criterion for regression problems was the Root Mean Square Error (RMSE).

2.2 Synthetic Datasets

Using real-world datasets has the following drawbacks: 1) They may contain imperfections that require a cleaning effort. 2) Since the function that relates the data with the target to predict is unknown, it is very difficult to ascertain which is the model's best performance. 3) The actual nature of the possible categorical attributes is unknown (nominal, ordinal, high cardinality). To avoid these drawbacks, we propose to use synthetic data. The purpose is to artificially generate datasets via a known function. The advantages of including this type of data in the comparative study are: 1) It is possible to create an arbitrary number of records, numeric attributes, and categorical attributes. 2) A better evaluation of the models' performance may be achieved since the function that relates the independent variables with the dependent variable is known a priori. 3) Possibly complex data-cleaning tasks are avoided. 4) Create arbitrary classification problems: linear, nonlinear, with class balanced, unbalanced, etc.

In this work, the solution of 3 synthetic problems is included. The functions used to generate the corresponding data were:

$$f(x, y) = x^2 + 100y \tag{1}$$

$$f(x, y) = x^2 + y^2 \tag{2}$$

$$f(x, y) = x^3 + y^3 \tag{3}$$

These functions were selected arbitrarily and as an initial proposal. We plan in future work to explore the inclusion of more complex functions with categorical attributes of a different nature. The values of the integer variables x and y were randomly generated in $[-500, 500]$. The total number of tuples was 10,000. The functions were evaluated for each pair (x, y) and the numerical result was stored in a so-called 'target' column, which was the value to be predicted by the ML algorithms.

To transform the regression problem into a classification one, we used the following two approaches: 1) Linear classification. A single cut-off line was set up taking the average value in the target, in such a way that all the values greater than the average were set to *true*, and all the values less than the average were set to *false*; 2) Nonlinear classification. Two lines were set up, the numerical values within the two lines were set to *true* and all others were set to *false*. In both cases, the cut-off lines were selected in such a way that the class distribution was balanced (roughly 50% true – 50% false). Figure 1 illustrates the two mentioned approaches.

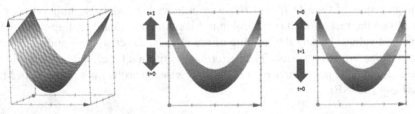

Fig. 1. *Left*: plot of function $x^2 + 100y$. *Center*: linear threshold for classification problem. *Right*: nonlinear classification problem

Since we want to better understand the effectiveness of the investigated encoders, it is necessary to map the numerical values of attributes x and y to a categorical value. This was done by adding the suffix '_encode' to every numerical value in the column to map (see Fig. 2). The resulting categorical attributes are ordinal and with relatively high cardinality (~1000 instances).

x	y	target		x	y	target
608	966	True		608	966_encoded	True
113	421	False	Encode	113	421_encoded	False
564	362	False	y	564	362_encoded	False
158	860	False		158	860_encoded	False
409	776	False		409	776_encoded	False
983	167	True		983	167_encoded	True

Fig. 2. Mapping of a numerical attribute into a categorical attribute

The same exercise was applied for the variable x but without coding y. This mapping task was done for the 3 selected functions, resulting in 6 datasets that were included in the comparative study.

3 Experimental Results

3.1 Real-World Dataset

Figures 3, 4, 5, 6, 7, 8 and 9 illustrate the results obtained for each dataset. Each figure contains the model-CET combinations that had the highest performance (left plot) and lowest performance (center plot), as well as the difference between these two values (right plot). This difference quantifies how widely CET can affect the same ML algorithm. In the case of the *Employee* (Fig. 5) and *Car Auction* (Fig. 6) datasets, the techniques that increase the size of the dataset were excluded due to the high cardinality in the categorical attributes.

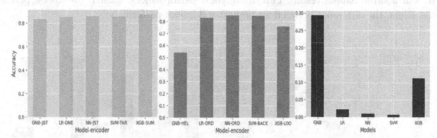

Fig. 3. Adult dataset results. Left: the highest performances. Center: the lowest performances. Right: the difference between the highest and lowest performance.

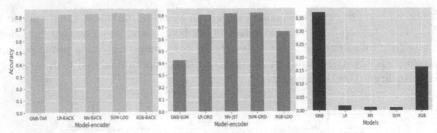

Fig. 4. Titanic dataset results. Left: the highest performances. Center: the lowest performances. Right: the difference the between the highest and lowest performance.

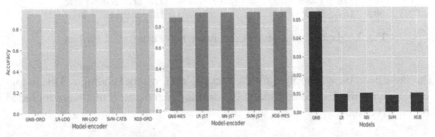

Fig. 5. Employee dataset results. Left: the highest performances. Center: the lowest performances. Right: the difference between the highest and lowest performance.

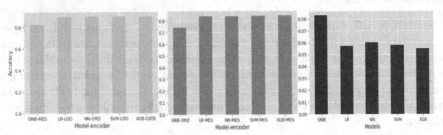

Fig. 6. Car auction dataset results. Left: the highest performances. Center: the lowest performances. Right: the difference between the highest and lowest performance.

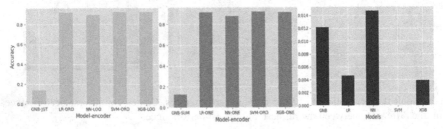

Fig. 7. Credit dataset results. Left: the highest performances. Center: the lowest performances. Right: the difference between the highest and lowest performance.

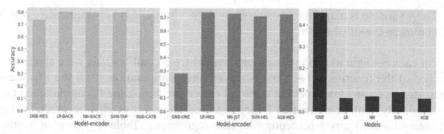

Fig. 8. Telco-churn dataset results. Left: the highest performances. Center: the lowest performances. Right: the difference between the highest and lowest performance.

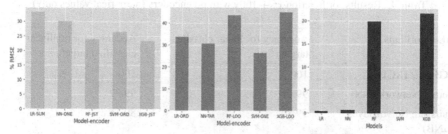

Fig. 9. Housing California dataset results. Left: the highest performances. Center: the lowest performances. Right: the difference between the highest and lowest performance.

It can be observed that CETs influence either positively or negatively the performance of ML models. One of the underlying behavior found is that multilayer perceptron (NN) are the models least affected by the different CETs. In contrast, the Naive-Bayes model is the most negatively affected, especially by those methods that add columns for each instance. Similar to the multilayer perceptron, the support vector machine and the logistic regression models are slightly affected regardless of the encoding technique. Furthermore, it was found that the CETs which appeared more frequently in the combinations with the worst performances were the Ordinal encoder (Figs. 3, 4, 6, 7 and 9) and the M-estimate encoder (Figs. 5, 6 and 8).

3.2 Synthetic-Datasets

The next results were obtained from synthetic data. Each dataset was solved as a classification problem in 6 different configurations:

1) Numerical values only and linear-classified target (reference values).
2) Numerical values only and nonlinear-classified target (reference values).
3) The x variable is mapped to a category, and the y variable is kept numerical, with a linear-classified target.
4) The y variable is mapped to a category, and the x variable is kept numerical, with a linear-classified target.
5) The x variable is mapped to a category, and the y variable is kept numerical, with a nonlinear-classified target.

6) The y variable is mapped to a category, and the x variable is kept numerical, with a nonlinear-classified target.

For each configuration, the 50 different model-CET combinations were applied. In the case of the function $x^2 + 100y$, when solving configuration 1, all the ML algorithms displayed an almost perfect accuracy (when predicting the linear-classified target). Similarly, with configuration 2 they achieved around over 90% accuracy for the nonlinear-classified target (excepting logistic regression). Table 4 shows the results obtained.

Table 4. Results for function $x^2 + 100y$ with no encoders (numerical values only)

ML algorithms	Accuracy for linear classified target (%)	Accuracy for nonlinear classified target (%)
XGBoost (XGB)	100	100
Neural Network (NN)	99.91	99.37
Gaussian Naïve-Bayes (GNB)	99.80	89.16
Logistic Regression (LR)	99.79	61.51
Support Vector Machine (SVC)	99.74	98.85

We denote these performances as *reference values*. It is expected that after mapping numerical values to a category (adding the prefix '*_encoded*') the obtained performances will be equal or lower than the reference values. In Tables 5 and 6 the results for configurations 3 and 4 are shown. As in real-word dataset results, for each ML model, we recorded the CET that got the highest and the lowest performance, as well as the difference between these two values. When a model got the same performance regardless of the CET, we use the word "*same*", and it means that the model was not affected at all by the encoding technique.

From Table 5, it can be observed that the ordinal encoder got the best performance. This could be because the variable x, before being mapped into a categorical attribute, had a numerical order. Therefore, probably ordinal encoder technique matched its integer assignments with the original order of variable x.

Similarly, Tables 7 and 8 show the results for configurations 6 and 7 (nonlinear-classified target).

The same measures were recorded for the functions $f(x, y) = x^2 + y^2$ and $f(x, y) = x^3 + y^3$ (see appendix). From the obtained results, it can be observed that the SVM, GNB, and NN models were negatively affected by those CETs which increase the size of the training dataset. In contrast, we found that the LR model, for the configurations with a nonlinear-classified target, performed better with those CETs that increase the size of data, even in some cases beat the references values.

Table 5. Results for $x^2 + 100y$ with x mapped to a category and linear-classified target

Model	CET with the best performance	Best accuracy	CET with the lowest performance	Lowest accuracy	Diff.
NN	ORD	100	HEL	95.29	4.71
XGB	– Same –	100	– Same –	100	0
LR	ORD	99.79	HEL	95.39	4.4
SVC	ORD	99.64	HEL	89.49	10.15
GNB	ORD	98.34	BACK	56.5	41.84

Table 6. Results for $x^2 + 100y$ with y mapped to a category and linear-classified target

Model	CET with the best performance	Best accuracy	CET with the lowest performance	Lowest accuracy	Diff.
XGB	– Same –	100	– Same –	100	0
NN	CAT	99.98	SUM	93.82	6.15
LR	CAT	99.81	HEL	95.96	3.85
SVC	CAT	99.57	ONE	75.57	24
GNB	CAT	98.84	SUM	45.57	53.26

Table 7. Results for function $x^2 + 100y$ with x mapped and nonlinear-classified target

Model	CET with the best performance	Best accuracy	CET with the lowest performance	Lowest accuracy	Diff.
XGB	ORD	100	JST	99.98	0.02
NN	ME	99.39	ONE	88.55	10.84
SVC	ORD	98.61	BACK	63.34	35.28
GNB	ORD	91.15	BACK	53.06	38.09
LR	HEL	79.34	ORD	61.56	17.78

Tables 9 and 10 summarize which CET obtained more frequently the highest and lowest performances in the different combinations. Catboost encoder [12] topped the list of the best performances, while backward difference encoder [11] did it for the lowest performances.

Table 8. Results for function $x^2 + 100y$ with y mapped and nonlinear-classified target

Model	CET with the best performance	Best accuracy	CET with the lowest performance	Lowest accuracy	Diff.
XGB	– Same –	100	– Same –	100	0
NN	CAT	99.59	ONE	80.87	18.71
SVC	ORD	98.71	BACK	53.65	45.06
GNB	CAT	89.15	SUM	53.12	36.03
LR	ONE	59.17	CAT	56.55	2.63

4   Conclusions

In this paper, we quantified how the different encoding techniques may affect machine learning algorithms. Although this effect can vary depending on the nature of the categorical attribute (nominal, ordinal) and its cardinality, some machine learning models are more susceptible to the application of certain categorical encoders. From the real-world data, we found that MLPs and SVM were the models least affected either positively or negatively. In contrast, the Naive Bayes model was severely affected by the categorical encoders that add attributes for each instance (performance decreased around 30%). We also observed that leave-one-out encoder negatively affected decision-tree-based models, such as XGBoost and Random Forest.

From the synthetic data, we found that, when predicting nonlinear-classified targets, the logistic regression model can have a better performance when one-hot encoder (or similar) is applied. It even outperformed the reference values. In addition, the XGBoost model was the least affected by the different encoding techniques. The encoding technique that reflected the best performance was the catboost encoder. On the other hand, the backward difference encoder appeared more frequently in combinations with the lowest performance. The ordinal encoder can reach good results, as long as the nature of the categorical attribute is ordinal and the order is known. Otherwise, if not applied carefully, a non-existent pattern will be induced which will affect negatively the performance of the model. Regarding the categorical attributes with high cardinality, it becomes impractical to apply those techniques which increase importantly the number of columns (like one-hot encoding).

Finally, the use of synthetic data allows us to better understand the effectiveness of the investigated encoders in a more controlled and faster way since is known the function that relates the dataset in predicting tasks. Moreover, it allows us to save time in cleaning and preprocessing tasks. Comparing with real-world datasets, the implementation of synthetic datasets requires less effort to explore more broadly the proposed research question. The comparative study will be expanded with the inclusion of more complex functions and with categorical attributes of different nature. One question which arises

Table 9. Summary of categorical encoders with the best performance obtained from synthetic data

CET with the highest performances	Count of high performing cases (total 55)	% of high performing cases
CatBoost	15	27.27
Leave one out	14	25.45
Ordinal	11	20.00
One hot encoder	5	9.09
Helmert	4	7.27
Sum	2	3.63
M estimate	2	3.63
James Stein	1	1.81
Backward difference	1	1.81

Table 10. Summary of categorical encoders with the worst performance obtained from synthetic data

CET with the lowest performance	Count of low performing cases (total 55)	% of low performing cases
Backward difference	22	40.00
Ordinal	13	23.63
Sum	8	14.54
Helmert	5	9.09
One hot encoder	4	7.27
CatBoost	2	3.63
James Stein	1	1.81

naturally is whether there is a way to tackle the problem of categorical encoding without the restrictions we have identified in all the methods analyzed herein. Our contention is that, indeed, at least one such method exists. We have called this the CESAMO algorithm and it will be the main subject of a paper to appear shortly.

Appendix

Results from synthetic data for the functions $x^2 + y^2$ (2) and $x^3 + y^3$ (3) (Tables 11, 12, 13, 14, 15, 16, 17, 18, 19 and 20).

Table 11. Results for the function $x^2 + y^2$ with no encoders (numerical values)

Model	Accuracy for linear classified target	Accuracy for nonlinear classified target
Support Vector Machine (SVC)	99.48	96.18
XGBoost (XGB)	98.88	97.43
Neural Network (NN)	98.84	94.68
Gaussian Naïve-Bayes (GNB)	95.78	56.09
Logistic Regression (LR)	52.58	52.31

Table 12. Results for the function $x^2 + y^2$ with x mapped and linear-classified target

Model	Encoder with the best performance	Best accuracy	Encoder with the lowest performance	Lowest accuracy	Diff.
NN	LOO	90.82	ORD	75.86	14.96
SVC	LOO	90.80	BACK	55.75	35.05
XGB	CAT	90.15	SUM	81.46	8.69
GNB	CAT	85.79	BACK	50.34	35.45
LR	LOO	73.66	ORD	54.45	19.21

Table 13. Results for the function $x^2 + y^2$ with y mapped and linear-classified target

Model	Encoder with the best performance	Best accuracy	Encoder with the lowest performance	Lowest accuracy	Diff.
NN	LOO	91.08	ORD	76.39	14.69
SVC	CAT	90.95	BACK	55.14	35.81
XGB	ME	90.95	SUM	77.19	13.76
GNB	CAT	85.35	BACK	50.4	34.95
LR	LOO	72.46	ORD	52.26	20.2

Table 14. Results for the function $x^2 + y^2$ with x mapped and nonlinear-classified target

Model	Encoder with the best performance	Best accuracy	Encoder with the lowest performance	Lowest accuracy	Diff.
NN	SUM	73.36	BACK	54.21	19.15
SVC	LOO	64.77	BACK	52.5	12.27
XGB	HEL	64.44	CAT	62.51	1.93
LR	CAT	56.17	ORD	51.89	4.29
GNB	JST	55.85	BACK	50.08	5.77

Table 15. Results for the function $x^2 + y^2$ with y mapped and nonlinear-classified target

Model	Encoder with the best performance	Best accuracy	Encoder with the lowest performance	Lowest accuracy	Diff.
NN	HEL	74.41	BACK	54.43	19.99
XGB	ORD	66.55	SUM	63.65	2.9
SVC	LOO	65.58	BACK	52.82	12.75
LR	BACK	57	ORD	51.61	5.39
GNB	LOO	56.67	BACK	49.86	6.81

Table 16. Results for the function $x^3 + y^3$ with no encoders (numerical values)

Model	Accuracy in linear classified target (%)	Accuracy in nonlinear classified target (%)
Neural Network (NN)	99.45	99.53
Support Vector Machine (SVC)	98.98	99.06
XGBoost (XGB)	98.84	99.94
Logistic Regression (LR)	98.27	48.25
Gaussian Naïve-Bayes (GNB)	97.97	94.88

Table 17. Results for the function $x^3 + y^3$ with x mapped and linear-classified target

Model	Encoder with the best performance	Best accuracy	Encoder with the lowest performance	Lowest accuracy	Diff.
LR	ONE	91.48	ORD	75.21	16.26
NN	HEL	90.08	ORD	75.39	14.69
SVC	CAT	87.2	BACK	75.44	11.76
GNB	LOO	86.71	BACK	52.89	33.83
XGB	LOO	85.89	SUM	78.39	7.5

Table 18. Results for the function $x^3 + y^3$ with y mapped and linear-classified target

Model	Encoder with the best performance	Best accuracy	Encoder with the lowest performance	Lowest accuracy	Diff.
LR	ONE	91.99	ORD	75.59	16.4
NN	SUM	89.76	ORD	75.87	13.89
SVC	LOO	88.13	BACK	76.05	12.08
GNB	LOO	86.8	BACK	52.2	34.6
XGB	LOO	86.63	SUM	79.15	7.48

Table 19. Results for the function $x^3 + y^3$ with x mapped and nonlinear-classified target

Model	Encoder with the best performance	Best accuracy	Encoder with the lowest performance	Lowest accuracy	Diff.
XGB	– Same –	99.95	– Same –	99.95	0
NN	CAT	99.59	HEL	83.4	16.19
SVC	CAT	98.49	BACK	53.15	45.34
GNB	ORD	95.99	BACK	51.69	44.3
LR	ONE	63.99	ORD	48.63	15.36

Table 20. Results for the function $x^3 + y^3$ with y mapped and nonlinear-classified target

Model	Encoder with the best performance	Best accuracy	Encoder with the lowest performance	Lowest accuracy	Diff.
XGB	– Same –	99.95	– Same –	99.95	0
NN	LOO	99.53	ONE	84.45	15.07
SVC	CAT	98.68	BACK	53.54	45.14
GNB	ORD	95.61	BACK	52.07	43.54
LR	ONE	66.01	ORD	47.84	18.17

References

1. Zheng, A., Casari, A.: Feature Engineering for Machine Learning: Principles and Techniques for Data Scientists. O'Reilly Media, Inc. (2018)
2. Agresti, A.: An Introduction to Categorical Data Analysis, pp. 1–10. John Wiley & Sons (2018)
3. Potdar, K., Pardawala, T.S., Pai, C.D.: A comparative study of categorical variable encoding techniques for neural network classifiers. Int. J. Comput. Appl. **175**(4), 7–9 (2017)
4. Hancock, J.T., Khoshgoftaar, T.M.: Survey on categorical data for neural networks. J. Big Data **7**(1), 1–41 (2020)
5. Pargent, F., Bischl, B., Thomas, J.: A benchmark experiment on how to encode categorical features in predictive modeling (Doctoral dissertation, M.Sc. Thesis), p. 12. Ludwig-Maximilians–Universitat Munchen (2019)
6. Alkharusi, H.: Categorical variables in regression analysis: a comparison of dummy and effect coding. Int. J. Educ. **4**(2), 202 (2012)
7. Bolón-Canedo, V., Sánchez-Maroño, N., Alonso-Betanzos, A.: A review of feature selection methods on synthetic data. Knowl. Inf. Syst. **34**(3), 483–519 (2013)
8. Pedregosa, F., et al.: Scikit-learn: machine learning in Python. J. Mach. Learn. Res. **12**, 2825–2830 (2011)
9. Abadi, M., et al.: Tensorflow: a system for large-scale machine learning. In: 12th USENIX symposium on operating systems design and implementation (OSDI 16), pp. 265–283 (2016)
10. McGinnis, W.D., Siu, C., Andre, S., Huang, H.: Category encoders: a scikit-learn-contrib package of transformers for encoding categorical data. J. Open Source Softw. **3**(21), 501 (2018)
11. UCLA Statistical Consulting Group: R Library: Contrast Coding Systems for Categorical Variables (2011)
12. Dorogush, A.V., Ershov, V., Gulin, A.: CatBoost: gradient boosting with categorical features support. arXiv preprint arXiv:1810.11363 (2018)

Long-Term Exploration in Persistent MDPs

Leonid Ugadiarov[1], Alexey Skrynnik[1,2]([✉]), and Aleksandr I. Panov[1,2]

[1] Moscow Institute of Physics and Technology, Moscow, Russia
[2] Artificial Intelligence Research Institute FRC CSC RAS, Moscow, Russia
skrynnik@isa.ru

Abstract. Exploration is an essential part of reinforcement learning, which restricts the quality of learned policy. Hard-exploration environments are defined by huge state space and sparse rewards. In such conditions, an exhaustive exploration of the environment is often impossible, and the successful training of an agent requires a lot of interaction steps. In this paper, we propose an exploration method called Rollback-Explore (RbExplore), which utilizes the concept of the persistent Markov decision process, in which agents during training can roll back to visited states. We test our algorithm in the hard-exploration Prince of Persia game, without rewards and domain knowledge. At all used levels of the game, our agent outperforms or shows comparable results with state-of-the-art curiosity methods with knowledge-based intrinsic motivation: ICM and RND. An implementation of RbExplore can be found at https://github.com/cds-mipt/RbExplore.

Keywords: Reinforcement learning · Curiosity based exploration · State space clustering

1 Introduction

Exploration is an essential component of reinforcement learning (RL). During training, agents have to choose between exploiting the current policy and exploring the environment. On the one hand, exploration can make the training process more efficient and improve the current policy. On the other hand, excessive exploration may waste computing resources visiting task-irrelevant regions of the environment [4,6].

Exploration is essential to solving sparse-reward tasks in environments with high dimensional state space. In this case, an exhaustive exploration of the environment is impossible in practice. A considerable amount of interaction data is required to train an effective policy due to the sparseness of the reward. A common approach is to use knowledge-based or competence-based intrinsic motivation [10]. In the first more commonly used approach, it is proposed to augment an extrinsic reward with the additional dense intrinsic reward that encourages exploration [2,3,15]. Another approach is to separate an exploration phase from a learning phase [6]. As noted by the authors of [6], the disadvantage of the first approach

© Springer Nature Switzerland AG 2021
I. Batyrshin et al. (Eds.): MICAI 2021, LNAI 13067, pp. 108–120, 2021.
https://doi.org/10.1007/978-3-030-89817-5_8

is that an intrinsic reward is a non-renewable resource. After exploring an area and consuming the intrinsic reward, the agent likely will never return to the area to continue exploration due to catastrophic forgetting and inability to rediscover the path because it has already consumed the intrinsic reward that could lead to the area.

Implementing a mechanism that reliably returns the agent to the neighborhood of known states from which further exploration might be most effective is a challenging task for both approaches. In the case of resettable environments (e.g., Atari games or some robotic simulators), it is possible to save the current state of the simulator and restore it in the future. Many real-world RL applications are inherently reset-free and require a non-episodic learning process. Examples of this class of problems include robotics problems in real-world settings and problems in domains where effective simulators are not available and agents have to learn directly in the real world. Recent work has focused on reset-free setting [17,18]. On the other hand, for many domains, simulators are available and widely used at least in the pretraining phase (e.g., robotics simulators [9]). Specific properties of resettable environments make it possible to reliably return to previously visited states and increase exploration efficiency by reducing the required number of interactions with the environment. Therefore, exploration algorithms should effectively visit all states of an environment. However, factoring in the high dimension of the state space, it is intractable in practice to store all the visited states. Therefore, effective exploration of the environment remains a difficult problem, even for resettable environments.

In this paper, we propose to formalize the interaction with resettable environments as a persistent Markov decision process (pMDP). We introduce the RbExplore algorithm, which combines the properties of pMDP with clustering of the state space based on similarity of states to approach long-term exploration problems. The distance between states in trajectories is used as a feature for clustering. The states located close to each other are considered similar. The states distant from each other are considered dissimilar. Clusters are organized into a directed graph where vertices correspond to clusters, and arcs correspond to possible transitions between states belonging to different clusters. RbExplore uses a novelty detection module as a filter of perspective states. We introduce the Prince of Persia game environment as a hard-exploration benchmark suitable for comparing various exploration methods. The percentage coverage metric of the game's levels is proposed to evaluate exploration. RbExplore outperforms or shows comparable performance with state-of-the-art curiosity methods ICM and RND on different levels of the Prince of Persia environment.

2   Related Work

Three types of exploration policies can be indicated. Exploration policies of the first type use an intrinsic reward as an exploration bonus. Exploration strategies of the second type are specific to multi-goal RL settings where exploration is driven by selecting sub-goals. Exploration policy of the third type use clustered representation of the set of visited states.

In recent works [3,4,11,15], the curiosity-driven exploration of deep RL agents is investigated. The exploration methods proposed by these works can be attributed to the first type. The extrinsic sparse reward is replaced or augmented by a dense intrinsic reward measuring the curiosity or uncertainty of the agent at a given state. In this way, the agent is encouraged to explore unseen scenarios and unexplored regions of the environment. It has been shown that such a curiosity-driven policy can improve learning efficiency, overcome the sparse reward problem to some extent, and successfully learn challenging tasks in no-reward settings.

Another line of recent work focuses on multi-goal RL and can be attributed to the second type. Algorithm HER [1] augments trajectories in the memory buffer by replacing the original goals with the actually achieved goals. It helps to get a positive reward for the initially unsuccessful experience, makes reward signal denser, and learning more efficient especially in sparse-reward environments. A number of RL methods [7,13] focus on developing a better policy for selecting sub-goals for augmentation of failure trajectories in order to improve HER. These policies ensure that the distribution of the selected goals adaptively changes throughout training. The distribution should have greater variance in the early stages of training and direct the agent to the original goal in the latter stages. Other works [8,12,16] propose methods to generate goals that are feasible, and their complexity corresponds to the quality of the agent's policy. The distribution of generated goals changes adaptively to support sufficient variance ensuring exploration in goal space.

The Go-Explore [6] algorithm could be attributed to the third type of exploration policy. It builds a clustered lower-dimensional representation of a set of visited states in the form of an archive of cells. Two types of representation are proposed for Montezuma's Revenge environment: with domain knowledge based on discretized agent coordinates, room number, collected items, and without domain knowledge based on compressed grayscale images with discretized pixel intensity into eight levels.

Exploration of the state space is implemented as an iterative process. At each iteration, a cell is sampled from the archive, its state is restored in the environment, and the agent starts exploration with stochastic exploration policy. If the agent visits new cells during the run, they are added to the archive. The statistic of visits is updated for existing cells in each iteration. For both types of representation, the cell stores the highest score that the agent had when it visited the cell. A cell is sampled from the archive by heuristic, preferring more promising cells.

Exploiting domain-specific knowledge makes it difficult to use Go-Explore in a new environment. In our work, we use the idea of clustering of a set of visited states and propose to use a supervised learning model to perform clustering based on the similarity of states. We use a reachability network from the Episodic Curiosity Module [15] as a similarity model predicting similarity score for a pair of states. The clusters are organized into a graph using connectivity information between their states in a similar way as the Memory graph [14] is built. RND module [4] is used to detect novel states. Our approach does not exploit domain knowledge, which allows us to apply RbExplore to the Prince of Persia environment without feature handcrafting.

3   Background

3.1   Markov Decision Processes

A Markov Decision Process (MDP) for a fully observable environment is considered as a model for interaction of an agent with an environment:

$$\mathcal{U} = (\mathcal{S}, \mathcal{A}, p, r, \gamma, s_{init}),\qquad(1)$$

$\mathcal{S}$—a state space, $\mathcal{A}$—an action space, $p : \mathcal{S} \times \mathcal{A} \times \mathcal{S} \to \mathbb{R}$—a state transition distribution, $r : \mathcal{S} \times \mathcal{A} \times \mathcal{S} \to \mathbb{R}$—a reward function, $\gamma \in [0; 1]$—a discount factor, and $s_{init} \in \mathcal{S}$—an initial state of the environment.

An episode starts in the state s_0. Each step t the agent samples an action a_t based on the current state s_t: $a_t \sim \pi(\cdot|s_t)$, where $\pi : \mathcal{S} \times \mathcal{A} \to \mathbb{R}$—a stochastic policy, which defines the conditional distribution over the action space. The environment responds with a reward $r_t = r(s_t, a_t, s_{t+1})$ and moves into a new state $s_{t+1} \sim p(\cdot|s_t, a_t)$. The result of the episode is a return R_0—a discounted sum of the rewards obtained by the agent during the episode, where $R_t = \sum_{i=t}^{T} \gamma^{i-t} r_i$. Action-value function Q^π is defined as the expected return for using action a_t in a certain state s_t: $Q^\pi(s_t, a_t) = \mathbb{E}_{s_{t+1} \sim p(\cdot|s_t,a_t), a_{t+1} \sim \pi(\cdot|s_{t+1})} [R_t|s_t, a_t]$. State-value function V^π can be defined via action-value function Q^π: $V^\pi(s) = \max_a Q^\pi(s, a)$. The goal of reinforcement learning is to find the optimal policy π^*:

$$\pi^* = \mathrm{argmax}_\pi Q^\pi(s, a) \quad \forall s \in \mathcal{S}, \forall a \in \mathcal{A}\qquad(2)$$

3.2   Persistent MDPs

The persistent data structure allows access to any version of it at any time [5]. Inspired by that structures, we propose persistent MDPs for RL. We consider an MDP to have a persistence property if for any state $s_v \in \mathcal{S}$ exists policy $\pi_{s_v}^p$, which transits agent from the initial state s_{init} to state s_v, in a finite number of timesteps T. Thus, a persistent MDP is expressed as:

$$\mathcal{U}^p = (\mathcal{S}, \mathcal{A}, p, r, \gamma, s_{init}, \pi^p),\qquad(3)$$

However, the way of returning to visited states can differ. For example, instead of policy $\pi_{s_v}^p$, it could be an environment property, that allows one to save and load states.

4   Exploration via State Space Clustering

In this paper, we propose the RbExplore algorithm that uses similarity of states to build clustered representation of a set of visited states. There are two essential components of the algorithm: a similarity model, which predicts a similarity measure for a pair of states, and a graph of clusters, which is a clustered representation of a set of visited states organized as a graph. The scheme of the algorithm is shown in Fig. 1.

A high-level overview of one iteration of the RbExplore algorithm:

1. Generate exploration trajectories: sample M clusters from the graph of clusters $\mathbf{G}$ based on cluster visits statistics (e.g., preferring the least visited clusters), roll back to corresponding states, and run exploration.
2. Generate training data for the similarity model R from the exploration trajectories and additional trajectories starting from novel states filtered by the novelty detection module. Full trajectory prefixes are used to generate negative examples.
3. Train the similarity model R.
4. Update the graph $\mathbf{G}$ with states from the exploration trajectories and merge its clusters. A state is added to the graph $\mathbf{G}$ and forms new clusters if it is dissimilar to states which are already in the graph $\mathbf{G}$. The similarity model is used to select such states.
5. Train the novelty detection module on the states from the exploration trajectories.

As a result of one iteration, novel states are added to the graph $\mathbf{G}$, the statistics of visits to existing clusters are updated, the similarity model and the novelty detection module are trained on the data collected during the current iteration.

4.1   Similarity Model

As a feature for clustering, it is proposed to use the distance between states in trajectories. The states located close to each other are considered similar, the states distant from each other are considered dissimilar. A supervised model is used to estimate the similarity measure between states $R : \mathcal{S} \times \mathcal{S} \to [0, 1]$. It takes a pair of states as input and outputs a similarity measure between them. The training dataset is produced by labeling pairs of states for the same trajectory $\{\tau^k = s_1^k, \ldots, s_T^k\}$: triples (s_i^k, s_j^k, y_{ij}^k) are constructed, where y_{ij}^k is a class label. States s_i^k, s_j^k are considered similar ($y_{ij}^k = 1$) if the distance between them in the trajectory τ^k is less than n steps: $|i - j| < n$. Negative examples ($y_{ij}^k = 0$) are obtained from pairs of states that are more than N steps apart from each other: $|i - j| > N$. The model R is trained as a binary classifier predicting whether two states are close in the trajectory (class 1) or not (class 0).

Figure 2 illustrates the training data generation procedure. A neural network model is used as a similarity model R as the experiments are performed in environments with high-dimensional state spaces. The network R with parameters

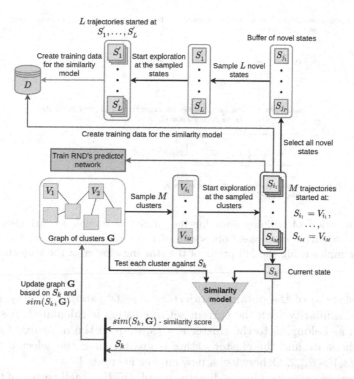

Fig. 1. Scheme of the RbExplore algorithm: M exploration trajectories are generated by running exploration from clusters sampled from $\mathbf{G}$. L additional trajectories are generated by running exploration from novel states selected from the exploration trajectories by the novelty detection module. Training data is generated for the similarity model from the exploration trajectories and the additional trajectories. The similarity model is trained on the generated data. $\mathbf{G}$ is updated based on the states from the exploration trajectories and their similarity score with clusters of $\mathbf{G}$. The novelty detection module is trained on the states from the exploration trajectories.

w is trained on the training data set D using binary cross-entropy as a loss function:

$$\mathbb{E}_{(s_1,s_2,y)\sim D}\left[-y\log R_w(s_1,s_2) - (1-y)\log\left(1 - R_w(s_1,s_2)\right)\right] \to \min_w$$

4.2   Graph of Clusters

The clustering of the state space $\mathcal{S}$ is an iterative process using the similarity model R and the chosen stochastic exploration policy $\pi_{explore}(a|s)$ (e.g. uniform distribution over actions). A cluster $v = (s, snap)$ is a pair of state s—the center of the cluster and the corresponding snapshot of the simulator $snap$. At each iteration, a cluster is selected from which exploration will be continued. The state of the selected cluster is restored in the environment using the corresponding snapshot, and the agent starts exploration with stochastic exploration policy $\pi_{explore}$.

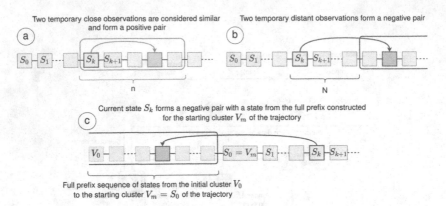

Fig. 2. Generation of training data for the similarity model. **a), b)** Generation of positive and negative examples from states of the same trajectory. **c)** Generation of negative examples using the full prefix of the starting cluster of the trajectory.

For each state s_i of the obtained trajectory $\tau = (s_1, snap_1), \dots, (s_T, snap_T)$ a measure of similarity with the current set of clusters is calculated. A state s_i is considered as belonging to the cluster $v = (s, snap)$ if the measure of similarity between the state and the cluster's state is greater than the selected threshold θ_{sim}: $R(s, s_i) > \theta_{sim}$. Otherwise, a new cluster is created.

Clusters are organized into a directed graph (V, E). Each vertex of the graph corresponds to a cluster. If two successive states s_k and s_{k+1} in the same trajectory $\tau = s_1, \dots, s_T$ belong to different clusters v_i and v_j, an arc between those clusters (v_i, v_j) is added to the graph. Cluster visit statistics and arc visit statistics are updated each iteration. The graph is initialized with the initial state of the environment $(s_{init}, snap_{init})$. The cluster is selected from the graph for exploration using sampling strategy σ that can take into account the structure of the graph and the collected statistics (e.g., the probability of sampling a cluster is inversely proportional to the number of visits). Each iteration of the graph building procedure can be alternated with training the similarity model R on the obtained trajectories. The search for a cluster to which the current state s_i of the trajectory belongs can be accelerated by considering first those vertices which are adjacent to the cluster to which the previous state s_{i-1} was assigned.

In order to improve the quality of the similarity model R on states from novel regions of the state space $\mathcal{S}$, an RND module is used. For each state, the RND module outputs an intrinsic reward, which is used as a measure of the state's novelty. The state is considered novel if the intrinsic reward is greater than $\beta_{intrinsic}$. At each iteration, all states from the trajectory τ, which are detected by the RND module as a novel, are placed into the buffer of novel states B. The buffer B is used to generate additional training data for the similarity model R which includes states from novel regions of the state space $\mathcal{S}$. A set of states $\{(s_i^*, snap_i^*)\} \sim B$ is randomly sampled from the buffer B, and the agent starts an exploration with an exploration strategy by restoring the sampled simulator states. The resulting trajectories are used solely to generate additional training data for the similarity model R.

When an exploration trajectory is processed and a new cluster v is added to the graph G a new arc from v to the parent cluster from which v was created is added to a set of arcs to parent cluster E_{parent}. A prefix—a sequence of states from the parent cluster to v is stored along with the arc. Thus, for any cluster in the graph G, it is possible to construct a sequence of states that leads to the initial cluster $(s_{init}, snap_{init})$. This property is used to add negative pairs (s_1, s_2) to the training data set of the similarity model R such that $s_1 \in \tau_1$ and $s_2 \in \tau_2$ are distant from each other and $\tau_1 \neq \tau_2$. If the exploration trajectory τ started from a cluster $(s, snap)$, a full prefix $\tau^{prefix} = s_{init}, \dots, s$ for the trajectory $\tau = (s, snap), \dots$ is constructed to obtain additional negative examples. For a state $s_1 \in \tau$, a sufficiently distant state $s_2 \in \tau^{prefix}$ is randomly selected to form a negative example (s_1, s_2). Figure 2(c) illustrates this procedure.

Redundant clusters are created at each iteration due to the inaccuracy of the similarity model R. A cluster merge procedure is proposed to mitigate the issue. It tests all pairs of clusters in the graph G and merges the pair $v_1 = (s_1, snap_1), v_2 = (s_2, snap_2)$ into a new cluster if the similarity measure of their states is greater than the selected threshold θ_{merge}: $R(s_1, s_2) > \theta_{merge}$. The new cluster is incident to any arc that was incident to the two original clusters. Cluster visits statistics are summarized during merging. As a state and a snapshot of the new cluster, the states and the snapshot of the cluster that was added to the graph G earlier are selected.

5   The Prince of Persia Domain

We evaluate our algorithm on the challenging Prince of Persia game environment, which has over ten complex levels. Figure 3 shows the first level of the game. To pass it, the prince needs to: find a sword, avoiding traps; return to the starting point; defeat the guard, and end the level. In most cases, the agent goes to the next level when he passes the final door, which he also needs to open somehow.

The input of the agent is a 96×96 grayscale image. The agent chooses from seven possible joystick actions no-op, left, right, up, down, A, B. The same action may work differently depending on the situation in the game. For example, the agent can jump forward for a different distance, depending on the take-off run. The agent can jump after using action A or strike with a sword if in combat. Also, the agent can interact with various objects: ledges, pressing plates, jugs, breakable plates.

The environment is difficult for RL algorithms because it has the action space with changing causal relationships and requires mastery in many game aspects, such as fighting. Also, the first reward is thousands of steps apart from the initial agent position.

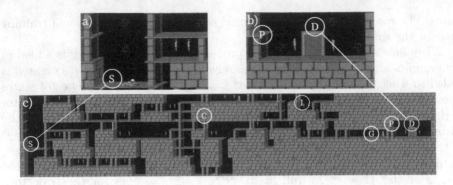

Fig. 3. The Prince of Persia environment. **a)**, **b)** Examples of environment observations from the agent's view. **c)** The complete map of the first level of the game. The agent's task is to get from the initial location (I) to the final door (D). To solve this problem, the agent needs to: pick up a sword (S), go back and defeat a guard (G), stand on the pressure plate (P) to open the door, proceed to the exit. The environment has many obstacles, such as cell doors (C) and various traps.

We use the percentage coverage metric $\% \text{ Cov} = \frac{|U_{visited}|}{|U_{full}|}$, i.e. the ratio of coverage to the max coverage in the corresponding level, where $U_{visited}$ is a set of the visited units and U_{full} full coverage set. We consider the minimum unit to be the area that roughly corresponds to space above the plate. For example, for the first level in the room with the sword, the full area has 36 units, but the agent can visit only 34 of them.

6   Experiments

We evaluate the exploration performance of our RbExplore algorithm on the first three levels of the Prince of Persia environment alongside state-of-the-art curiosity methods ICM and RND.

6.1   Experimental Setup

Raw environment frames are preprocessed by applying a frameskip of 4 frames, converting into grayscale, and downsampling to 96×96. Frame pixels are rescaled to the 0–1 range. The neural network R consists of two subnetworks: ResNet-18 network and a four-layer fully connected neural network. ResNet-18 accepts a frame as input and produces its embedding of dimension 512. Embeddings of the pair's frames are concatenated and fed into the fully connected network that performs binary classification of the pair of embeddings.

The same algorithm parameters are used for all levels. The maximum number of frames per exploration trajectory is 1,500. Episodes are terminated on the loss of life. Training data for the similarity model R is generated with parameters $n = 5$ and $N = 25$. The similarity threshold $\theta_{similarity} = 0.5$. The similarity threshold for the cluster merging $\theta_{merge} = 0.5$. Merge is run every 15 iteration. For exploration, clusters are sampled from the graph G with probabilities inversely proportional to the number of visits. Every iteration $M = 30$ clusters are sampled from the graph. The uniform distribution overall actions are used as the exploration policy $\pi_{explore}$. RND module intrinsic reward threshold for detecting novel states $beta_{intrinsic} = 2.5$. $L = 1000$ states are sampled from the buffer of novel states.

To prevent the formation of a large number of clusters at the early stages due to the low quality of the similarity model R, the similarity model is pretrained for 500,000 steps with a gradual increase of the value of the similarity threshold from 0 to $\theta_{similarity}$. At the same time, the necessary normalization parameters of the RND module are initialized. After pretraining the graph G is reset to the environment's initial state $(s_{init}, snap_{init})$ and RbExplore is restarted with the fixed similar threshold $\theta_{similairty}$.

6.2   Exploring the Prince of Persia Environment

By design of the Prince of Persia environment, the agent's observation does not always contain information about whether the agent carries a sword. To get around this issue the agent starts the first level with the sword. Also, the agent is placed at the point where the sword is located. This location is far enough from the final door, so reaching the final door is still a challenging task. For the other levels, we did not make any changes to the initial state.

Evaluation of RbExplore, ICM, and RND on the first three levels of the Prince of Persia environment is shown in Fig. 5. On the first level, RbExplore performed significantly outperforms ICM and RND, and also have visited all possible rooms of the level. The visualization of the coverage is shown in Fig. 4.

On levels two and three, none of the algorithms were able to visit all the rooms in 15 million steps. On level two RbExpore shows slightly worse results than RND and ICM. We explain it by the fact that the learning process in RND and ICM is driven by an exploration bonus, which helps them to explore local areas inside rooms more accurately. Each of the algorithms was able to visit only seven rooms at the very beginning of the level. On level three, RbExplore shows slightly better coverage than RND, and both of them outperform ICM.

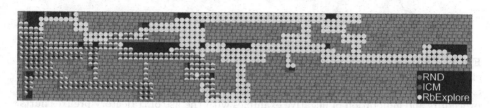

Fig. 4. The visualization of the first level coverage for the best RND, ICM, and RbExplore runs. RbExplore significantly outperforms RND and ICM.

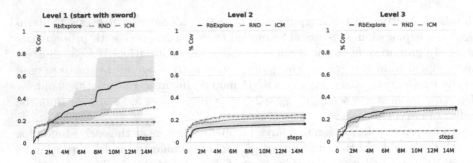

Fig. 5. Performance of RbExplore, RND, and ICM for the first three levels of the Prince of Persia environment. **Left:** Level 1—RbExplore significantly outperforms RND and ICM; **Center:** Level 2—All methods resulted in visits to seven rooms. RND and ICM outperform RbExplore as they deal better with local explorations and explore rooms more thoroughly; **Right:** Level 3—RbExplore significantly outperforms ICM and show comparable performance with RND; curves are averaged over three runs. The shading indicates the min-max range.

6.3   Ablation Study

In order to evaluate the contribution of each component of the RbExplore algorithm, we perform an ablation study. Experiments were run on level one with two versions of RbExplore. The first version does not merge clusters in the graph G. The second one does not build the full trajectory prefix when generating negative examples for the training data of the similarity model; thus, states of

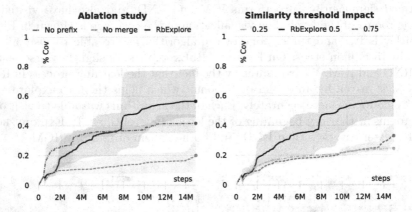

Fig. 6. Left: Level 1 - Comparison of RbExplore with its versions that do not merge clusters (No merge) and do not build the full trajectory prefix to generate a negative example for the training data for the similarity model (No prefix). Disabling merge procedure and generation of negative examples with the use of trajectory prefixes hurts performance of the RbExplore algorithm; **Right:** Level 1—Comparison of RbExplore with similarity thresholds for merging $\theta_{merge} \in \{0.25, 0.5, 0.75\}$. The performance of RbExplore with $\theta_{merge} \in \{0.25, 0.75\}$ is worse than that of RbExplore with $\theta_{merge} = 0.5$; curves are averaged over three runs. The shading indicates the min-max range.

negative pairs are sampled only from the same trajectory. Figure 6 shows that disabling these components hurts the performance of the algorithm.

Additional experiments were run on level one to study the impact of the value of θ_{merge} parameter on the performance. Figure 6 shows that the performance of RbExplore with $\theta_{merge} \in \{0.25, 0.75\}$ is worse than that of RbExplore with $\theta_{merge} = 0.5$.

7   Conclusion

In this paper, we introduce a pure exploration algorithm RbExplore that uses the formalized version of a resettable environment named persistent MDP. The experiments showed that RbExplore coupled with a simple exploration policy, which is a uniform distribution over actions, demonstrates performance comparable with RND and ICM methods in the hard-exploration environment of the Prince of Persia game in a no-reward setting. RbExplore, ICM, and RND got stuck on the second and third levels roughly in the same locations where the agent must perform a very specific sequence of actions over a long-time horizon to go further. The combining of RbExplore exploration and exploitation of RL approaches, which also utilize pMDPs, is an important direction for future work to resolve this problem.

Acknowledgements. This work was supported by the Russian Science Foundation (Project No. 20-71-10116).

References

1. Andrychowicz, M., et al. (eds.): Advances in Neural Information Processing Systems, vol. 30, pp. 5048–5058. Curran Associates, Inc. (2017). https://proceedings.neurips.cc/paper/2017/file/453fadbd8a1a3af50a9df4df899537b5-Paper.pdf
2. Bellemare, M.G., Srinivasan, S., Ostrovski, G., Schaul, T., Saxton, D., Munos, R.: Unifying count-based exploration and intrinsic motivation. In: Lee, D.D., Sugiyama, M., von Luxburg, U., Guyon, I., Garnett, R. (eds.) Advances in Neural Information Processing Systems 29: Annual Conference on Neural Information Processing Systems 2016, Barcelona, Spain, 5–10 December 2016, pp. 1471–1479 (2016). https://proceedings.neurips.cc/paper/2016/hash/afda332245e2af431fb7b672a68b659d-Abstract.html
3. Burda, Y., Edwards, H., Pathak, D., Storkey, A., Darrell, T., Efros, A.A.: Large-scale study of curiosity-driven learning. In: ICLR (2019)
4. Burda, Y., Edwards, H., Storkey, A., Klimov, O.: Exploration by random network distillation. In: International Conference on Learning Representations (2019). https://openreview.net/forum?id=H1lJJnR5Ym
5. Driscoll, J.R., Sarnak, N., Sleator, D.D., Tarjan, R.E.: Making data structures persistent. J. Comput. Syst. Sci. **38**(1), 86–124 (1989)
6. Ecoffet, A., Huizinga, J., Lehman, J., Stanley, K.O., Clune, J.: Go-explore: a new approach for hard-exploration problems (2021)

7. Fang, M., Zhou, T., Du, Y., Han, L., Zhang, Z.: Curriculum-guided hindsight experience replay. In: Wallach, H., Larochelle, H., Beygelzimer, A., d'Alché-Buc, F., Fox, E., Garnett, R. (eds.) Advances in Neural Information Processing Systems, vol. 32, pp. 12623–12634. Curran Associates, Inc. (2019). https://proceedings.neurips.cc/paper/2019/file/83715fd4755b33f9c3958e1a9ee221e1-Paper.pdf
8. Florensa, C., Held, D., Geng, X., Abbeel, P.: Automatic goal generation for reinforcement learning agents. In: Dy, J., Krause, A. (eds.) Proceedings of the 35th International Conference on Machine Learning. Proceedings of Machine Learning Research, vol. 80, pp. 1515–1528. PMLR, Stockholmsmässan, Stockholm, July 2018. http://proceedings.mlr.press/v80/florensa18a.html
9. OpenAI, et al.: Learning dexterous in-hand manipulation. Int. J. Robot. Res. **39**(1), 3–20 (2020)
10. Oudeyer, P.Y., Kaplan, F.: How can we define intrinsic motivation? In: Proceedings of the 8th International Conference on Epigenetic Robotics: Modeling Cognitive Development in Robotic Systems (2008)
11. Pathak, D., Agrawal, P., Efros, A.A., Darrell, T.: Curiosity-driven exploration by self-supervised prediction. In: ICML (2017)
12. Racaniere, S., Lampinen, A., Santoro, A., Reichert, D., Firoiu, V., Lillicrap, T.: Automated curriculum generation through setter-solver interactions. In: International Conference on Learning Representations (2020). https://openreview.net/forum?id=H1e0Wp4KvH
13. Ren, Z., Dong, K., Zhou, Y., Liu, Q., Peng, J.: Exploration via hindsight goal generation. In: Wallach, H., Larochelle, H., Beygelzimer, A., d'Alché-Buc, F., Fox, E., Garnett, R. (eds.) Advances in Neural Information Processing Systems, vol. 32, pp. 13485–13496. Curran Associates, Inc. (2019). https://proceedings.neurips.cc/paper/2019/file/57db7d68d5335b52d5153a4e01adaa6b-Paper.pdf
14. Savinov, N., Dosovitskiy, A., Koltun, V.: Semi-parametric topological memory for navigation. In: International Conference on Learning Representations (2018). https://openreview.net/forum?id=SygwwGbRW
15. Savinov, N., Raichuk, A., Vincent, D., Marinier, R., Pollefeys, M., Lillicrap, T., Gelly, S.: Episodic curiosity through reachability. In: International Conference on Learning Representations (2019). https://openreview.net/forum?id=SkeK3s0qKQ
16. Alexey, S., Panov, A.I.: Hierarchical reinforcement learning with clustering abstract machines. In: Kuznetsov, S.O., Panov, A.I. (eds.) RCAI 2019. CCIS, vol. 1093, pp. 30–43. Springer, Cham (2019). https://doi.org/10.1007/978-3-030-30763-9_3
17. Xu, K., Verma, S., Finn, C., Levine, S.: Continual learning of control primitives: skill discovery via reset-games (2020)
18. Zhu, H., et al.: The ingredients of real world robotic reinforcement learning. In: International Conference on Learning Representations (2020). https://openreview.net/forum?id=rJe2syrtvS

Source Task Selection in Time Series via Performance Prediction

Jesús García-Ramírez[✉] , Eduardo Morales , and Hugo Jair Escalante

Instituto Nacional de Astrofísica Óptica y Electrónica (INAOE),
Sta. Maria Tonantzintla, 72840 Puebla, Mexico
{gr_jesus,emorales,hugojair}@inaoep.mx

Abstract. Deep Learning has shown high performance in different domains, however, they need large computational resources and training data sets. Transfer learning offers an alternative to reduce both aspects but requires an effective way for selecting relevant pre-trained models for the current task. In this work, we propose a meta-learning formulation using the entropies of the feature maps produced by a pre-trained model for predicting its performance in a new target task via a regression model. Our method is tested in the time-series domain, where it obtains a better top-1 precision and a better position of a useful source task than state-of-the-art methods in 85 datasets from the UCR archive.

Keywords: Source task selection · Meta-learning · Time series

1 Introduction

Deep learning has shown impressive performance in difficult tasks such as image classification [6], games (e.g., Atari games [8] and Go games [13,14]), and natural language processing [2], among others. Nevertheless, long training times and the number of instances to learn a new task are the main limitations of this approach. Transfer Learning (TL) offers an alternative to reduce these limitations, by reusing knowledge from one or more source tasks task to learn, more efficiently, a new similar target task.

Yang et al. [16] discuss three important questions to take into account when TL methods are developed: (i) *When to transfer?*, this question is related to the selection of a source tasks (pre-trained models), the aim is to avoid a negative transfer, where the performance of the training without transfer is better than training with a transferred pre-trained model; (ii) *What to transfer?* this question refers to determining the part of the knowledge that will be transferred to a new task, in some cases transferring part of a model can obtain similar performance compared to transferring the entire model; (iii) *How to transfer?* refers to the method used to transfer elements to the new task. In this work, we approach the first question in the context of deep learning: we want to select a pre-trained model that will be useful (positive transfer) for the target task, in time series analysis domain.

© Springer Nature Switzerland AG 2021
I. Batyrshin et al. (Eds.): MICAI 2021, LNAI 13067, pp. 121–130, 2021.
https://doi.org/10.1007/978-3-030-89817-5_9

A time series $X = [x_1, x_2, \ldots, x_T]$ is an ordered set of real values[1], that is $x_i \in \mathbb{R}$. The length of X is equal to the number of real values T [5]. Some previous works that have tried to answer the question of *when to transfer?* in time series domain. Fawaz et al. [5] propose a method based on a reduction of the class prototypes. Then, they search the minimum distance between classes using dynamic time warping [9], they transfer the pre-trained models between each available task, and provide the results of these experiments. Meiseles et al. [7] propose to use the mean silhouette coefficient to find the separation between the classes of a dataset using a hidden layer of a pre-trained model. Differently to them, a method based on a regression model used to predict the performance of a pre-trained model is proposed. We use the output of the regression model (estimation of the performance transferring the pre-trained model to the new task) to rank the pre-trained models and select one that obtains positive or neutral transfer when we fine-tune it in the target task.

In this work, we propose a meta-learning approach that uses the entropies of the feature maps as input to a regression model that predicts the performance of a pre-trained model in a new task, specifically we use convolutional neural networks (CNN) to test our approach. We observe that the pre-trained models that produce outputs with diverse values are more *"transferables"* (useful when transferred) than those that produce uniform outputs. This variability of values can be estimated by the entropy. A limitation of our method is that it needs some (meta) examples of TL experiments to train the model. To alleviate this, the datasets with the smallest number of instances, in their training sets, are used to train the regression model. Also, our method does not require knowledge of the source data and the classes of the target dataset as other state-of-the-art methods, consequently, our method can be applied in a semi-supervised learning framework or in deep reinforcement learning, where the actions of the optimal policy are not available at the beginning of training. The code and additional material of this work are publicity available[2].

We evaluated our approach in the UCR archive [4], which contains 85 datasets of time series with different data such as images, real values, or signals. According to the results, our method obtains a better top-1 precision than other state-of-the-art methods, also the top-1 pre-trained models are placed in better positions. Also, we obtain a useful model in better positions (at least the 30th. position), when compared to those of other state-of-the-art methods (Fawaz method 37th and Miseles 53th).

The remainder of the paper is organized as follows. In Sect. 2, background information on transfer learning is presented. In Sect. 3, we provide a description of state-of-the-art methods for source task selection in time series domains. Details of the proposed method are given in Sect. 4. In Sect. 5, we report the experimental results. Finally, in Sect. 6, we summarize the conclusions and give possible directions for future work.

---

[1] Please note that timse series can be multi variate as well (i.e., $x_i \in \mathbb{R}^d$).
[2] https://github.com/gr-jesus/source_task_time_series.

2 Transfer Learning

In this section, we present background information on Transfer Learning (TL). TL offers an alternative to reduce training times, and is useful when the number of instances is insufficient to train a model for a specific problem. TL uses previously obtained knowledge from one or more source tasks to learn a new target task. An interesting problem of TL is to answer the question of *when to transfer?* [16], this problem is approached in this work. Specifically, we want to select, from a set of source tasks (or pre-trained models), one that will be useful for a target task.

TL uses a source task with sufficient labeled examples to approximate a conditional distribution $P(Y_s|X_s)$, it could be seen as a function $f(\cdot)$), using a marginal distribution $P(X_s)$, where $X_s = \{x_1, x_2, \ldots, x_n\}$ are the instances and $Y_s = y_1, \ldots, y_k$ are the labels for the source task. The target task may have few examples, insufficient to approximate the conditional distribution $P(X_t|Y_t)$. TL can approximate the conditional probability distribution of the target task using $P(X_s)$, $P(X_s|Y_s)$, and $P(X_t)$ [10]. One potential problem of TL is negative transfer, that appears when training with information from a source task can worsen the results obtained when training the agent from scratch.

In the traditional process of TL in DL models (specifically in CNNs), the weights of a pre-trained model are used to initialize the training process for the target task. Some experimental results of transferring different layers in deep models are shown in [17], they conclude that the first layers obtain general features, and these are more *transferable* than deeper ones. The main limitation of TL is negative transfer. This appears when the training using TL, is worse than train from scratch. In DL the use of rectified linear unit produce values lower than zero, then this units are difficult to optimize.

3 Related Work

In this section, we review the state-of-the-art on source task selection for transfer learning in time series domains. The methods proposed in this domain are based on finding a distance between two tasks using their classes [5], or using the separation between the classes, using different coefficients to select an appropriate pre-trained model [7]. In the next paragraphs we present the details of these works.

The work proposed by Fawaz et al. [5] reports experiments transferring a pre-trained model of the 85 datasets in the UCR archive [4]. They first train a model using the architecture proposed by Wang et al. [15], then the pre-trained models are transferred to each dataset until the penultimate layer, and the last layer is initialized with random weights. For training the target task, the model is fine-tuned using the same hyperparameters used in the source tasks for 2,000 epochs, the number of experiments is 7,140, and they used a computer with 60 GPUs for less than one week. The method of Fawaz consists of two stages. First, they use a method for reducing the instances to one prototype for each class [11], then they use dynamic time wrapping to find the distance between the classes prototypes. Finally, the minimum distance between the classes is used as the distance between the tasks.

On the other hand, Meiseles et al. [7] use a mean silhouette coefficient (MSC) to find the separation between the classes of the dataset. Different from the method of Fawaz, they use the output of the global average pooling layer to find the MSC, which can be seen as a similarity measure, and they select the task with the highest value.

Our method uses the entropies produced by the outputs of a pre-trained CNN. We hypothesize that models with higher entropies will obtain better performance than those that produce low entropies. The output of the regression model is the predicted performance of a pre-trained model using the features previously described. Then, we rank all the pre-trained models according to their predicted performance, and select the best one to apply fine-tuning. Table 1 shows a qualitative comparison of these methods with the one proposed in this paper. Column two briefly describes the main technique, and the next three columns correspond to the information required to apply the method (the source data, the pre-trained model, and the classes of the dataset). Contrary to the method proposed by Fawaz et al. [5], we only use information from the pre-trained models. Our method does not need data from the source tasks or information of their class values.

Table 1. Comparison of the method of the state of the art

Reference	Description	Source data	Pre-trained model	Classes
Fawaz et al. [5]	Dynamic Time Warping to find the most closer classes	●	○	●
Meiseles et al. [7]	MSC to find the separation of the classes	○	●	●
Ours	Regression model for predicting the performance of a pre-trained model	○	●	○

4   Performance Prediction for Source Task Selection

In this section, we present the proposed method for source task selection. The proposed method uses the entropies produced by the outputs of the hidden layers in a pre-trained model using the target training set to estimate its performance in the target task. Our method consists of three stages: (i) first, we use a subset of the available tasks with the lowest instances in their datasets to build a training set, then a CNN is trained for each task, and their models are transferred between each task; (ii) then, we extract the entropies on each hidden layer, and we use them as features of the regression model; (iii) finally, we train a regression model, and it is evaluated with new datasets. Figure 1 shows a flowchart of our proposal.

In the next subsections, we present details of the proposed methodology. First, we present the characterization of the pre-trained models for predicting the performance using the entropies produced by the pre-trained models. Then, we describe how to train the regression model, in this stage we use the features extracted in the characterization stage. Finally, a selection of the higher predicted performance pre-trained model is applied.

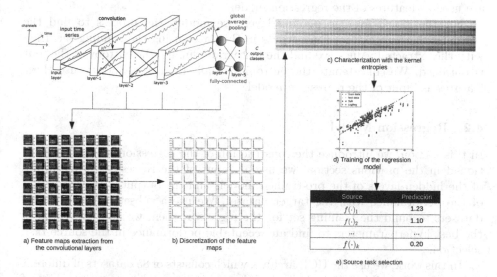

Fig. 1. Proposed method for selecting the pre-trained model: a) feature maps extraction; b) discretization of the feature maps; c) feature maps characterization; d) model regression training; e) selection of the pre-trained models.

4.1 Feature Extraction

In this section, we introduce the characterization of the pre-trained model to train a regressor. Such a regression model will then be used to predict the expected performance of the source tasks when approaching an unseen task. We observe that pre-trained models that produce outputs with a diversity of values in the feature maps are more useful to transfer than those that produce uniform outputs. This is because, in most cases, the uniform outputs produce outputs with zeros and these units are difficult to adjust in a fine-tuning process. On the other hand, if the units produce outputs with diverse values, the pre-trained model can obtain better performance. Taking into account this, we use the entropy values of the feature maps to characterize this diversity; if the feature map contains diverse values, the entropy value will be higher than those that contain uniform values.

In the first part of Fig. 1, the method for the extraction of the characterization is presented. We build a training set using some performance samples obtained by transferring the pre-trained models between a subset of the available datasets. First, the feature maps of the hidden layers using the training dataset are extracted, then the histogram of the feature maps is obtained, we use 100 bins to characterize the distribution of the feature maps, then we obtain the entropy value of the histogram [12]. Finally, the mean of the found entropies are used as features of the regression model.

For the global average pooling layer, we modify the process to find the entropies, here we extract the outputs in this layer using the training set, then with the extracted values, we find the histogram; finally, the entropy of the unit is obtained. We concatenate the entropies of all the layers, and we use these features as input of the regression model.

4.2   Regression Model

In this section, we introduce the construction of the regression model. As mentioned in the previous section, we use as input of the regressor the entropies of the hidden layers of the pre-trained models and the output is the prediction of the performance with the target dataset. We use a subset of the available datasets to build the training set for the regressor; then, we apply a search of the best hyperparameters to find an acceptable performance in the source task selection.

In this work, we use the UCR archive, which consists of 85 datasets of different time series. To build the training set of the regression model, we use a subset of these datasets to transfer the model to each selected task. We decide to use the datasets with the smallest number of instances in their training set. Then, for the evaluation, we use the entire archive with their pre-trained model.

After the selection of the datasets, we tune the hyperparameters for the regressor. We use Random Forest [1] as the regression model because of the algorithm performance in regression tasks; some of its hyper-parameters are tuned, in particular, the numbers of trees and the quality measure. The best regression model is selected according to the evaluation metrics (mean square error, precision in the first, second, and third positions). In the next section, we present the experimental results of the proposed method.

5   Experimental Results

In this section, we introduce the experimental results of the proposed method. The aim of the experiments is to show that training a regressor with few examples is enough to generalize over more datasets. In order to obtain a comparison according to the selected source tasks, we use a more comprehensive set of metrics than those proposed in the state-of-the-art: Top-1 precision, is the precision when the optimal model is selected in the first position; Top-1 mean, is the mean position where the optimal model is obtained; Maximum Useful, is the maximum position where a useful model (positive or neutral transfer) is obtained; Mean Useful is the mean position where a useful model is obtained; finally we use the precision at the first, second and third position, similarly to a learning to rank evaluation.

To find the best regression model, we use Random Forest because of its good performance in regression tasks, we use the implementation of SciKitLearn [3] and we varied the number of trees $\in [10–150]$ and the quality measure (mean square error and mean absolute error, MSE and MAE respectively), the rest of hyperparameters are those proposed by the library. We use 15 datasets with the smallest number of instances in their training set to build the training set for our regressor. The best hyperparameters found during the search are shown in Table 2, the most stable set using the proposed metrics are when 18 trees and MSE are used for training the regressor. The prediction matrix is shown in Fig. 2, this is the performance prediction of the regression model between the datasets, the columns corresponds to the target tasks and the rows to the source task, the lighter values correspond to low predicted performance, and darker ones are the higher predicted performance, we do not take into account the main diagonal because this corresponds to the transfer between the same tasks.

Table 2. Best hypeparameters (QM is the quality measure) for random forest experiments, the hyperparameters correspond to the number of trees and the quality measure.

Trees	QM	Top-1	Top-1 mean	Useful mean	Useful max	p@1	p@2	p@3
18	MSE	0.1058	33.3765	2.9294	30	0.5176	0.5412	0.5414
79	MSE	0.1647	33.8700	3.4000	49	0.5647	0.5235	0.5137
83	MSE	0.1176	31.5882	3.3529	68	0.6118	0.5588	0.5098
15	MSE	0.0823	32.4118	2.4118	23	0.5647	0.5588	0.5373
22	MAE	0.1176	29.3294	3.4588	48	0.5294	0.5471	0.5490

We also compare our results with state-of-the-art methods. Table 2 shows the results of the proposed metrics with the best hyper-parameters of the proposed method. We can see that the proposed method obtains better results than the proposed method by Meiseles et al. [7] in all the metrics. When compared with the Fawaz' method, it can be seen that they obtain better performance in the first, second, and third position precision; also, they obtained a useful model in a better position than the proposed method. Nevertheless, we obtain a better Top-1 precision, select the optimal model in a lower position than Fawaz's method, and obtain a useful model in the worst case at 30th. position while they have it in the 37th. position. We should also emphasize the high computational resources needed by the Fawaz method (Table 1).

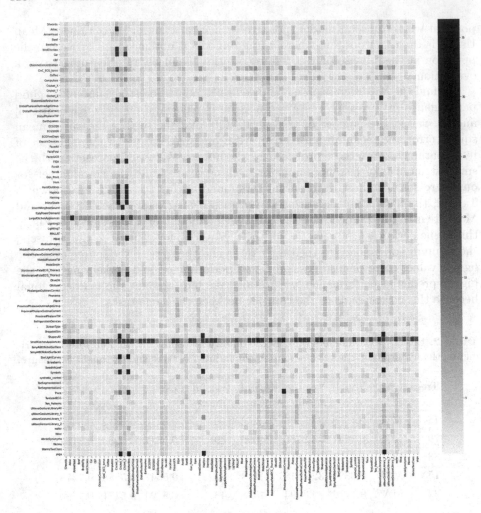

Fig. 2. Prediction matrix using the regression model with the best hyperparameters. Lighter values corresponds to low predicted performance, darker ones are high predicted performance. Main diagonal corresponds to the transfer between the same taks, because of that we do not take them into account.

Table 3. Comparison using the proposed metrics with the state of the art.

Work	Top-1	Top-1 mean	Useful mean	Useful max	p@1	p@2	p@3
Fawaz et al. [5]	0.0941	44	**2.5764**	37	**0.6588**	**0.6176**	**0.6313**
Meiseles et al. [7]	0.0706	55	3.1059	53	0.4471	0.5000	0.5137
Ours	**0.1058**	**33.3765**	2.9294	**30**	0.5176	0.5412	0.5412

6   Conclusions and Future Work

In this work, we present a method for selecting a pre-trained model that is useful to train a new task using a meta-learning formulation. Our main contribution is a novel method to predict the performance of a pre-trained model using the entropies produced by the output of a pre-trained model using a regression algorithm. We use the predicted performance to rank the pre-trained models and select a useful source task in order to avoid negative transfer. According to the results, our method selects the optimal model in more datasets than state-of-the-art methods, also we guarantee to obtain a useful model in a better position than other proposed methods for time series domain. Contrary to the other methods, our method does not need the class values, and consequently it can be applied to semi-supervised domains or in deep reinforcement learning tasks.

Nevertheless, our method need some examples of the transfer learning performance in the datasets, which is the main limitation of our work. To alleviate this problem we use the datasets with fewer instances in their training set.

For future work we will test other set of features in order to obtain better performance in the prediction of the pre-trained models. Also, we will applied the proposed method to other domains for testing its robustness.

Acknowledgements. The authors thankfully acknowledge computer resources, technical advice and support provided by Laboratorio Nacional de Supercómputo del Sureste de México (LNS), a member of CONACYT national laboratories with projects No. 201901047C and 202002030c. Jesús García-Ramírez acknowledges CONACYT for the scholarship that supports his PhD studies associated to CVU number 701191. This work was partially supported by project grant CONACYT CB-S-26314.

References

1. Breiman, L.: Random forests. Mach. Learn. **45**(1), 5–32 (2001). https://doi.org/10.1023/A:1010933404324
2. Brown, T.B., et al.: Language models are few-shot learners. arXiv preprint arXiv:2005.14165 (2020)
3. Buitinck, L., et al.: API design for machine learning software: experiences from the scikit-learn project. In: ECML PKDD Workshop: Languages for Data Mining and Machine Learning, pp. 108–122 (2013)
4. Chen, Y., et al.: The UCR time series classification archive, July 2015
5. Fawaz, H.I., Forestier, G., Weber, J., Idoumghar, L., Muller, P.A.: Transfer learning for time series classification. In: 2018 IEEE International Conference on Big Data (Big Data), pp. 1367–1376. IEEE (2018)
6. Krizhevsky, A., Sutskever, I., Hinton, G.E.: ImageNet classification with deep convolutional neural networks. In: Advances in Neural Information Processing Systems, pp. 1097–1105 (2012)
7. Meiseles, A., Rokach, L.: Source model selection for deep learning in the time series domain. IEEE Access **8**, 6190–6200 (2020)
8. Mnih, V., et al.: Human-level control through deep reinforcement learning. Nature **518**(7540), 529 (2015)

9. Müller, M.: Dynamic time warping. In: Müller, M. (ed.) Information Retrieval for Music and Motion, pp. 69–84. Springer, Heidelberg (2007). https://doi.org/10.1007/978-3-540-74048-3_4

10. Pan, S.J., Yang, Q.: A survey on transfer learning. IEEE Trans. Knowl. Data Eng. **22**(10), 1345–1359 (2009)

11. Petitjean, F., Gançarski, P.: Summarizing a set of time series by averaging: from Steiner sequence to compact multiple alignment. Theor. Comput. Sci. **414**(1), 76–91 (2012)

12. Shannon, C.E.: A mathematical theory of communication. Bell Syst. Tech. J. **27**(3), 379–423 (1948)

13. Silver, D., et al.: Mastering the game of go with deep neural networks and tree search. Nature **529**(7587), 484–489 (2016)

14. Vinyals, O., et al.: Grandmaster level in StarCraft II using multi-agent reinforcement learning. Nature **575**(7782), 350–354 (2019)

15. Wang, Z., Yan, W., Oates, T.: Time series classification from scratch with deep neural networks: a strong baseline. In: 2017 International Joint Conference on Neural Networks (IJCNN), pp. 1578–1585. IEEE (2017)

16. Yang, Q., Zhang, Y., Dai, W., Pan, S.J.: Transfer Learning. Cambridge University Press, Cambridge (2020)

17. Yosinski, J., Clune, J., Bengio, Y., Lipson, H.: How transferable are features in deep neural networks? arXiv preprint arXiv:1411.1792 (2014)

Finding Significant Features for Few-Shot Learning Using Dimensionality Reduction

Mauricio Mendez-Ruiz[1], Ivan Garcia[2], Jorge Gonzalez-Zapata[2],
Gilberto Ochoa-Ruiz[1]([✉]), and Andres Mendez-Vazquez[2]

[1] School of Engineering and Sciences, Tecnologico de Monterrey, Monterrey, Mexico
{A00812794,gilberto.ochoa}@tec.mx
[2] CINVESTAV Unidad Guadalajara, Zapopan, Mexico

Abstract. Few-shot learning is a relatively new technique that specializes in problems where we have little amounts of data. The goal of these methods is to classify categories that have not been seen before with just a handful of samples. Recent approaches, such as metric learning, adopt the meta-learning strategy in which we have episodic tasks conformed by support (training) data and query (test) data. Metric learning methods have demonstrated that simple models can achieve good performance by learning a similarity function to compare the support and the query data. However, the feature space learned by a given metric learning approach may not exploit the information given by a specific few-shot task. In this work, we explore the use of dimension reduction techniques as a way to find task-significant features helping to make better predictions. We measure the performance of the reduced features by assigning a score based on the intra-class and inter-class distance, and selecting a feature reduction method in which instances of different classes are far away and instances of the same class are close. This module helps to improve the accuracy performance by allowing the similarity function, given by the metric learning method, to have more discriminative features for the classification. Our method outperforms the metric learning baselines in the miniImageNet dataset by around 2% in accuracy performance.

Keywords: Few-shot learning · Image classification · Metric learning

1 Introduction

In recent years, we have witnessed the great progress of successful deep learning models and architectures [7], and the application in real-world problems. For example, in cases in Computer Vision, Natural Language Processing (NLP), speech synthesis, strategic games, etc. The new performance levels achieved by such deep architectures, have revolutionized those fields. However, despite such advances in deep learning, the standard supervised learning does not offers a satisfactory solution for learning from small datasets (Few shot problem). This

Currently under review for the Mexican Conference on AI (MICAI 2021).

© Springer Nature Switzerland AG 2021
I. Batyrshin et al. (Eds.): MICAI 2021, LNAI 13067, pp. 131–142, 2021.
https://doi.org/10.1007/978-3-030-89817-5_10

is due to the overfitting problem that Deep Learning incurs when a small dataset is used reducing their generalization capabilities. Furthermore, there are many problem domains, such as health and medical settings, where obtaining labeled data can be very difficult or the amount of work required to obtain the ground truth representations is very large, time consuming and costly.

An interesting phenomena is how humans deal with data scarcity and are able to make generalizations with few samples. Thus, it is desirable to reproduce these abilities in our Artificial Intelligence systems. In the case of Machine Learning, Few-Shot learning (FSL) methods has been proposed [5,9,18,19,21] to imitate this ability by classifying unseen data from a few new categories. There are two main FSL approaches: The first one is Meta-learning based methods [1,4,5,12], where the basic idea is to learn from diverse tasks and datasets and adapt the learned algorithm to novel datasets. The second are Metric-learning based methods [9,22], where the objective is to learn a pairwise similarity metric such that the score is high for similar samples and dissimilar samples get a low score. Later on, these metric learning methods started to adopt the meta learning policy to learn across tasks [18,19,21].

The main objective of these methods is to learn an effective embedding network in order to extract useful features of the task and discriminate on the classes which we are trying to predict. From this basic learning setting, many extensions have been proposed to improve the performance of metric learning methods. Some of these works focus on pre-training the embedding network [2], others introduce task attention modules [3,11,23], whereas other try to optimize the embeddings [10] and yet others try to use a variety of loss functions [23].

In this work, we focus on finding task-significant features by applying different feature reduction techniques and assigning the reduced features a score based on the inter and intra class separability. We believe that finding those relevant features for each task is important, as we can better discriminate between classes and obtain a better inference.

The rest of this paper is organized as follows. In Sect. 2 we introduce the related work and explain the problem setting. In Sect. 3 we introduce our proposed model, with the ICNN module which helps us to choose the best dimensionality reduction technique. Then, in Sect. 4 we give details on how we implemented our model, the design choices taken based on experiments and the comparison with baselines and state-of-the-art models. Finally, in Sect. 5 we summarize our work and discuss future directions.

2   Materials and Methods

2.1   Meta-learning Tasks

The few-shot meta-learning setup consists of episodic tasks, which can be seen as batches in traditional deep learning. A task is made up of support data and query data. The support set contains k previously unseen classes and n instances for each class, and the objective is to classify q queries using the support data.

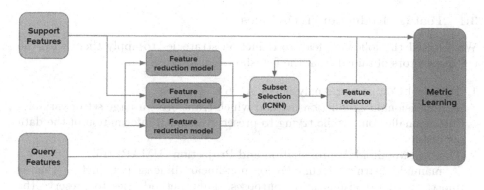

Fig. 1. The components of the (ICNNnet). After obtaining the features from the feature extractor, a number of feature reduction models are applied to the support embeddings. The features obtained from these models are measured via the ICNN Score to select the best feature reductor. Finally, this reductor is applied to the support and query embeddings to continue with the metric learning inference.

This setting is also known as k-way n-shot (e.g. 5-way 1-shot or 5-way 5-shot). As described in [16,21], the model is trained using an episodic mechanism, where each episode is loaded with a new random task taken from the training data.

2.2 MiniImageNet Dataset

For the experimental results we used MiniImageNet [21], which is a subset of ImageNet version of ILSVRC-2012 [17] and is used as a benchmark for the evaluation of few-shot learning methods. This subset is comprised of 100 classes, each one containing 600 images, making up a total of 60,000 images. We follow the split proposed by Ravi and Larochelle [16], dividing the dataset into 64 classes for training, 16 classes for validation and 20 classes for testing.

3 Proposed Model

For our proposed model (see Fig. 1), we adopt a feature selection strategy based on dimensionality reduction assisted by an inter and intra class nearest neighbors distance score. As illustrated in Fig. 1, after obtaining the embeddings from the feature extractor we are left with a number of feature vectors, each one representing a task sample. From these features, we want to obtain the more relevant for the given task. We apply different feature reduction methods, and obtain an intra-class and inter-class score for each one. These scores are used to select the method which helps us to obtain the best dimensions for the current task. The obtained features are then used by a metric learner to produce a classification.

3.1 Feature Reduction Techniques

We selected the following feature reduction strategies to apply them with the feature vectors obtained from the few-shot learning task:

1. Principal Component Analysis (PCA) [6]:
   A dimensionality reduction method which transforms a large set of variables into a smaller one, while trying to preserve as much information of the data as possible.
2. Uniform Manifold Approximation and Projection (UMAP) [15]:
   A manifold learning technique for non-linear dimension reduction that is mostly used for visualization purposes. This method tries to preserve the global structure of the data.
3. Isometric mapping (Isomap) [20]:
   A non-linear dimensionality reduction which seeks a low dimensional embedding which preserves the geodesic distances between the data points. The three stages of this algorithm are to create a neighborhood network, use a shortest path graph search to calculate the geodesic distance between all pair of points, and finally find the low dimensional embedding through eigenvalue decomposition.

These methods cover the main attempts for feature generation from Linear Algebra, kernel methods and the use of random graphs for learning manifolds. Although other methods exist, as the t-SNE [14], they can be seen as particular applications of the previous methods. For example, the t-SNE defines a series of probability based on Gaussian kernels. Thus, using Kullback-Leiber a minimization procedure is executed to obtain a manifold based on those kernels. These types of minimization also happen in the PCA when using lagrange multipliers to obtain the eigenvectors and eigenvalues of the data covariance matrix. Later on Sect. 4, we show experiments made with more methods (Umap, PCA, Isomap, Kernel PCA, Truncated SVD, Feature Agglomeration, Fast ICA and Non-Negative Matrix Factorization) to prove that we do not need more methods than the three mentioned above.

3.2 Inter and Intra Class Nearest Neighbors Score (ICNN Score)

The ICNN Score [8] was proposed to aid the feature selection based on supervised dimensionality reduction with subset evaluation. This measure improves the performance of dimensionality reduction techniques based on manifold algorithms by removing noisy features. There are two main concepts for the ICNN feature selection technique: Inter-class distance and Intra-class distance. Inter-class distance refers to the distance between points of different classes, and Intra-class distance refers to the distance between points of the same class. The idea for a successful feature selection approach is to choose those which increase the inter-class distance and reduce the intra-class distance, in order to allow the task to be differentiated.

The ICNN Score is a measure that combines the distance and variance of the inter-intra k-nearest neighbors of each instance in the data. The formula to calculate this score is the following:

$$ICNN(X) = \frac{1}{|X|} \sum_{x_i \in X} \lambda(X_i)^{\frac{1}{p}} \omega(X_i)^{\frac{1}{q}} \gamma(X_i)^{\frac{1}{r}}, \tag{1}$$

where p, q and r are control constants.

Here, λ is a function that penalizes the neighbors of X_i with the same class based on how distant they are, and the neighbors of different classes based on how close they are:

$$\lambda(X_i) = \frac{\sum_{p \in K_{\tilde{x}_i}} \frac{d(X_i,p) - \theta(X_i)}{\alpha(X_i) - \theta(X_i)} + \sum_{q \in K_{x_i}} 1 - \frac{d(X_i,q) - \theta(X_i)}{\alpha(X_i) - \theta(X_i)}}{|K_{x_i}| + |K_{\tilde{x}_i}|}, \tag{2}$$

where $K_{x_i} = KNN(x_i) \in y_i$ are the set of k-nearest neighbors of x_i that have the same class. $K_{\tilde{x}_i} = KNN(x_i) \in y_j \neq y_i$ are the set of k-nearest neighbors of x_i that has different class. $d(a,b)$ is a distance function, which in this case is the euclidean distance. $\alpha(X_i)$ and $\theta(X_i)$ are the maximum distance and the minimum distance of the x_i neighbors, respectively.

In the ideal scenario, the neighbor's distance of the same class are close to 0 and the distance with different classes are close to 1.

Now, ω is a function that penalizes the distance variance of neighbors (Eq. 3):

$$\omega(X_i) = 1 - \left(Var\left(\sum_{p \in K_{\tilde{x}_i}} \frac{d(X_i,p) - \theta(X_i)}{\alpha(X_i) - \theta(X_i)} \right) \right.$$
$$\left. + Var\left(\sum_{q \in K_{x_i}} 1 - \frac{d(X_i,q) - \theta(X_i)}{\alpha(X_i) - \theta(X_i)} \right) \right), \tag{3}$$

where a high variance is penalized because it increases the possibility of overlapping classes.

Finally, the γ function describes the ratio of the neighbor's classes:

$$\gamma(x_i) = \frac{|K_{x_i}|}{|K_{x_i}| + |K_{\tilde{x}_i}|}, \tag{4}$$

where each instance is penalized based on the neighbors in the same class of x_i. Each of the three functions (λ, ω and γ) have an output with a range between 0 and 1. In Fig. 2 and 3, we can visualize how some few-shot tasks scenarios would be rated using the ICNN score.

Using this metric, we can evaluate each feature reduction technique, as well as the original feature vector, to choose the best selection of features that are relevant for the current task.

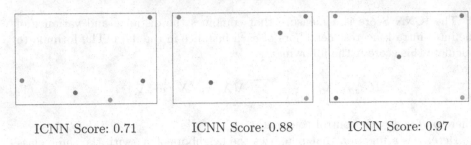

ICNN Score: 0.71   ICNN Score: 0.88   ICNN Score: 0.97

Fig. 2. ICNN score behaviour on different 5-way 1-shot scenarios. The hyperparameters for calculating the score are set to: $k = 3$, $p = 2$, $q = 2$, $r = 2$

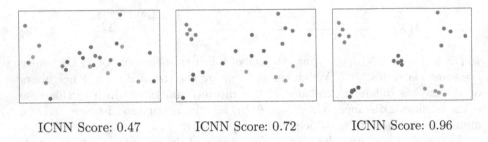

ICNN Score: 0.47   ICNN Score: 0.72   ICNN Score: 0.96

Fig. 3. ICNN score behaviour on different 5-way 5-shot scenarios. The hyperparameters for calculating the score are set to: $k = 5$, $p = 2$, $q = 2$, $r = 2$

4   Experiments

Evaluation Metric. Following most of the metric learning methods [18,19,21], we report our results on the mean accuracy (%) over 1,000 test episodes with 95% of confidence intervals.

4.1   Implementation Details

Feature Extractor. For the feature extractor, we test with two different backbones: ConvNet and ResNet-12. The ConvNet architecture follows the setting used by Vinyals et al. [21], with 4 layers of convolutional blocks. Each block is composed of a 3 × 3 convolution with 64 filters, followed by Batch Normalization and a ReLU layer. This network is optimized with Adam optimizer with an initial learning rate of 10^{-3}. For the ResNet-12, following recent work [10,11], the network is pre-trained using the SGD optimizer with momentum of 0.9 and learning rate of 0.1 over 100 epochs with a batch size of 128. Then, we apply the meta-training using SGD optimizer with momentum of 0.9 and learning rate of 0.0001.

Meta-learning Setup. In order to compare against the baselines, our experiments are made under the 5-way 1-shot and 5-way 5-shot setting with 15 query

Table 1. Test accuracies of the design choices experiments for 5-way tasks. See Sect. 4.2 for more details

(# of components; type; model)	BackBone	1shot	5shot
Prototypical Networks (Paper)	ConvNet	49.52	68.20
(1) 6 Components; Support&Query; Base	ConvNet	52.37	**69.08**
(2) Multiple Components; Support&Query; Base	ConvNet	**52.43**	68.77
(3) 6 Components; Support; Base	ConvNet	46.88	38.24
(4) 6 Components; Support&Query; All	ConvNet	51.44	67.26
(5) Multiple Components; Support& Query; All	ConvNet	51.34	67.08
Prototypical Networks (our implementation)	ResNet-12	61.13	76.21
(1) 6 Components; Support&Query; Base	ResNet-12	63.03	**78.14**
(2) Multiple Components; Support& Query; Base	ResNet-12	62.30	78.12
(3) 6 Components; Support; Base	ResNet-12	56.67	77.37
(4) 6 Components; Support&Query; All	ResNet-12	**63.81**	76.68
(5) Multiple Components; Support&Query; All	ResNet-12	63.19	76.93

images for each class in the task. All the input images are resized to 84 × 84. On the training phase, we randomly construct 100 tasks over 200 epochs and apply validation over 500 tasks after every epoch. We train the network and obtain the cross-entropy loss. The initial learning rate is reduced by half every 20 epochs. For the testing phase, we randomly construct 1,000 tasks and measure the mean accuracy with 95% confidence intervals.

4.2 Model Design Choices

Having the setup for the model extension stated above, we now discuss the ICNN hyper-parameters and feature reduction techniques settings that we chose to test in order to find the best combination that will give us better accuracy performance.

The first set of ablation studies is carried out in order to find the best hyper-parameters for the ICNN score. There are four constants that we need to choose for the algorithm, k, p, q and r. We decided to give the same weight for λ, ω and γ. For this, we take the decision to use the same value for those three constants and set its value to 2. For the k (k - Nearest Neighbors), we decide to assign it a value related to the few-shot task. Since in the given task we have c classes and n shots, we decide to set to k the value of n. In this way, we can ensure that, for each point, it always appears a nearest neighbor with different class. By having always a nearest neighbor of different class, we can obtain a better estimation in the λ function, since we have now a perspective of the neighbors in the same class in relation with those of other classes. For the case of the 1-shot setting, we decide to set k to 3, since all the other data points are of different class and this allows us to better understand the inter-class distance.

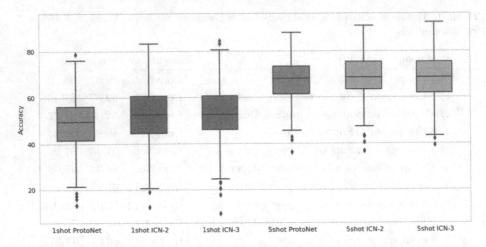

Fig. 4. Accuracy results over 1000 test episodes for the Prototypical Networks and our best proposed models using a ConvNet as a Feature Extractor, on the 1-shot and 5-shot setting. The ICN model number corresponds to the design choice showed on Table 1.

For the second set of ablation studies for setting the design choices, we study the effect of the variations in the feature dimensionality reduction techniques. There are three main concepts that we test related to the dimension reduction strategies:

1. The number of components to which we reduce the feature embedding vectors. In this case, we started by testing the reduction to 4, 6, 8 and 10 components. We found out that it was giving almost the same results, but it was a little better with 6 components. For the next experiments, we tested reducing to multiple components by halfs depending on the feature extractor. For the ConvNet, the feature embedding has 64 dimensions, and the reduced components were 32, 16, 8 and 4. For the ResNet-12, the feature embedding has 512 dimensions, and the reduced components were 256, 128, 64, 32, 16 and 8.
2. The set of points used for the reduction (support/support & query). We added these experiments to tests if only the support data was being useful for obtaining a good reduction, or we could aid the reduction by adding the query data.
3. The set of feature reduction techniques used. Here we tested two different settings: using only PCA, Isomap and Umap (Base), or using all the models stated in Sect. 3.1 (All). The base setting was obtained by testing the model on 1000 episodes and keeping the three feature reduction techniques that were chosen the most.

There are some findings obtained from these experiments. We found that using UMAP in our feature reduction models, the training phase execution time greatly increased. For this reason, we decided to remove UMAP from the methods used in training, and use it only on the testing phase.

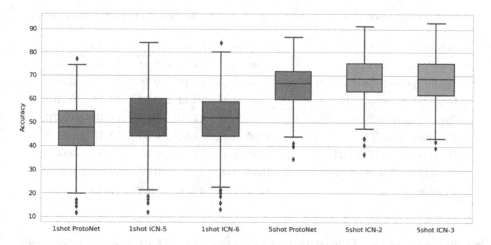

Fig. 5. Accuracy results over 1000 test episodes for the Prototypical Networks and our best proposed models using a ResNet-12 as a Feature Extractor, on the 1-shot and 5-shot setting. The ICN model number corresponds to the design choice showed on Table 1.

By testing the number of components, we first found that 6 components gave us slightly better results than reducing to 4, 8 and 10 components. Then, we also tested the model by reducing dimensions to multiple components. We found that, in most cases, reducing only to 6 components was giving better results. We believe that 6 components are good to keep most of the main information from the task while giving a better representation of the data.

With the experiments on the set used for reducing the dimensions, we found that using only the support data give us bad results. This proves that, on the few-shot settings tested (5-way 1-shot and 5-way 5-shot), the support data is not enough to obtain a good representation of the data after reducing the dimensions with the different feature reduction techniques. Using the support and query data allowed the feature reduction methods to better interpret the structure of the data, thus obtaining a better ICNN score.

As for the feature reduction techniques, we found that using the base models (PCA, Isomap and Umap) give us better results than using all the reduction models, on most of the experiments. This could be happening because using the worst behaving models may be resulting in wrong decisions for the prototypes. Further work can focus on visualizing the features obtained by each feature reduction technique.

The results of these experiments are summarized in Table 1. We also provide a visualization for the behaviour of the best models obtained with the experiments stated above. In Fig. 4, we can visualize the accuracies from 1000 episodes obtained using the original prototypical networks and the best models obtained adding the ICNN module, all of these using the ConvNet as feature extractor and with the 5-way 1-shot and 5-way 5-shot setting. We can see improvements

Table 2. Comparison of test accuracies for 5-way tasks. See Sect. 4.3 for more details.

Model	Network	1-shot	5-shot
Matching Networks [21]	ConvNet	43.56 ± 0.84	55.31 ± 0.73
Prototypical Networks [18]	ConvNet	49.42 ± 0.78	68.20 ± 0.66
Relation Networks [19]	ConvNet	50.44 ± 0.82	65.32 ± 0.70
ProtoNets + ICNN	ConvNet	**52.43 ± 0.73**	**69.08 ± 0.76**
K-tuplet Nets [13]	ResNet-12	58.30 ± 0.84	72.37 ± 0.63
ProtoNets + CTM [11]	ResNet-12	59.34 ± 0.55	77.95 ± 0.06
Principal Characteristic Nets [23]	ResNet-12	63.29 ± 0.76	77.08 ± 0.68
ProtoNets + ICNN	ResNet-12	**63.81 ± 0.71**	**78.14 ± 0.50**

on the first and fourth quartile, and in some cases we have a smaller box from the second and third quartile, which means that the classification become a little more robust.

The same visualization is showed in Fig. 5, but these accuracies are obtained using a ResNet-12 as feature extractor. We can see here that the first and second quartile are improving, with a better improvement on the first quartile. We can also see that, while the 1-shot setting gets a better improvement on the mean accuracy, the 5-shot setting obtain a better improvement on the first quartile.

4.3 Comparison with Baselines

To validate the effectiveness of our new module, we compare it against similar state-of-the-art models following the standard few-shot learning setting. First, we compare it against the three main metric learning methods: (1) Prototypical Networks [18], (2) Matching Networks [21] and (3) Relation Networks [19]. These three models use a ConvNet as feature extractor, which obtains a feature embedding of size 64. We also compare against Category Traversal Module (CTM) [11], a model with the same idea of looking for task-relevant features, Principal Characteristic Network [23] and K-Tuplets Network [13]. These three models use a ResNet-12 as feature extractor, with 512 dimensions in the output feature embedding.

Table 2 illustrate the comparison of all the previous mentioned models with our method. We obtained an improvement of around 2% for the 5-way 1-shot setting on the test set using Prototypical Networks. We also achieved a better performance than Matching Nets and Relation Nets on 5-way 5-shot setting by a large margin, but obtained a little improvement of around 1% compared with Prototypical Nets. As for the other three models with ResNet-12 as feature extractor, we obtained an improvement of around 1% on both settings of 1-shot and 5-shot.

5   Conclusion

In this paper, we propose a new module with the purpose of finding task-significant features by using dimensionality reduction techniques and a discriminant score based on intra and inter class nearest neighbors. The performance of the proposed model improves the accuracy compared to the metric learning baselines. We also compare the results with state-of-the-art models with deeper backbone and obtain a gain in accuracy performance.

Our experiments are based on the combination of Prototypical Networks and ICNN but, as this method is proposed to obtain better features, any other metric learning technique (Matching Networks, Prototypical Networks) is expected to improve. The experimentation of our ICNN module with these other techniques are left for future work.

References

1. Andrychowicz, M., et al.: Learning to learn by gradient descent by gradient descent. CoRR abs/1606.04474 (2016). http://arxiv.org/abs/1606.04474
2. Chen, D., Chen, Y., Li, Y., Mao, F., He, Y., Xue, H.: Self-supervised learning for few-shot image classification. CoRR abs/1911.06045 (2019). http://arxiv.org/abs/1911.06045
3. Chen, H., Li, H., Li, Y., Chen, C.: Multi-scale adaptive task attention network for few-shot learning (2020)
4. Chen, Y., et al.: Learning to learn without gradient descent by gradient descent. In: International Conference on Machine Learning, pp. 748–756. PMLR (2017)
5. Finn, C., Abbeel, P., Levine, S.: Model-agnostic meta-learning for fast adaptation of deep networks. CoRR abs/1703.03400 (2017). http://arxiv.org/abs/1703.03400
6. Pearson, K.: LIII. on lines and planes of closest fit to systems of points in space. London Edinburgh Philos. Mag. J. Sci. **2**(11), 559–572 (1901). https://doi.org/10.1080/14786440109462720
7. Ganatra, N., Patel, A.: A comprehensive study of deep learning architectures, applications and tools. Int. J. Comput. Sci. Eng. **6**, 701–705 (2018). https://doi.org/10.26438/ijcse/v6i12.701705
8. García Ramírez, I.: Supervised feature space reduction for multi-label classification. Ph.D. dissertation, CINVESTAV Unidad Guadalajara (2021)
9. Koch, G., Zemel, R., Salakhutdinov, R.: Siamese neural networks for one-shot image recognition (2015)
10. Lee, K., Maji, S., Ravichandran, A., Soatto, S.: Meta-learning with differentiable convex optimization. CoRR abs/1904.03758 (2019). http://arxiv.org/abs/1904.03758
11. Li, H., Eigen, D., Dodge, S., Zeiler, M., Wang, X.: Finding task-relevant features for few-shot learning by category traversal. CoRR abs/1905.11116 (2019). http://arxiv.org/abs/1905.11116
12. Li, K., Malik, J.: Learning to optimize. CoRR abs/1606.01885 (2016). http://arxiv.org/abs/1606.01885
13. Li, X., Yu, L., Fu, C., Fang, M., Heng, P.: Revisiting metric learning for few-shot image classification. CoRR abs/1907.03123 (2019). http://arxiv.org/abs/1907.03123

14. van der Maaten, L., Hinton, G.: Visualizing data using t-SNE. J. Mach. Learn. Res. **9**(86), 2579–2605 (2008). http://jmlr.org/papers/v9/vandermaaten08a.html
15. McInnes, L., Healy, J., Melville, J.: UMAP: uniform manifold approximation and projection for dimension reduction. arXiv preprint arXiv:1802.03426 (2018)
16. Ravi, S., Larochelle, H.: Optimization as a model for few-shot learning. In: ICLR (2017)
17. Russakovsky, O., et al.: ImageNet large scale visual recognition challenge. CoRR abs/1409.0575 (2014). http://arxiv.org/abs/1409.0575
18. Snell, J., Swersky, K., Zemel, R.: Prototypical networks for few-shot learning. In: Guyon, I. (eds.) Advances in Neural Information Processing Systems, vol. 30, pp. 4077–4087. Curran Associates, Inc. (2017). http://papers.nips.cc/paper/6996-prototypical-networks-for-few-shot-learning.pdf
19. Sung, F., Yang, Y., Zhang, L., Xiang, T., Torr, P.H.S., Hospedales, T.M.: Learning to compare: relation network for few-shot learning. CoRR abs/1711.06025 (2017). http://arxiv.org/abs/1711.06025
20. Trosset, M.W., Buyukbas, G.: Rehabilitating isomap: euclidean representation of geodesic structure. arXiv preprint arXiv:2006.10858 (2020)
21. Vinyals, O., Blundell, C., Lillicrap, T., Kavukcuoglu, k., Wierstra, D.: Matching networks for one shot learning. In: Lee, D.D., Sugiyama, M., Luxburg, U.V., Guyon, I., Garnett, R. (eds.) Advances in Neural Information Processing Systems, vol. 29, pp. 3630–3638. Curran Associates, Inc. (2016). http://papers.nips.cc/paper/6385-matching-networks-for-one-shot-learning.pdf
22. Xing, E., Jordan, M., Russell, S.J., Ng, A.: Distance metric learning with application to clustering with side-information. In: Becker, S., Thrun, S., Obermayer, K. (eds.) Advances in Neural Information Processing Systems, vol. 15. MIT Press (2003). https://proceedings.neurips.cc/paper/2002/file/c3e4035af2a1cde9 f21e1ae1951ac80b-Paper.pdf
23. Zheng, Y., Wang, R., Yang, J., Xue, L., Hu, M.: Principal characteristic networks for few-shot learning. J. Vis. Commun. Image Represent. **59**, 563–573 (2019)

Seasonality Atlas of Solar Radiation in Mexico

Mónica Borunda[1,2(✉)], Adrián Ramírez[2], Nayeli Liprandi[2], Miriam Rodríguez[2], and Alejandro Sánchez[2]

[1] Conacyt - Instituto Nacional de Electricidad y Energías Limpias, Cuernavaca, Mexico
monica.borunda@ineel.mx
[2] Universidad Nacional Autónoma de México, Mexico City, Mexico
{felos,ditesli,miriamrc,alex8379}@ciencias.unam.mx

Abstract. Due to the imminent climate-change emergency, it is urgent to boost the exploitation of renewable resources to produce clean energy, being solar energy one of the most promising ones. However, one of the greatest challenges that solar energy faces is its intermittency. Thus, to get the biggest benefit from this resource, especially for photovoltaic generation, it is required to predict its availability to estimate variations in energy production. As the first step for solar radiation forecasting, a seasonality analysis is mandatory to obtain better results. In this work, we perform a seasonality analysis of solar radiation in Mexico using Machine Learning. Specifically, we accomplish a cluster analysis of solar radiation data in locations representative of the different climate conditions in Mexico to obtain a seasonality atlas of the solar resource. Cluster analysis is performed with two algorithms, k-means and k-medoids. Finally, the Silhouette method is used to validate the results.

Keywords: Solar radiation · Seasonality · Cluster analysis · Machine learning

1 Introduction

The climate change that planet Earth is experiencing places humanity at a point where it is mandatory to take actions to mitigate it before reaching disastrous and irreversible consequences [1]. One of the main actions in this direction is to reduce the use of fossil fuels while increasing the exploitation of renewable energies. Solar energy is one of the most promising renewable energy sources, with an estimated global potential of 4.4×10^{14} to 1.4×10^{16} kWh [2].

Solar energy can be used for either electricity generation or heat production. In the last decades, technology for both applications has been developed and improved reducing production costs and increasing efficiencies. However, the main drawback of this energy source is its intermittent nature. Solar radiation reaching a site depends on many factors such as the site's location, the day of the year, the time of the day, cloud cover, local meteorological conditions. Therefore, in order to obtain the maximum benefit from it, in particular for photovoltaic applications, it becomes extremely useful to forecast it. Forecast horizon goes from short, medium and long-term timescales, depending on the application, which covers network operations, voltage and frequency regulations, unit commitment, and the electricity market, among others. Forecasting models can be

© Springer Nature Switzerland AG 2021
I. Batyrshin et al. (Eds.): MICAI 2021, LNAI 13067, pp. 143–157, 2021.
https://doi.org/10.1007/978-3-030-89817-5_11

classified in statistical models, conventional (autoregressive models) or with Artificial Intelligence (AI) techniques (Artificial Neural Networks ANN), cloud imagery models (from either satellite or ground-based images), Numerical Weather Prediction (NWP) models, and hybrid models. Depending on the forecast horizon they are better or less suited for the prediction [3].

Given the intermittency nature of solar energy, many authors look for the seasonality of the resource at the site and apply their forecast method to each season to obtain better predictions [4]. Lan et al. constructed a hybrid model for day-ahead solar radiation forecasting and applied it to the four seasons of the year [5]. Ghofrani et al. proposed a soft computing method for short-term prediction and its performance was evaluated against seasonality variations [6]. Wang et al. proposed a model using support vector machine for one hour ahead forecasting and carried out experiments diving the data into seasonally sets [7]. Bigdeli et al. presented a hybrid model combining different techniques of AI for one hour ahead forecast and divided their data into two seasons corresponding to sunny and cloudy days [8]. Voyant et al. proposed a model for hourly forecasting consisting of an ARMA model for spring and summer and an ANN for the rest of the year [9]. Yacef et al. constructed a combined method using empirical models and Bayesian neural networks and tested it on two seasons corresponding to winter and summer [10]. Akarslan and Hocaoglu introduced an hourly solar radiation forecasting model using a linear method for winter and summer and an empirical model for spring and autumn obtaining results with good accuracy [11].

Solar radiation seasonality depends on the location, and deciding whether there are no seasonality or 2, 3 or 4 seasons may not be an obvious issue. At some sites, seasonality may coincide with the four seasons according to the Gregorian calendar, namely, spring, summer, autumn and winter. However, there is also a six-season model for template climate regions that are prevernal, vernal, estival, serotinal, autumnal and hibernal. Some tropical regions have a wet season and a dry season, and other locations have a third season, which could be mild or Harmattan season. Usually, seasonality is analyzed by statistical inspection of the data but, in this work, we are interested in defining a methodology, using Machine Learning (ML). In particular, we perform a cluster analysis [12, 13], which is validated by the Silhouette method [14], to determine the optimal number of seasons of the solar radiation data.

This work is organized as follows. The following Subsect. 1.1, presents a summary of related work. Section 2 introduces basic facts about the solar resource in Mexico. Section 3 presents the proposed methodology to find the seasonality for solar radiation data. The results and the seasonality atlas of solar radiation in Mexico are presented in Sect. 4. A discussion of the results is provided in Sect. 5. Finally, the conclusions of this work are outlined in the last section.

1.1 Related Work

There are several published works analyzing the seasonality of solar radiation in different sites, some of the most relevant are described as follows. Boland remarked two approaches for characterizing seasonality: the multiplicative approach that consists on dividing the solar radiation by the extraterrestrial radiation to produce the clearness index and the additive approach that consists on subtracting a mean function from the solar radiation. He concluded that the additive approach, using Fourier series, is better than the other one and is compatible with the seasonality of other climate variables [15]. Lima et al. studied the solar irradiation variability and trends based on a statistical analysis of the measured data in weather stations in the Northeastern Brazilian region. They performed a cluster analysis and found five regional patterns with temporal regimes [16]. Vindel et al. analyzed the temporal variability patterns for solar radiation estimations in Zambia using principal component analysis finding two patterns associated with global climate characteristics and to a local climate effect respectively. In addition, they performed a cluster analysis and found a spatial distribution of the radiation [17]. Fernández et al. use a cluster analysis to select representative days corresponding to typical weather data to evaluate the performance of a concentrating solar thermal power plant [18]. Bessafi et al. performed cluster analysis of 2-year irradiance data for prediction purposes. They used two approaches and compared the results [19]. Govender et al. presented a day-ahead forecast model using clustering of solar irradiance and cloud cover forecasts from NWP. They grouped the data according to diurnal patterns based on sunny and cloudy conditions and, based on cloudy patterns finding four groups [20].

2 Mexico's Solar Radiation

Solar radiation reaching Earth's surface depends on many factors, being the latitude and longitude the main ones. In particular, the annual behavior of solar radiation may be very different for locations on the equator or in the northern or southern hemispheres. Figure 1 exemplifies this fact by showing the annual solar radiation in some cities in the northern and southern hemispheres and close to the Equator.

Fig. 1. Mean irradiance throughout the year in New York and Vancover (northern hemisphere), Antofagasta and Mar del Plata (southern hemisphere) and, Guayaquil and Manaos (Equator).

Mexico is located between $15°$ and $35°$ latitude, which is a privileged location with a mean annual irradiance around 5.5 kWh/m^2 per day. In particular, Fig. 2 shows the mean daily irradiance in the country. This graph is obtained by calculating the mean daily irradiance from a 20-year dataset of 30-min radiation data averaged over 24 locations distributed throughout the country.

Moreover, Mexico counts with a diversity of climate zones, as shown in Fig. 3, ranging from warm climate with four varieties, temperate climate with six, dry climate with seven, to cold climate. Therefore, a first guess could be that solar radiation seasonality strongly depends on weather conditions. In the next section, we describe the methodology to find the seasonality of solar radiation throughout the country.

Fig. 2. Mean irradiance in Mexico throughout the year.

Fig. 3. Different climate types in Mexico.

3 Methodology

3.1 Designing the Atlas

Climate zones shown in Fig. 3 can be grouped into four main groups, as shown in Fig. 4, corresponding to cold, dry, mild and tropical climates. We select six locations for each group such that they are spread over the country, represented by the blue dots in Fig. 4. These locations are our reference points to build the atlas. It is important to remark that all the chosen locations correspond to communities with a strong energy need and the exploitation of the solar resource could be of great benefit to them.

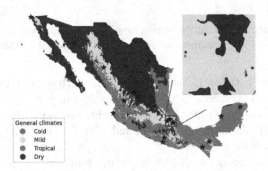

Fig. 4. Main climate groups in Mexico and selected locations.

Next, we obtain data for the selected communities. The National Renewable Energy Laboratory (NREL) counts with a free, open-access meteorological database for many locations in the northern hemisphere [21]. We use 30-min solar radiation data from 2000 to 2020 of the selected locations to perform the cluster analysis outlined in the next section.

3.2 Cluster Analysis

Our objective is to determine the solar radiation seasonality in the sites of interest. To achieve this we consider the incident solar energy per unit area called Solar Irradiation (SI) [22]. We analyze the temporal variability of the solar radiation at the sites by performing a cluster analysis of the SI. First, we search for groups with similar daily SI along the year, performing a one-dimensional cluster analysis with the mean daily SI at each site. Next, we perform a two-dimensional cluster analysis considering the Standard Deviation (SD) of the daily SI and the daily SI.

The daily SI is calculated by

$$SI = \sum_{i=0}^{n} \Delta t_i GHI_i, \tag{1}$$

where GHI_i is the global horizontal irradiance in the time interval Δt_i and the index i runs from 0 to n where n stands for the maximum number of measurements in a day, for example, if the sunlight during the day lasted 8 h in one site and we count with 30 min data, n = 16. The GHI is the flux of total radiation from the Sun on a horizontal surface on Earth.

The mean daily SI over the 20 years is calculated by

$$\overline{SI} = \frac{1}{20} \sum_{i=1}^{20} SI_i, \tag{2}$$

The SD of the is calculated by

$$SD = \sqrt{\frac{\sum_{i=1}^{n} \left(SI_i - \overline{SI}\right)^2}{n-1}}, \tag{3}$$

where $n = 20$ since we are considering the database of 20 years.

First, we calculate the mean daily solar energy during 20 years on the sites of interest using the 30-min solar radiation data. Figure 5 shows a representative behavior of the radiation in the regions corresponding to the four groups with the main climates by taking the average date of the six locations of each group.

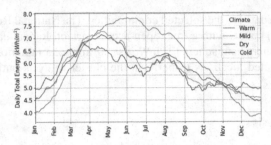

Fig. 5. Representative behavior of the solar radiation in the main climates in Mexico.

3.3 Algorithms

There are many algorithms to group data objects. First, we use the unsupervised k-means algorithm that aims to partition a set of n observations $(x_1, x_2, \ldots, x_n)$, where each observation is a d-dimensional vector, into $k (\leq n)$ sets or *clusters* $S = \{S_1, S_2, \ldots, S_k\}$ such that the Euclidean distance between the objects and the mean of the points μ_i in S_i (the centers of the cluster) is minimized [23].

$$\arg min \sum_{i=1}^{k} \sum_{x \in S_i} \|x - \mu_i\|^2. \tag{4}$$

To obtain a comparative analysis we also use the k-medoids algorithm [24]. This algorithm also groups the data by minimizing the sum of dissimilarities between points and the center of the cluster, medoid, which in this case corresponds to a data from the dataset, in contrast with k-means.

These algorithms are the most used in cluster analysis and one is better than the other in some respects and viceversa. For instance, k-medoids is more robust than k-means but k-means reduces noise and outliers. Thus, we use both algorithms to complement our analysis.

3.4 Validation of the Results

The number of clusters k, as well as the goodness of the clustering, are evaluated with the Silhouette method. Even though, there are several approaches to appraise clustering, the Silhouette method is the best suited for assessing the clustering of a dataset which given its nature there exists an intrinsic natural number of clusters [14]. The goodness of the clustering is determined with the Silhouette value that measures how similar an object is to other objects in the same cluster compared to objects in other clusters. Let i be a data point in the i-th cluster C_i, $i \in C_i$, then the mean distance between i and all other data points in the same cluster is defined by

$$a(i) = \frac{1}{C_i - 1} \sum_{j \in C_i, i \neq j} d(i, j), \tag{5}$$

where $d(i, j)$ is the distance between the points i and j in C_i. Also, the smallest mean distance of i to all points in other cluster is defined by

$$b(i) = \min \frac{1}{|C_k|} \sum_{j \in C_k} d(i, j). \tag{6}$$

Then, whether or not the data i is appropriately grouped is given by

$$s(i) = \begin{cases} \frac{b(i) - a(i)}{\max\{a(i), b(i)\}}, & \text{if } C_i > 1, \\ 0, & \text{if } C_i = 1, \end{cases} \tag{7}$$

where $-1 \leq s(i) \leq 1$ and $s(i) = 1$ means the data is well grouped, and $s(i) = -1$ means the data should be in the neighboring cluster. Thus, the Silhouette score SS is the mean $\tilde{s}(i)$ over all points of the dataset is a measure of the goodness of the clustering. In particular, the silhouette coefficient, SC, is the maximum value of the mean $s(i)$ over all data and provides the best k

$$SC = \max \tilde{s}(k), \tag{8}$$

where $\tilde{s}(k)$ represents the mean $s(i)$ over all data of the entire dataset for a specific number of clusters k. The next section shows the results of the cluster analysis.

4 Results

4.1 1-Dimensional Cluster Analysis

As mentioned in Sect. 3, we perform a one-dimensional cluster analysis considering the mean daily SI at each site. We group the data in k groups, with k running from 2 to 6 with k-means and k-medoids algorithms for all sites. In order to find the best k we calculate the Silhouette score. It turns out that for this 1-d cluster analysis, the results obtained with both algorithms are very similar. Figure 6 shows the value of SS as a function of k for the sites of the four climate zones using both algorithms. The best k corresponds to the one with highest SS, in particular, as SS approaches 1, the goodness of the clustering increases. As seen in Fig. 6, $k = 2$ is the best one for all locations corresponding to cold and warm climates. On the other hand, $k = 2$ is dominant for all sites for dry and mild climates except for $k = 3$ and $k = 4$, for Los Cerritos and Tuxpan de Bolaños, respectively.

Table 1 summarizes the best k, and its corresponding SS, using both algorithms for all sites. From this analysis, one infers that there solar radiation exhibits two seasons in most of the selected sites.

4.2 2-Dimensional Cluster Analysis

Given the fact that for forecasting applications, not only the intensity of solar radiation is relevant but also its variation, we perform a 2-d cluster analysis considering the mean daily SI and its variability, given by the SD. As in the previous section, we group the

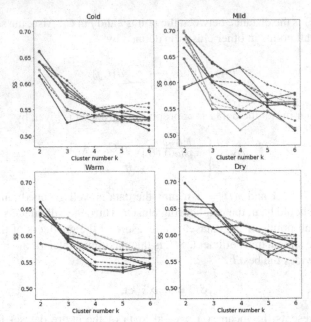

Fig. 6. SS for 1-d cluster analysis for locations of the different climate zones. Solid and dashed lines correspond to results obtained with k-medoids and k-means algorithms, respectively.

Table 1. Number of clusters for each location according to k-means and k-medoids algorithms for the 1-d cluster analysis.

Climate	Site	k-means		k-medoids	
		k	SS	k	SS
Cold	Guadalupe de Córdoba, Tlaxcala	2	0.66	2	0.66
	San Salvador el Verde, Puebla	2	0.64	2	0.64
	Atlauta, Estado de México	2	0.64	2	0.64
	Pico de Orizaba, Puebla	2	0.63	2	0.63
	Alpatláhuac, Veracruz	2	0.62	2	0.61
	Timimilco, Puebla	2	0.66	2	0.66
Dry	Hermosillo, Sonora	2	0.63	2	0.63
	Tijuana, Baja California	2	0.70	2	0.70
	Conejos, Durango	2	0.66	2	0.66
	Valladares, Coahuila	2	0.65	2	0.65
	Ojo Caliente, Zacatecas	2	0.64	2	0.65
	Los Cerritos, Baja California Sur	3	0.64	3	0.64

(*continued*)

Table 1. (*continued*)

Climate	Site	k-means		k-medoids	
		k	SS	k	SS
Mild	Tuxpan de Bolaños, Jalisco	4	0.63	4	0.63
	Llano de San Francisco, Querétaro	2	0.70	2	0.70
	Chanal, Chiapas	2	0.65	2	0.65
	Las Yerbitas, Chihuahua	2	0.70	2	0.70
	Cochoapa el Grande, Guerrero	2	0.70	2	0.70
	Zoquitlán, Puebla	2	0.67	2	0.67
Warm	Tixcancal, Yucatán	2	0.65	2	0.65
	Juchitán, Guerrero	2	0.65	2	0.65
	Francisco Murguía, Chiapas	2	0.58	2	0.58
	Campeche, Campeche	2	0.66	2	0.66
	Llano Grande de Ipala, Jalisco	2	0.63	3	0.63
	Pánuco, Zacatecas	2	0.64	2	0.64

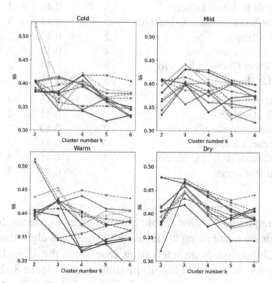

Fig. 7. SS for 2-d cluster analysis for locations of the different climate zones. Solid and dashed lines correspond to results obtained with k-medoids and k-means algorithms, respectively.

data in k groups, with k running from 2 to 6 using k-means and k-medoids algorithms for all sites and, to find the best k, we compute the Silhouette score. Figure 7 shows the results.

In this case, the number of clusters differs from the ones obtained in the 1-d analysis. Depending on the climate zones, the optimum number of groups can be 2, 3, or even 4. We also note in some cases, the number of clusters is different according to the used algorithm. Additionally, the Silhouette scores for both algorithms exhibit a slightly

Table 2. Number of clusters for each location according to k-means and k-medoids algorithm for the 2-d cluster analysis.

Climate	Site	k-means		k-medoids	
		k	SS	k	SS
Cold	Guadalupe de Córdoba, Tlaxcala	3	0.41	3	0.41
	San Salvador el Verde, Puebla	4	0.41	4	0.41
	Atlauta, Estado de México	4	0.40	2	0.38
	Pico de Orizaba, Puebla	2	0.52	2	0.52
	Alpatláhuac, Veracruz	2	0.40	2	0.40
	Timimilco, Puebla	4	0.40	4	0.39
Dry	Hermosillo, Sonora	3	0.46	3	0.47
	Tijuana, Baja California	2	0.48	2	0.48
	Conejos, Durango	3	0.43	3	0.42
	Valladares, Coahuila	3	0.46	3	0.46
	Ojo Caliente, Zacatecas	3	0.45	3	0.44
	Los Cerritos, Baja California Sur	3	0.45	3	0.45
Mild	Tuxpan de Bolaños, Jalisco	3	0.40	3	0.40
	Llano de San Francisco, Querétaro	3	0.43	3	0.43
	Chanal, Chiapas	2	0.41	2	0.41
	Las Yerbitas, Chihuahua	4	0.40	3	0.40
	Cochoapa el Grande, Guerrero	3	0.44	3	0.44
	Zoquitlán, Puebla	3	0.41	2	0.41
Warm	Tixcancal, Yucatán	3	0.43	3	0.42
	Juchitán, Guerrero	2	0.51	2	0.51
	Francisco Murguía, Chiapas	2	0.40	2	0.40
	Campeche, Campeche	3	0.41	2	0.40
	Llano Grande de Ipala, Jalisco	3	0.45	3	0.45
	Pánuco, Zacatecas	3	0.43	3	0.42

bigger difference than in the 1-d case. Table 2 summarizes the results for the best k, and its respective SS, for each site using k-means and k-medoids algorithms.

Las Yerbitas, Chihuahua, is an interesting case since the $k = 4$ or 3, depending on the algorithm and it is useful to highlight some facts. Figure 8 shows the clusters obtained with both algorithms as well as the behavior of the SS. First thing to notice is that, from the orange dashed-line, the value of SS for $k = 3$ is the biggest one, implying that three groups is the best clustering according to k-medoids. However, from the blue dashed-line, we note that even though $k = 4$ is the largest value of k, $k = 3$ is very close to it. Moreover, looking at the clusters and their respective centroids, which are represented by the red stars, we realize that the centroid of the 3rd cluster obtained by k-means, yellow cluster, does not correspond to data from the dataset. This was to be expected, since k-means calculates the best centroid, whereas k-medoids uses the best data from the dataset to choose the centroid. Thus, given the physical interpretation, we consider it is more appropriate to use k-medoids for the purposes of this work.

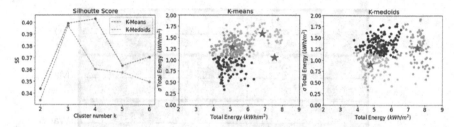

Fig. 8. Clustering results according to k-means and k-medoids for Las Yerbitas, Chihuahua.

Figure 9 provides a summary of the seasons found. Dry and mild climates are characterized by the following seasonality:

a) the first one, corresponding to $k = 2$, presents two seasons: high/low incident solar energy corresponding to the months colored with orange/blue;

b) the second one, corresponding to $k = 3$, presents three seasons: high/low incident solar energy corresponding to the months colored with orange/blue, and a third with high variability of the solar irradiance and average incident solar energy corresponding to the months colored with green.

Mild climate presents a very similar seasonality but the time of the year in which the change of season occurs is slightly different. Finally, cold climate exhibits a different solar radiation characterization:

a) the first one, corresponding to $k = 2$, presents two seasons: high/low incident solar energy corresponding to the months colored with orange/blue;

b) the second one, corresponding to $k = 3$, with three seasons: high/low incident solar energy corresponding to the months colored with orange/blue, and a third with average incident solar energy and high variability corresponding to the months colored with green;

c) the third one, corresponding to $k = 4$, with four seasons: high and low incident solar energy and high and low variability with average incident solar energy corresponding to months colored with green and purple, respectively.

These results are shown in the respective localities on the map in Fig. 10. The colored dots in the map correspond to the different types of seasonality described above and match with the colors on the left hand side of Fig. 9. Each dot is located at the corresponding site in agreement with Fig. 4. Thus, this represents the first step towards a seasonality atlas for solar radiation in Mexico. Indeed, the more points we have the resolution of the atlas becomes better, as mentioned in the last section.

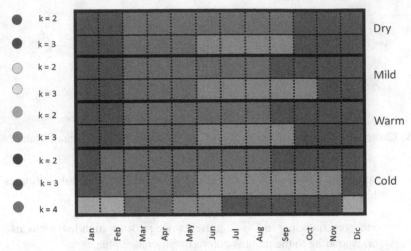

Fig. 9. Types of seasonality in the main climates in Mexico according to k-medoids results. The four climate zones are shown with the corresponding solar radiation seasons. The colored squares colored correspond to: blue and orange for low and high values of incident solar energy, respectively, whereas green and purple correspond to high and low variability in the incident radiation (Color figure online).

Fig. 10. Types of seasonality in the main climates in Mexico according to k-medoids results.

5 Discussion

In this work, we perform two different cluster analyses:

a) A one-dimensional clustering problem, which corresponds to the search for clusters that classify the incident solar radiation in accordance to their similarity.
b) A two-dimensional clustering problem, which consists of characterizing the solar radiation by groups considering the incident solar radiation and its variability, standard deviation, according to their likeness.

Since the values of the Silhouette score of the first analysis turn out to be higher than the second analysis, one could infer that it is better to use that approach to get conclusive results. However, by examining the results, it turns out that the 2-dimensional approach captures better the rainy season in Mexico, from June to November, corresponding to the high variability with average solar radiation. Figure 11 exemplifies this fact by presenting the clustering results in two locations, Ojo Caliente y las Yerbitas. The figures are done such that each color represents a cluster and the months of the year are distributed along the circumference. The first two figures correspond to the best clustering ($k = 2$) according to the 1-dimensional analysis. Notice that the rainy season is not reflected in it. Moreover, if one forces the number of clusters to be three, as shown in the middle graphs, the grouping continues without reflecting the rainy season. Finally, the graphs below show the clustering for the 2-dimensional analysis and the rainy season, as it is reflected from the grouping corresponding to the months of June to November.

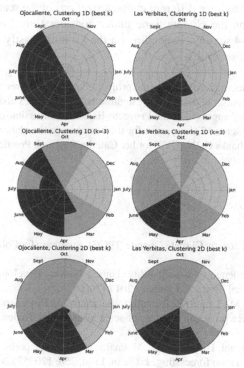

Fig. 11. Clustering results represented along the year for two locations done with the one and two-dimensional approaches. The pink/purple colors correspond to low/high incident solar energy, whereas orange corresponds to average incident solar energy with high variability.

Finally, it is worth to mention that this is a rough estimation of a seasonality atlas for Mexico. In future work, we will improve this work by constructing a mesh covering the country and calculating the average radiation values on each element. Then, the procedure performed in this paper can be followed.

6 Conclusions

The identification of solar seasonality in a given place is very useful to construct a solar radiation forecasting model. Solar seasonality contains the information for dividing and treating data, resulting in better forecasting results. This work performs a cluster analysis to determine the seasonality of the solar radiation of different locations. We use k-means and k-medoids algorithms, and even though both of them are partitioning algorithms, we end up preferring k-medoids to find the seasonality since the centroids of the clusters belong to data from the dataset and therefore a straightforward interpretation is generated.

We performed one and two-dimensional cluster analysis considering the incident solar energy and its variability. We conclude that the two-dimensional analysis better captures the nature of the solar radiation seasonality.

Finally, we find up to three different characterizations of the seasonality of the solar radiation depending on the climate zone under consideration. Each characterization presents either 2, 3 or 4 seasons, depending on the location. Lastly, we present an atlas of Mexico showing these results.

Acknowledgments. This work arises from the project "Predicción del recurso solar usando imágenes satelitales para impulsar el desarrollo sostenible en comunidades aisladas con energía asequible y no contaminante" approved in the Proyecto Espacial Universitario (PEU) from UNAM. The authors wish to thank the PEU program for their support in the elaboration and publication of this work. M.B. also thanks CONACYT for her Catedra Research Position with ID 71557, and to INEEL for its hospitality.

References

1. Intergovernmental Panel on Climate Change (IPCC). Summary for Policymakers. IPCC AR5 WG1 (2013).
2. Energy and the challenge of sustainability. United Nations Development Programme and World Energy Council. Retrieved January 2017 (2000)
3. Diagne, M., David, M., Lauret, P., Boland, J., Schmutz, N.: Review of solar irradiance forecasting methods and a proposition for small-scale insular grids. Renew. Sustain. Energy Rev. **27**, 65–76 (2013)
4. Guermoui, M., Melgani, F., Gairaa, K., Mekhalfi, M.L.: A comprehensive review of hybrid models for solar radiation forecasting. J. Clean. Prod. **258**, 120357 (2020)
5. Lan, H., Zhang, C., Hong, Y.Y., He, Y., Wen, S.: Day-ahead spatiotemporal solar irradiation forecasting using frequency-based hybrid principal component analysis and neural network. Appl. Energy **247**, 389–402 (2019)
6. Ghofrani, M., Ghayekhloo, M., Azimi, R.: A novel soft computing framework for solar radiation forecasting. Appl. Soft Comput. **48**, 207–216 (2016)
7. Wang, S.Y., Qiu, J., Li, F.F.: Hybrid decomposition-reconfiguration models for long-term solar radiation prediction only using historical radiation records. Energies **11**(6), 1376 (2018)
8. Bigdeli, N., Borujeni, M.S., Afshar, K.: Time series analysis and short-term forecasting of solar irradiation, a new hybrid approach. Swarm. Evol. Comput. **34**, 75–88 (2017)
9. Voyant, C., Muselli, M., Paoli, C., Nivet, M.L.: Hybrid methodology for hourly global radiation forecasting in Mediterranean area. Renew. Energy **53**, 1–11 (2013)

10. Yacef, R., Mellit, A., Belaid, S., Sen, Z.: New combined models for estimating daily global solar radiation from measured air temperature in semi-arid climates: application in Ghardaïa, Algeria. Energy Convers. Manag. **79**, 606–615 (2014)

11. Akarslan, E., Hocaoglu, F.O.: A novel adaptive approach for hourly solar radiation forecasting. Renew. Energy **87**, 628–633 (2016)

12. Kaufman, L., Rousseeuw, P.J.: Finding Groups in Data: An Introduction to Cluster Analysis. John Wiley & Sons (2005)

13. King, R.S.: Cluster Analysis and Data Mining An Introduction. Mercury Learning and Information LLC (2015)

14. Rousseeuw, P.J.: Silhouettes: a graphical aid to the interpretation and validation of cluster analysis. Comput. Appl. Math. **20**, 53–65 (1987)

15. Boland, J.: Characterising seasonality of solar radiation and solar farm output. Energies **13**, 471 (2020)

16. Lima, F.J.L., Martins, F.R., Costa, R.S., Goncalves, A.R., Santos, A.P.P., Pereira, E.B.: The seasonal variability and trends for the surface solar irradiation in northeastern region of Brazil. Sustainable Energy Technol. Assess. **23**, 335–346 (2019)

17. Vindel, J.M., Valenzuela, R.X., Navarro, A.A., Polo, J.: Temporal and spatial variability analysis of the solar radiation in a región affected by the intertropical convergence zone. Meteorol. Appli. **27**, e1824 (2020)

18. Fernández, P.C.M., García, B.J., Guisado, M.V., Gastón, M.: A clustering approach for the analysis of solar energy yields: A case study for concentrating solar thermal power plants. AIP Conf. Proc. **1734**, 070008 (2016)

19. Bessafi, M., et al.: Clustering of solar irradiance. In: Lausen, B.., Krolak-Schwerdt, S.., Böhmer, M.. (eds.) Data Science, Learning by Latent Structures, and Knowledge Discovery. SCDAKO, pp. 43–53. Springer, Heidelberg (2015). https://doi.org/10.1007/978-3-662-449 83-7_4

20. Govender, P., Brooks, M., Matthews, A.P.: Cluster analysis for classification and forecasting of solar irradiance in Durban, South Africa. J. Energy South. Afr. **29**(2), 63–76 (2018)

21. https://nsrdb.nrel.gov/

22. Al-Waeli, A.H.A., Kazem, H.A., Chaichan, M.T., Sopian, K.: Photovoltaic Thermal System: Principles, Design and Applications, 1st edn. Springer Nature (2020)

23. Pelleg, D., Moore, A.: Accelerating exact k-means algorithms with geometric reasoning. In: Proceedings of the Fifth ACM SIGKDD International Conference on Knowledge Discovery and Data Mining – KDD'99. California (1999).

24. Kaufman, L., Rousseeuw, P.J.: Partitioning Around Medoids (Program PAM). Wiley Series in Probability and Statistic. John Wiley & Sons, Hoboken, NJ, USA (1990)

Best Paper Award, Third Place

Comparing Machine Learning Based Segmentation Models on Jet Fire Radiation Zones

Carmina Pérez-Guerrero[1]($\boxtimes$), Adriana Palacios[2]($\boxtimes$), Gilberto Ochoa-Ruiz[1]($\boxtimes$),
Christian Mata[3], Miguel Gonzalez-Mendoza[1],
and Luis Eduardo Falcón-Morales[1]

[1] School of Engineering and Sciences, Tecnologico de Monterrey, Monterrey, Mexico
carmina.perez@neodatus.mx, gilberto.ochoa@tec.mx
[2] Department of Chemical, Food and Environmental Engineering, Universidad de las
Americas Puebla, 72810 Puebla, Mexico
adriana.palacios@udlap.mx
[3] EEBE, Universitat Politècnica de Catalunya, Eduard Maristany 16,
08019 Barcelona, Catalonia, Spain

Abstract. Risk assessment is relevant in any workplace, however, there
is a degree of unpredictability when dealing with flammable or hazardous
materials so that detection of fire accidents by itself may not be enough.
An example of this is the impingement of jet fires, where the heat fluxes
of the flame could reach nearby equipment and dramatically increase the
probability of a domino effect with catastrophic results. Because of this,
the characterization of such fire accidents is important from a risk management point of view. One such characterization would be the segmentation of different radiation zones within the flame, so this paper presents
exploratory research regarding several traditional computer vision and
Deep Learning segmentation approaches to solving this specific problem. A data set of propane jet fires is used to train and evaluate the
different approaches and given the difference in the distribution of the
zones and background of the images, different loss functions, that seek to
alleviate data imbalance, are also explored. Additionally, different metrics are correlated to a manual ranking performed by experts to make
an evaluation that closely resembles the experts criteria. The Hausdorff
Distance and Adjusted Rand Index were the metrics with the highest correlation and the best results were obtained from the UNet architecture
with a Weighted Cross-Entropy Loss. These results can be used in future
research to extract more geometric information from the segmentation
masks or could even be implemented on other types of fire accidents.

Keywords: Semantic segmentation · Deep learning · Computer
vision · Jet fires

I. Batyrshin et al. (Eds.): MICAI 2021, LNAI 13067, pp. 161–172, 2021.
https://doi.org/10.1007/978-3-030-89817-5_12

1   Introduction

In the industrial environment, there are certain activities such as the storage of fuels or the transportation of hazardous material, that can be involved in severe accidents that affect the industrial plant or activity border, as well as external factors like human health, environmental damage, property damage, and others. Overall knowledge of the features and characteristics of major accidents is required to prevent and manage them or, in the worst case scenario, take action to reduce and control their severity and aftermath. Some accidents are relatively well known and researched, however, other accidents have not been broadly explored.

Sometimes the detection of fire, in an industrial setting, is not enough to make the correct decisions when managing certain fire accidents. An example of this, is the impingement of jet flames on nearby pipes, depending on the heat fluxes of the flame, the pipe wall could heat up quickly and reach dangerous temperatures that lead to severe domino effect sequences. These heat fluxes can be defined as three main areas of interest within the flame [6], and the definition of their geometric characteristics and localization becomes valuable information in risk management.

Semantic segmentation could be a useful approach for this kind of flame characterization, and the same procedure could even be employed in other fire-related accidents. To explore this proposal, different segmentation methodologies are evaluated on a set of images from real jet fires of propane to accurately segment the radiation zones within the flames. These different zones are illustrated in Fig. 1 and are defined across this paper as the Central Zone, the Middle Zone, and the Outer Zone.

The rest of this paper is organized as follows. Section 2 describes the previous work present in the literature related to the problem and different semantic segmentation methods. Section 3 describes the approaches used to perform the exploration research. Section 4 explains the data set used for the experiments and the pre-processing methods applied. Section 5 describes the different evaluation metrics and loss functions explored during the experiments. Section 6 contains the training protocol used for the experiments with Deep Learning architectures. Section 7 explains the testing procedure applied to the segmentation methods. Section 8 presents the results of the exploration research. This section also offers a discussion of the future work that can stem from the knowledge obtained. Finally, Sect. 9 summarizes the major findings of the presented work.

2   State of the Art

There has been previous research work regarding the use of Computer Vision and image processing for the detection and monitoring of flares in the context of industrial security. For example, Rodrigues and Yan [13] used imaging sensors combined with digital image processing to determine the size, shape, and region of the flare, successfully characterizing the dynamism of the fire.

Another example is the work done in Janssen and Sepasian [7], where, by separating the flare from the background using temperature thresholding, and

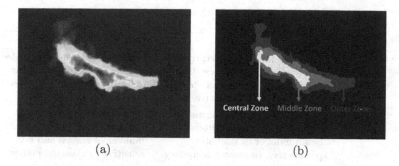

(a) (b)

Fig. 1. Image (a) is an infrared visualization of a horizontal propane jet flame. Image (b) is the corresponding ground truth segmentation, with the segment names indicated. Modified from [6].

adding false colors to represent different temperature regions, a system was created, capable of tracking the flare size for automated event signaling.

2.1 Deep Learning Architectures

Deep learning algorithms, such as Convolutional Neural Networks, have shown outstanding performance in many complex tasks, such as image recognition, object detection, and semantic and instance segmentation [9]. Advancements on these methods have been increasing rapidly, with constant research being published regarding better and more robust algorithms. Some important architectures present in the literature are summarized in Table 1 with their advantages and disadvantages described. In general, these architectures were selected due to their efficiency during segmentation and their capacity to accurately portray the shape of such dynamic figures obtained from the flames.

2.2 Traditional Computer Vision Methods

Different traditional segmentation methods are included in the analysis as a baseline for the results given by the Deep Learning architectures, these methods are Gaussian Mixture Model (GMM), K-means clustering, Thresholding, and Chan-Vese segmentation.

- GMM is one of the main methods applied to fire and smoke image segmentation. It offers a clear definition of their dynamic shape, but sometimes missed pixels from the inner parts of the fire and smoke [1].
- K-means clustering provides shape-based image segmentation [18] and has been previously used for fire segmentation [1,15]. However, this method does not guarantee continuous areas [18].
- Thresholding is the simplest method of image segmentation and has a fast operation speed. It is most commonly used in region-based segmentation, but it is sensitive to noise and gray scale unevenness [17].
- The Chan-Vese segmentation [3] belongs to the group of active contour models, which present some advantages for infrared image segmentation,

Table 1. Summary of the selected architectures.

Source	Architecture	Pros	Cons
[4]	DeepLabv3	Recovers detailed structures lost due to spatial invariance. Efficient approximations via probabilistic inference. Wider receptive fields.	Low accuracy on small scaled objects. Has trouble capturing delicate boundaries of objects.
[2]	SegNet	Efficient inference in terms of memory and computational time. Small number of trainable parameters. Improved boundary delineation.	Input image must be fixed. Performance drops with a large number of classes and is sensitive to class imbalance.
[14]	UNet	Can handle inputs of arbitrary sizes. Smaller model weight size. Precise localization of regions.	Size of network comparable to the size of features. Significant amount of time to train. High GPU memory footprint in larger images.
[10]	Attention UNet	Avoids the use of multiple similar feature maps. Focuses on the most informative features without additional supervision. Enhances the results of the UNet architecture.	Adds more weight parameters The time for training increases, especially for long sequences.

since the edges obtained are smooth and are represented with closed curves. However, this kind of segmentation is very sensitive to noise and depends heavily on the location of the initial contour, so it needs to be manually placed near the image of interest [19].

3 Proposed Approach

To explore the semantic segmentation of radiation zones within the flames as characterization of fire incidents, a group of 4 traditional segmentation methods and 4 deep learning architectures are explored. These segmentation methods are trained and tested using a data set of jet fire images obtained from videos of an experiment performed in an open field.

To properly evaluate and compare the different segmentation approaches, several metrics with different evaluation methods are correlated to manual rankings performed by two experts in the field, this is to make sure that the evaluation is the most representative of a fire engineer's perception of good segmentation.

The best model found from this exploratory research will then be used in future work, not included in this paper, to extract other geometric characteristics from the resulting segmentation masks.

4 Data Set

Investigators from Universitat Politècnica de Catalunya performed an experiment to produce horizontal jet fires at subsonic and sonic gas exit rates.

The experiment was filmed using an infrared thermographic camera, more specifically, an AGEMA 570 by Flir Systems. The video was saved in four frames per second, resulting in a total of 201 images with a resolution of 640 × 480 pixels. After obtaining the infrared visualizations of the flames, segmentation of the three radiation zones within the fire was performed, the results were validated by experts in the field and the result was a ground truth for each of the infrared images. The images were saved as Matlab files that contain a temperature matrix corresponding to the temperature values detected by the camera for the infrared images, and the label matrix corresponding to the segments for the ground truth segmentations. These files were then exported as PNG files to be used by the different segmentation algorithms described in Sect. 2.1.

4.1   Image Processing

To enhance the characteristics of the jet fires represented in the infrared images, and to reduce their variance, a process of image normalization was employed, which also helps in the convergence of the Deep Learning methods. The ground truth images were transformed into labeled images, where a label id was used instead of the original RGB values. Given the small number of samples of the data set, and to increase the variability of the input during the training of the models, data augmentation techniques were applied. This processing of the images can increase the performance of the models, can avoid the over-fitting to the training samples, and can help the models to also perform well for instances that may not be present in the original data set. Horizontal flipping, random cropping, and random scaling were applied in parallel to the training workflow, so for each iteration of training, a different augmented image was inputted. The probability of horizontal flipping was set to 50%, and random scaling had values that ranged between 0.7 and 2.0.

5   Metrics and Loss Functions

5.1   Metrics

An analysis was performed to select evaluation metrics that were more representative of an expert's evaluation of the segmentation. The metrics analysed are separated into groups that describe their evaluation method. This diversity is important because the consideration of particular properties could prevent the discovery of other particular errors or lead to over or underestimating them. To compare the metrics enlisted in Table 2, a group of images were evaluated with the metrics and two manual rankings performed by two experts in the field. A Pearson pairwise correlation is used to perform this comparison at segmentation level.

5.2   Loss Functions

The proportion of the different radiation zones within each flame is different. The Outer Zone tends to be the largest segment and the central zone is usually

Table 2. Summary of the metrics analysed in this paper. The "Group" column describes the method group that the metric belongs to. Based on [16].

Metric	Group
Jaccard Index	Spatial overlap based
F-measure	Spatial overlap based
Adjusted Rand Index	Pair counting based
Mutual Information	Information theoretic based
Cohen's Kappa	Probabilistic based
Hausdorff Distance	Spatial distance based
Mean Absolute Error	Performance based
Mean Square Error	Performance based
Peak Signal to Noise Ratio	Performance based

smaller than the Middle Zone, these differences could affect the overall segmentation obtained during training, which is why the loss functions employed in this research work were focused on dealing with this class imbalance. The implemented loss functions are summarized in Table 3.

Table 3. Summary of the loss functions implemented in this paper.

Source	Loss function	Description
[12]	Weighted cross-entropy loss	Combines log softmax and negative log likelihood. Useful for multiple classes that are unbalanced
[8]	Focal loss	Addresses foreground-background class imbalance. Training is focused on a sparse set of hard examples
[5]	Generalized Wasserstein Dice Loss (GWDL)	Semantically-informed generalization of the Dice score. Based on the Wasserstein distance on the probabilistic label space

6   Training

The PyTorch framework [12] was used for the implementation of the Deep Learning architectures. To maintain the weights of the Convolutional Neural Networks as small as possible, a weight decay strategy was used with L2 regularization. The learning rate had an initial value of 0.0001 and used an ADAM optimizer during training. The class weights used by the Weighted Cross-Entropy and Focal losses were computed according to the ENet custom class weighing scheme [11] and are defined as the following: 1.59 for background, 10.61 for Outer zone, 17.13 for Middle zone, and 22.25 for Central zone. The class distances used for the loss function of GWDL is defined as 1 between the background and the radiation zones, and as 0.5 between the zones themselves. The training was performed using an Nvidia DGX workstation that has 8 Nvidia GPUs and allows a batch size of 4 as maximum, which is the batch size used for the models. The data was split into 80% for training and 20% for testing and validation, resulting in a total of 161 images for training, 20 images for testing, and 20 images for validation. The models were trained for up to 5000 epochs with an Early Stopping strategy to avoid overfitting. The resulting loss values for the best models can be visualized in Fig. 2.

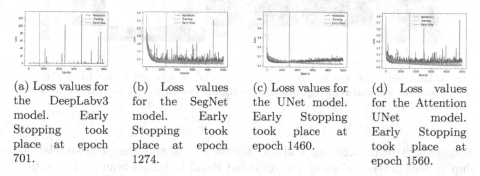

(a) Loss values for the DeepLabv3 model. Early Stopping took place at epoch 701.

(b) Loss values for the SegNet model. Early Stopping took place at epoch 1274.

(c) Loss values for the UNet model. Early Stopping took place at epoch 1460.

(d) Loss values for the Attention UNet model. Early Stopping took place at epoch 1560.

Fig. 2. The Weighted Cross-Entropy loss values for the best models obtained for each architecture explored.

7   Testing

Testing and Validation were performed on 20% of the data set, which represent a total of 40 images. The results are compared using the metrics with the highest correlation to the manual ranking of experts. The results for each Deep Learning architecture are first compared across all 3 different loss functions mentioned in Sect. 5.2. The best performing combination, for each architecture, is then compared to the results obtained from the other 4 traditional segmentation models. The time each method takes to segment the whole data set of 201 images is also taken into account. The goal of the comparison is to find the best overall model for the segmentation of the radiation zones within the flames.

8   Results and Discussion

8.1   Selected Metrics

The results of the correlation between the metrics, mentioned in Sect. 5, and the manual rankings performed by experts can be observed in Fig. 3.

The highest Pearson correlation value was given by the Hausdorff Distance, with a value of -0.32 to the first manual ranking, this negative relationship is because a smaller Hausdorff Distance is preferable and the manual ranking assigned higher values as the segmentation improved. The second highest correlations were given by the Adjusted Rand Index in both manual rankings, with a

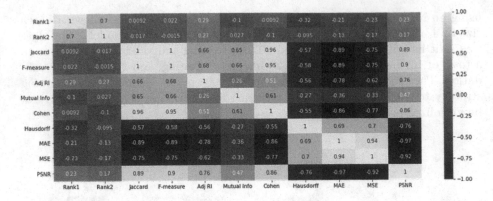

Fig. 3. Heatmap representing the Pearson Correlation Coefficient for the metrics in Sect. 5 and the manual rankings Rank1 and Rank2 done by experts in the field of the problem.

value of 0.29 for the first one and 0.27 for the second one, this positive relationship is because the values of the Adjusted Rand Index go from 0 to 1, with 1 being the best result, and showing similar behavior to the manual ranking that assigns higher values to better segmentation.

8.2   Best Loss Function

Each combination of Deep Learning architecture and Loss Function was evaluated using the Hausdorff Distance as the main metric. The best results across all architectures were obtained when using the Weighted Cross-Entropy Loss, this can be observed in Fig. 4. In general, the Focal Loss also showed good results, being close to the Weighted Cross-Entropy results and most of the times surpassing the GWDL results, with the only exception being the Attention UNet model, where the mean Hausdorff Distance with Focal Loss is larger than with GWDL, but has a much closer distribution. The most dramatic difference happened with the SegNet architecture, which showed very poor results with GWDL.

8.3   Traditional and Deep Learning Segmentation

A comparison was done between the traditional Computer Vision methods mentioned in Sect. 2.2 and the Deep Learning models mentioned in Sect. 2.1, using the Weighted Cross-Entropy loss. The Hausdorff Distance and the Adjusted Rand Index are used to evaluate the models across the testing set and the time each method takes to perform the segmentation on all the images from the data set is also taken into account. These results are summarized in Table 4 and the distribution of the results can be visualized in Fig. 5. Overall the best performing models are observed to be the UNet and Attention UNet models. Even if Attention UNet achieved a slightly better Adjusted Rand Index score, UNet obtained

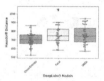 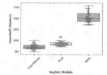 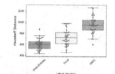 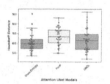

(a) Hausdorff Distance for all the DeepLabv3 models.
(b) Hausdorff Distance for all the SegNet models.
(c) Hausdorff Distance for all the UNet models.
(d) Hausdorff Distance for all the Attention UNet models.

Fig. 4. The Hausdorff Distance for all the model and loss function combinations across the validation set. For each model the order of loss functions is Weighted Cross-Entropy, Focal and GWDL from left to right.

a better Hausdorff distance and segmentation time, therefore we can say that the best model is the one that uses the UNet architecture.

The difference in the segmentation of each method can also be visualized in Fig. 6. It can be observed that the shape of the Outer and Middle zones are generally well represented in most of the segmentation models, however, the most important differences are found in the Central zone, where the Deep Learning architectures defined more clearly its shape. Similar results of the UNet and Attention UNet architectures are also observed in this sample segmentation masks, with close resemblance to the ground truth.

Table 4. Mean Hausdorff Distance and Adjusted Rand Index values for all the segmentation models across the testing set, as well as the time in seconds that each method takes to segment the whole data set. The best results are in bold.

Method	Hausdorff distance	Adjusted Rand Index	Time (s)
GMM	1288.10	0.9156	2723.8
K-means	1000.63	0.8855	3035.1
Thresholding	1029.08	0.9152	30.7
Chan-Vese	1031.90	0.8568	18177.5
DeepLabv3	784.86	0.9514	17.1
SegNet	692.73	0.9381	16.4
UNet	**586.46**	0.9504	**15.7**
Attention UNet	601.05	**0.9592**	17.7

8.4   Discussion

Overall, the Deep Leaning algorithms greatly outperformed the traditional Computer Vision methods and the best-proposed model would be a UNet architecture with a Weighted Cross-Entropy loss function. The segmentation masks obtained

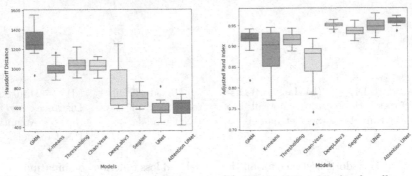

(a) Hausdorff distance for all segmentation models.

(b) Adjusted Rand Index for all segmentation models.

Fig. 5. The Hausdorff Distance and Adjusted Rand Index values for all the segmentation models across the testing set.

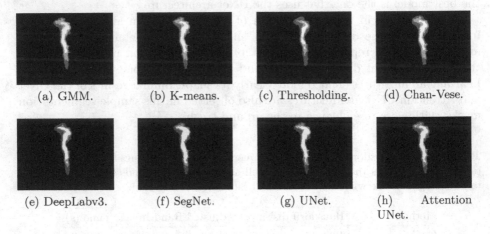

(a) GMM.   (b) K-means.   (c) Thresholding.   (d) Chan-Vese.

(e) DeepLabv3.   (f) SegNet.   (g) UNet.   (h)   Attention UNet.

Fig. 6. Sample segmentations of all the models.

from this model could be used in the future to further extract other geometric characteristics of the flame, such as length, area, and lift-off distance. This additional information would improve greatly the decision making process involved in the risk assessment and management of fire related accidents that can take place in an industrial setting. Furthermore, the metrics of Hausdorff Distance and Adjusted Rand Index can be used to evaluate other segmentation approaches that may try to solve similar problems in the future, having the certainty that the evaluation would be a close representation of an expert's opinion.

9   Conclusions

The semantic segmentation of radiation zones within the flames can be used to characterize fire accidents, such as jet fires, and the information obtained from the segmentation can prove to be critical when dealing with risk management in industrial settings. The exploratory research presented in this paper continued to show that Deep Learning architectures greatly outperform other traditional Computer Vision approaches. It was also found that for this specific problem, the best loss function to train a Deep Learning model is a Weighted Cross-Entropy Loss, the best architecture to be used is UNet, and the best evaluation metrics are both the Hausdorff Distance and the Adjusted Rand Index. All this knowledge can be later used in future research focused on extracting even more geometric information from the segmentation masks. This could bring about a more complete characterization analysis of jet fires and the methods applied to these types of fire accidents could then be used on other fire scenarios.

Acknowledgements. This research is supported in part by the Mexican National Council of Science and Technology (CONACYT). This research is part of the project 7817-2019 funded by the Jalisco State Council of Science and Technology (COECYT-JAL). The data set that supports the findings of this study are available upon reasonable request.

References

1. Ajith, M., Martínez-Ramón, M.: Unsupervised segmentation of fire and smoke from infra-red videos. IEEE Access **7**, 182381–182394 (2019). https://doi.org/10.1109/ACCESS.2019.2960209
2. Badrinarayanan, V., Kendall, A., Cipolla, R.: SegNet: a deep convolutional encoder-decoder architecture for image segmentation. IEEE Trans. Pattern Anal. Mach. Intell. **39**(12), 2481–2495 (2017). https://doi.org/10.1109/TPAMI.2016.2644615
3. Chan, T., Vese, L.: Active contours without edges. IEEE Trans. Image Process. **10**(2), 266–277 (2001). https://doi.org/10.1109/83.902291
4. Chen, L.C., Papandreou, G., Schroff, F., Adam, H.: Rethinking atrous convolution for semantic image segmentation (2017). arXiv: 1706.05587
5. Fidon, L., et al.: Generalised Wasserstein dice score for imbalanced multi-class segmentation using holistic convolutional networks. In: Crimi, A., Bakas, S., Kuijf, H., Menze, B., Reyes, M. (eds.) BrainLes 2017. LNCS, vol. 10670, pp. 64–76. Springer, Cham (2018). https://doi.org/10.1007/978-3-319-75238-9_6
6. Foroughi, V., et al.: Thermal effects of a sonic jet fire impingement on a pipe. J. Loss Prev. Process Ind. **71**, 104449 (2021). https://doi.org/10.1016/j.jlp.2021.104449
7. Janssen, R., Sepasian, N.: Automatic flare-stack monitoring. SPE Prod. Oper. **34**(01), 18–23 (2018). https://doi.org/10.2118/187257-PA
8. Lin, T., Goyal, P., Girshick, R., He, K., Dollar, P.: Focal loss for dense object detection. IEEE Trans. Pattern Anal. Mach. Intell. **42**(02), 318–327 (2020). https://doi.org/10.1109/TPAMI.2018.2858826

9. Litjens, G., et al.: A survey on deep learning in medical image analysis. Med. Image Anal. **42**, 60–88 (2017). https://doi.org/10.1016/j.media.2017.07.005
10. Oktay, O., et al.: Attention U-Net: learning where to look for the pancreas (2018). arXiv: 1804.03999
11. Paszke, A., Chaurasia, A., Kim, S., Culurciello, E.: ENet: a deep neural network architecture for real-time semantic segmentation (2016). arXiv: 1606.02147
12. Paszke, A., et al.: PyTorch: an imperative style, high-performance deep learning library. In: Wallach, H., Larochelle, H., Beygelzimer, A., d'Alché-Buc, F., Fox, E., Garnett, R. (eds.) Advances in Neural Information Processing Systems, vol. 32, pp. 8024–8035. Curran Associates, Inc. (2019)
13. Rodrigues, S.J., Yan, Y.: Application of digital imaging techniques to flare monitoring. J. Phys. Conf. Ser. **307**, 012048 (2011). https://doi.org/10.1088/1742-6596/307/1/012048
14. Ronneberger, O., Fischer, P., Brox, T.: U-Net: convolutional networks for biomedical image segmentation. In: Navab, N., Hornegger, J., Wells, W.M., Frangi, A.F. (eds.) MICCAI 2015. LNCS, vol. 9351, pp. 234–241. Springer, Cham (2015). https://doi.org/10.1007/978-3-319-24574-4_28
15. Rudz, S., Chetehouna, K., Hafiane, A., Laurent, H., Séro-Guillaume, O.: Investigation of a novel image segmentation method dedicated to forest fire applications*. Meas. Sci. Technol. **24**, 075403 (2013). https://doi.org/10.1088/0957-0233/24/7/075403
16. Taha, A.A., Hanbury, A.: Metrics for evaluating 3d medical image segmentation: analysis, selection, and tool. BMC Med. Imag. **15**(29) (2015). https://doi.org/10.1186/s12880-015-0068-x
17. Yuheng, S., Hao, Y.: Image segmentation algorithms overview (2017). arXiv:1707.02051
18. Zaitoun, N., Aqel, M.: Survey on image segmentation techniques. Procedia Comput. Sci. **65**, 797–806 (2015). https://doi.org/10.1016/j.procs.2015.09.027
19. Zhang, R., Zhu, S., Zhou, Q.: A novel gradient vector flow snake model based on convex function for infrared image segmentation. Sensors **16**, 1756 (2016). https://doi.org/10.3390/s16101756

A Machine Learning Approach for Modeling Safety Stock Optimization Equation in the Cosmetics and Beauty Industry

David Díaz$^{(\boxtimes)}$, Regina Marta, Germán Ortega, and Hiram Ponce

Facultad de Ingeniería, Universidad Panamericana, Augusto Rodin 498,
03920 Ciudad de México, Mexico
{0241060,0241058,0163470,hponce}@up.edu.mx

Abstract. Safety Stock is generally accepted as an appropriate inventory management strategy to deal with the uncertainty of demand and supply, as well as for limiting the risk of service loss and overproduction [6]. In particular, companies from the cosmetics and beauty industry face additional inventory management challenges derived from the strict regulatory standards applicable in different jurisdictions, in addition to the constantly changing trends, which highlight the importance of defining an accurate safety stock. In this paper, on the basis of the Linear Regression, Decision Trees, Support Vector Machine ("SVM") and Neural Network machine learning techniques, we modeled a general Safety Stock equation and one per product category for a multinational enterprise operating in the cosmetics and beauty industry. The results of our analysis indicate that the Linear Regression is the most accurate model to generate a reasonable and effective prediction of the company's Safety Stock.

Keywords: Stock optimization · Safety stock · Consumer industry · Inventory management · Linear Regression · Decision Trees · Support Vector Machine · Neural Networks

1 Introduction

The calculation of order points is a fundamental element of an optimized stock management; excessively high safety stocks raise costs, otherwise if they are too low, there is a greater risk of falling into stock outs [8]. The aim will always be balance. Facing such a changing and erratic panorama in the mass cosmetics and beauty consumer market, there is a great need for a better response capacity towards variations in demand, as well as the search for efficiency and flexibility at manufacturing for a make-to-stock production process.

This paper reviews the methodology currently used by a multinational company operating in the cosmetics and beauty industry for the calculation of the Safety Stock and propose an innovative alternative for its estimation. As of

© Springer Nature Switzerland AG 2021
I. Batyrshin et al. (Eds.): MICAI 2021, LNAI 13067, pp. 173–186, 2021.
https://doi.org/10.1007/978-3-030-89817-5_13

today, the methodology used by the company only reviews one picture of the main variables used for the calculation of reorder and Safety Stock points [7], the aforementioned may be troublesome since the said variables are dynamic and variate every day at each step of the supply chain.

In order to address the aforementioned, in this paper we propose the implementation of four different machine learning techniques for the calculation of an objective coverage (i.e. Safety Stock). Our aim is to achieve an optimal and adaptive coverage for series of the company's products. The results of the experimentation with diverse machine learning techniques will determine if any of the proposed models is capable of aiding the production validation process, with the main objective of optimizing the Safety Stock equation. The implementation of machine learning techniques for the estimation of Safety Stock will not only solve challenges related to accuracy and mass production for the company, but also optimize costs associated with high stock values and warehouse maintenance.

The rest of the paper is as follows. Section 2 of this paper contains a description of previous related work about the implementation of machine learning to inventory management strategy. Later, Sect. 3 consists of a description of our proposal, the current situation and methodology used for the calculation of an objective stock coverage. Further on, Sect. 4 describes the experimentation and methodology of the four different machine learning techniques implemented as an alternative for the calculation of an objective stock coverage. Afterwards, Sect. 5 presents the results and discussion of the machine learning techniques review and the selection of the optimal model for the estimation of an objective stock coverage. Finally, Sect. 6 concludes.

2 Related Work

Previous related work indicates that uncertainty is one of the major challenges companies face when defining their inventory optimization strategy. Throughout the weeks, key factors for inventory management, such as demand, production capacity and supply tend to fluctuate, causing disruptions in the companies' supply chain. In order to mitigate the risk derived from said fluctuations, companies define an extra level of inventory to hold in their warehouses (i.e. Safety Stock) to reduce the risk of an out-of-stock situation.

The importance of reducing the mentioned risk and improve inventory management is addressed in many published works; for instance, Anshuman, G and Costas M. [2] provide an overview of related works about demand uncertainty in midterm planning and multi site supply chain. In this article a stochastic model is described to model the planning process, which includes decisions of manufacturing and logistics to manage uncertainty. Also, the proposed model captures the trade-off between customer satisfaction level and production costs. Additionally, an application of the model through a study case is presented, where important features such as capacity constrained production equipment, carry-over of inventory and customer backlogs are covered. It concludes explaining that partitioning decision variables and constraints into manufacturing decisions and logistics decisions makes it possible to incorporate uncertainty in a framework. Also the authors suggest to use this framework to manage the risk of any company's assets and for retrieving valuable insights of the customer relationship aspects of supply chain. This paper was a valuable example of the construction of a model, definition of variables and management of uncertainty.

The paper by Syntetos et al. [3] is also helpful to understand the importance of Supply Chain Forecasting, since this work is a comprehensive review of literature in the field, which aims to discuss the process of supply chains relevant to forecasting and cover the existent gap between practice and theory of other related papers. This article starts explaining the Supply Chain process and its main features, raising some issues and highlighting the decisions that are to be made throughout the different steps, with this the importance of forecasting is explained. The authors analyzed the gap between theory and practice through a deep review of each dimension. The aforementioned aside of data and software issues that generate future challenges, making room for further investigation. This paper mentions that at the time it was written there was no Forecasting Support System commercially available; and finally, it discusses the importance of bridging the gap between statistical theory and operational practice, which can be achieved by software developers and through the application of big data technologies.

Moreover, from the review of related work, it was possible to identify research and modeling focused on determining the dimension of Safety Stock using machine learning modeling. For instance, Gonçalves et al. [6] made a literature review of 95 papers about methods and models for dimensioning safety stocks. Determining the dimension of the Safety Stock is an optimization problem, as the authors found in their review, "setting more Safety Stock increases the holding costs regardless the cycle time considered. In contrast, lower Safety Stock levels could lead to stock-outs when demands are volatile". Gonçalves et al. [6] review shows that the application of machine learning in determining Safety Stock still has a lot of opportunity to grow and be exploited. However, the authors highlight that a disadvantage of machine learning and artificial intelligence modeling is that if the values of Safety Stock used to train the model are not optimal, in terms of minimizing inventory costs while maximizing service level, the predicted safety stocks level may not be optimal as well. Finally, even though the review indicates that there is extensive research to understand the Safety Stock optimization problem, there is a pressing need for models that account for mechanisms to address data quality issues and report real-world case studies.

Zhong, W and Zhang, L. [10] use three different modeling techniques to predict the Safety Stock, starting from the assumption that when enterprises increase purchasing and production to meet the need of the production and sales, costs rise. In that sense, the forecasting of enterprise Safety Stock can reduce the inventory cost, enhance the market competitiveness, and meet the customers' needs better, which, according to the authors, has important practical significance. The authors found that SVM, Radial Basis Function neural network ("RBF") and a combination of those two ("GA") are appropriate approaches to solve the Safety Stock optimization problem. Moreover, for the empirical test of the models, Zhong, W and Zhang, L. [10] selected market demand, order delivery cycle, vendor products pass rate, storage costs, accessories ordering costs and stock losses as inputs for the Safety Stock optimization function. According to the authors' findings, the GA model has higher forecasting accuracy, since the values predicted using that approach were closer to the actual values, with a mean squared error of 0.74 and relative errors below 2%, concluding that the GA model is a "scientific, reasonable and effective method for safety stock forecasting".

Çolak et al. [8] start addressing that nowadays competition takes place among supply chains. In that sense, determination of Safety Stock levels is a key element to reach competitive advantage. In particular, the authors main motivation is to propose a scientific method to determine Safety Stock, providing an optimal combination of customer service level (i.e. the rate of responding to customer needs at the right time and amount) and stock levels. Çolak et al. [8] propose a methodology to determine Safety Stock for a company operating in the automotive sector with 1203 products, based on mean and standard deviations of quantitative changes and customer service level, constrained to the company's production capacity. The authors apply an ABC analysis to classify products based on their monetary values and group them through a XYZ analysis according to their order variations. Çolak et al. [8] conclude that their methodology is a comprehensive Safety Stock analysis because of product-based analyses and that it can be generalized to other companies with similar demand characteristics as the analyzed company.

Another approach for the application of artificial intelligence for Safety Stock forecasting is presented by Vasconcelos de Almeida, Y. and Tung W. [1]. In this work the authors use the Fuzzy Logic technique, which uses grades of uncertainty to determine whether an input variable belongs to one or more output variables. Therefore, Fuzzy Logic is used as a method to calculate the Safety Stock of a multinational company that, originally, used to calculate it in a subjective way considering its employees' experience. The favorable results of implementing the model were validated after some weeks of testing, confirming the hypothesis that an intelligent system to automate decisions of stock is needed to optimize this time consuming task.

In particular, Bertsimas et al. [4] detail the new opportunities for business decision making brought by data and machine learning. The authors focus on the applications of machine learning for inventory management, through the development of a model to determine the quantity of stock per product and location for a retail stores network. Similar to other publications described previously, Bertsimas et al. [4] consider a limiting factor (i.e. capacity) and other factors to determine the optimal quantity of stock, such as production costs, delivery costs, store location and uncertain demand for the product. A particularity of this paper is that the data used for the model was compiled from external and internal sources, for instance, the data used for the estimation of demand was obtained from Google Trends, while internal data consisted of sale and inventory records per product and store location. The authors' model was able to predict the optimal inventory with a coefficient of prescription of 88.

Aside from the previously works described, which use artificial intelligence, Li H, Jiang D [5] on their article propose solving the Safety Stock placement model as a scheduling problem, using a combination of constraint programming and genetic algorithm. The hybrid approach evaluated in this article with randomly generated instances, shows good results and performance for small-medium problems in networks between 10 and 80 nodes with moderately more computational time than constrain programming. Also, it is worth mentioning that this work applies a project network perspective, which allows the implementation of large sets of methods of the scheduling field to solve the stock placement problem, which opens new possibilities of study.

3   Methodology

As mentioned in Sect. 1, this paper reviews the application of four different machine learning techniques for the calculation of an objective stock coverage for a company that develops its business in the cosmetics and beauty industry. As of today, the company estimates the objective stock coverage using an Excel model, fed by a series of inputs obtained and/or estimated internally.

The equation below (1) is the Safety Stock main equation used by the company to calculate its objective stock coverage. Now, the said model has six inputs, which will be described in detail below.

$$Safety\_Stock = Safety\_Factor \times Future\_Demand \times \sqrt{(FE^2 \times Lead\_Time + LTE^2)} \quad (1)$$

Forecast Error
This input incorporates the Forecast Error calculation [7] which is represented by the equation shown in (2), it was obtained from proprietary documentation of the company and is used to make actual calculations; it is defined as the monthly ratio of the standard deviation of the expeditions (sales) and sales forecast, over the actual average sales forecast in line, expressed in percentage. The empiric experience indicates that this variable plays the most important role in the estimation of the objective stock coverage.

$$FE\% = \frac{\sigma(Expeditions - Forecast)}{Forecast} \quad (2)$$

Lead Time
This input is the amount of days from the start of production to the moment when the product arrives to IDC [7]. It is known from experience that the value of this variable depends on the product category. Moreover, for most of the categories lead time is 28 days of production and 3 days of goods receipt. In some cases it takes up to 6 days of transportation and quality inspections.

RLT Error (Lead Timer Error)
This input represents the error range that could happen to a product in terms of its lead time (31 days on average) [7]. It represents the probability of not having the product on IDC (distribution center) in the expected time and depends on the following variables:

- **Saturation (S)**. High amount of work load in production line.
- **Multi-Format (MF)**. The need of having product in multiple versions.
- **Supplier Accountability (SSF)**. Availability of raw materials or pieces necessary for production.
- **Subcontractor's reliability (SR)**. Delays in subcontractors delivery.
- **Quality (Q)**. Delays due to quality issues.

The actual documentation obtained from the company uses the factors mentioned above as shown in the Eq. 3; where LTE is the lead time error, expressed in percentage and the weights of each variable are fixed according to the company standars.

$$LTE = (0.3 * S + 0.1 * MF + 0.2 * SSF + 0.2 * SR + 0.2 * Q) \qquad (3)$$

On Table 1, we can observe how these variables are set according to the experience and supplier evaluation results ("MSC").

Table 1. LTE variables

	Saturation	Multi format	Supplier AC & SF	Subcontractor's reliability	Quality
0	<70% 0		Safety stock at factory 0	No subcontractor 0	X
1	70–80% 3	1 format 1	LT <15 days 2	Based on experience 2	No past problems 2
2	80–90% 6	2 formats 2	LT 15–30 days 4	Based on experience 4	Bulk past problems 4
3	>90% 9	3+ formats 3	LT >30 days 6	Based on experience 6	PF problems no solution 6

Service Level/Safety Factor

Service level is the probability that the amount of inventory on hand during the lead time is sufficient to meet expected demand [7], it can be seen as the probability that a stock out will not occur. The variance between the desired service level and the corresponding safety factor desired is normally distributed. This input depends on the category of the product and the type of product (i.e. Strategic, A, B, C or Launch).

The Safety Stock is the amount of units necessary to cover the demand and mitigate the risk of running out of stock; however, it can be converted to days of anticipation in the production.

Even though the calculation of the Safety Stock considers inputs estimated internally by the company, most of these inputs are fixed established values determined on the basis of previous experience. Moreover, the calculation of Safety Stock assumes that future demand will be almost equal to past demand. The aforementioned could be considered as a disadvantage, since most inputs are dynamic. Another disadvantage is that the company calculates Safety Stock only once a year, which turns it into a fixed period of past forecast that is not dynamic or updated throughout the year, which reduces its accuracy each month. Finally, as this model depends strongly on past demand and experience, it is less accurate for new products, since there is a lack of historic data, making it almost impossible to obtain a reliable Safety Stock estimation for new products.

4   Machine Learning Workflow

This section details the business understanding, data understanding and preparation, modeling and evaluation metrics, as well as the deployment of the applicable machine learning models and evaluation metrics that were used as explained in Fig. 1.

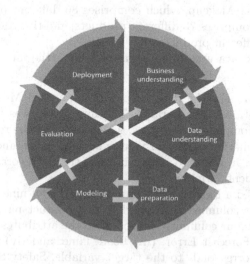

Fig. 1. CRISP reference model [9]

Business Understanding. In order to understand the processes and activities that the business currently performs to calculate the Safety Stock as presented in the previous Sect. 3. The importance of determining an appropriate level of Safety Stock is explained, in addition to the main equation and input parameters used to estimate the number of units necessary to avoid running out of stock. Moreover, the nature of these inputs values has been defined, from which it was noted that some of them are not updated or are based on data from previous years which makes it difficult to achieve a higher level of accuracy. Therefore, the aim of this project is to upgrade the current model with Machine Learning models and provide a more reliable Safety Stock value per category and a more accurate Safety Stock indicator for new products.

Data Understanding. The selection of variables, time frame and categories are described below:

- For the modeling of the Safety Stock equation, we considered four variables: (i) Forecast Error, (ii) Safety Factor, (iii) Lead Time and (iv) RLT Error. The estimation of each of these variables was described previously in Sect. 3.

- The values for these four variables were obtained directly from the company's internal databases.
- For the Forecast Error, we considered 24 months of data, from April 2019 to March 2021. While for the rest of the variables we considered annual fixed values.
- We were able to obtain data for 362 different products, which later were categorized in four groups: (i) Eye Makeup, which comprises 23 different products; (ii) Face Makeup, which comprises 86 different products; (iii) Lip Makeup, which comprises 73 different products; and (iv) Nail Makeup, which comprises 180 different products.
- The Safety Stock measured in days was elected as the target variable.

Data Preparation. For the data preparation we used Microsoft Excel and Pandas (Python data analysis library). The initial step was to clean the data obtained from the company's databases, which included the revision of missing values, the transformation of certain values, such as dates, and the elimination of unnecessary data in order to maintain only the data corresponding to the target and independent variables.

Later we prepared a Pandas' data frame consisting of nine columns and 362 rows. The first four columns correspond to the product name, type, life and category. The other four columns correspond to the attributes of our model: (i) Safety Factor; (ii) Forecast Error; (iii) Lead Time; and (iv) RLT Error. And the ninth column corresponds to the target variable, Safety Stock. Finally, we prepared a Pandas' data frame for each of the four categories by filtering the original data frame.

Table 2 shows an example of the data extracted from the company's databases, filtered for the Eye Makeup category. The columns corresponds to the inputs of our model: Safety Factor, Forecast Error, Lead Time and RLT Error.

Table 2. Data sample

SKU sample	Type	Life	Category	Safety factor	Forecast error M	Lead time (m)	RLT error (%)	Safety stock (days)
SKU100	HARMONIZED	50	Lip makeup	2.33	0.93	1.1	0.23	16.28
SKU103	HARMONIZED	50	Lip makeup	1.48	0.48	3.37	0.3	13.28
SKU104	HARMONIZED	50	Face makeup	2.33	0.44	1.03	0.3	20.94
SKU105	HARMONIZED	50	Face makeup	2.33	0.5	1.1	0.3	20.94

Modeling. For our modeling of the Safety Stock equation, we considered four different machine learning techniques: (i) Linear Regression; (ii) Decision Trees; (iii) SVM; and (iv) Neural Networks.

Moreover, for the modeling of the Safety Stock we used the scikit-learn Python library, which includes modules for different machine learning techniques, including those we selected for our modeling of the Safety Stock equation.

Our modeling process started from the Pandas' data frames described previously in this section. The first step was to define the dependent or target variable (i.e. Safety Stock) and the independent variables or attributes (i.e. Safety Factor, Forecast Error, Lead Time and RLT Error).

The second step consisted on the train and test split of the data, we decided to set 70% of the data for the training of the models and 30% for the testing part. For the data partition we used scikit-learn's "train_test_split" function.

The third step was to standardize the independent variables, which was done using scikit-learn's "StandardScaler" module. It was necessary to standardize our data since each attribute is measured in different units. For instance, RLT Error and Forecast Error are percentages, while Safety Factor and Lead Time are not.

Now, the fourth step is the implementation of each of the selected Machine Learning techniques for the estimation of the Safety Stock equation.

For each of the techniques we first trained the model, by using the training subset of data and then tested the trained model with the testing subset of data.

Evaluation. Since we implemented four different Machine Learning techniques, we evaluated each of our models using the Train/Test Split validation method. This validation method is based on the accuracy indicator. Accuracy indicates how many correct Safety Stock estimations were obtained using each model. Equation 4 sets forth the calculation of the accuracy. The application of this method consists of comparing the accuracy of the model using the training subset of data versus the accuracy of the model using the testing subset of data.

$$Accuracy = \frac{Correct\_Safety\_Stock\_Estimations}{Total\_Safety\_Stock\_Estimations} \tag{4}$$

5   Results and Discussion

This section describes the deployment of our modeling, using the previously described methodology and each of the four selected machine learning techniques.

1. Linear Regression. We ran five different regressions, a general one considering all the products and one for each makeup ("MU") categories. Results are shown in Table 3.

Table 3. Linear regression results

Parameter	General	Eye MU	Face MU	Lip MU	Nail MU
Intercept	17.92	13.78	17.8	14.88	20.16
Safety factor	2.75	2.36	2.41	2.57	2.75
Forecast error	−0.021	−0.153	−0.009	0.015	−0.006
Lead time	−0.05	−0.32	−0.08	−0.34	−0.05
RLT error	3.28	3.55	4.23	3.55	1.03
R^2	0.98	0.84	0.98	0.99	0.99
MSE	0.28	3.25	0.31	0.11	0.02
Accuracy train	0.9886	0.9954	0.991	0.9987	0.9983
Accuracy test	0.9874	0.8403	0.9806	0.9954	0.9981

2. Decision Trees. We ran five different decision trees, a general one considering all the products and one for each of the MU categories, as shown in Table 4.

Table 4. Decision trees results

Model decision tree	Accuracy train	Accuracy test
General	1	0.9999
Eye makeup category	1	0.7683
Face makeup category	1	0.8842
Lip makeup category	1	0.9227
Nail makeup category	1	1

For the modeling of the Decision Trees, the values of the hyperparameters used to obtain the results expressed in Table 3 were the following:

max_depth = None
max_features = None
max_leaf_nodes = None
min_samples_leaf = 1
min_weight_fraction_lead = 0.0
splitter = "best"

3. SVM (Support Vector Machines). We ran five different SVM models, accuracy results are shown at Table 5. A general one considering all the products and one for each of the MU categories.

Table 5. Support vector machines results

Model SVM	R^2	Accuracy train	Accuracy test
General	1	0.9793	0.9817
Eye makeup category	1	0.6247	0.6137
Face makeup category	1	0.9625	0.8484
Lip makeup category	1	0.9527	0.9327
Nail makeup category	1	0.9851	0.9806

For the modeling of the SVM, the values of the hyperparameters used to obtain the results expressed in Table 4 were the following:

C = 1.0
gamma = "scale"
kernel = "rbf"

4. Neural Networks We ran five different neural networks models, a general one considering all the products and one for each of the MU categories. Results are shown in Table 6.

Table 6. Neural networks results

Model neural N.	R^2	Accuracy train	Accuracy test
General	1	0.9791	0.8426
Eye makeup category	1	0.8209	0.5467
Face makeup category	1	0.9223	0.9089
Lip makeup category	1	0.9506	0.9288
Nail makeup category	1	0.8326	0.7943

For the modeling of the Neural Networks, the values of the hyperparameters used to obtain the results expressed in Table 5 were the following:

activation = "RELU"
hidden_layer_sizes=(20, 20, 20, 20, 20)
learning_rate = "constant"
momentum = 0.9
max_iter = 200
learning_rate_init = 0.001
batch_size = "auto"

Using the data and methodology described in previous sections we modeled the Safety Stock per category for a company from the cosmetics and beauty industry. To evaluate each modeling approach (i.e. Linear Regression, Decision

Tree, SVM and Neural Networks) we compared the accuracy of each model using training and testing data.

Table 7 contains a summary of the accuracy using train and test data for each of the models and a comparison of the said indicator. In that sense, the results obtained from the modeling with Linear Regression indicate that our general estimation of the Safety Stock is very close to the actual value. While the estimations of the Safety Stock per category have a similar accuracy, the estimation for the Eye Makeup category appears to have a lower predictive power than the rest. The aforementioned could be due to the lower number of observations for this category, in comparison with the observations available for the rest.

Table 7. Results comparison

Model	Category	AccuracyTrain	AccuracyTest	Accuracy train - Accuracy test
Linear Regression	General	0.9886	0.9874	0.0012
Linear Regression	Eye MakeupCategory	0.9954	0.8403	0.1551
Linear Regression	Face MakeupCategory	0.991	0.9806	0.0104
Linear Regression	Lip MakeupCategory	0.9987	0.9954	0.0033
Linear Regression	Nail MakeupCategory	0.9983	0.9981	0.0002
Decision Tree	General	1	0.9999	0.0001
Decision Tree	Eye MakeupCategory	1	0.7683	0.2317
Decision Tree	Face MakeupCategory	1	0.8842	0.1158
Decision Tree	Lip MakeupCategory	1	0.9227	0.0773
Decision Tree	Nail MakeupCategory	1	1	0
SVM	General	0.9793	0.9817	−0.0024
SVM	Eye MakeupCategory	0.6247	0.6137	0.011
SVM	Face MakeupCategory	0.9625	0.8484	0.1141
SVM	Lip MakeupCategory	0.9527	0.9327	0.02
SVM	Nail MakeupCategory	0.9851	0.9806	0.0045
Neural Networks	General	0.9791	0.8426	0.1365
Neural Networks	Eye MakeupCategory	0.8209	0.5467	0.2742
Neural Networks	Face MakeupCategory	0.9223	0.9089	0.0134
Neural Networks	Lip MakeupCategory	0.9506	0.9288	0.0218
Neural Networks	Nail MakeupCategory	0.8326	0.7943	0.0383

On the other hand, our estimation of the Safety Stock using Decision Trees seems to be over fitted since the accuracy using training data is 100%, while the accuracy of the models using testing data is lower for the general, Eye Makeup, Face Makeup and Lip Makeup Safety Stock estimations.

In regards to the application of the SVM, the training accuracy of our models seems to indicate that the models have high predictive power, nevertheless, when testing the model it is observed that accuracy is not as high as expected. Particularly, for the Face Makeup category for which the training accuracy is 96% and testing accuracy 84%.

Finally, the Neural Networks modeling also appears to be an accurate estimator for the Safety Stock, for the Face Makeup and Lip Makeup categories, as well as for the general scenario. Notwithstanding the aforementioned, the average testing accuracy of the Neural Networks models is only 80%.

Considering the above mentioned, the Linear Regression appears to be the best approach for modeling the general Safety Stock and for each MU category.

For further discussion, it is worth mentioning that our modeling of the Safety Stock equation per category could be used as an approximation of the Safety Stock when launching new products with similar characteristics. On the other hand, our general estimation of the Safety Stock could be used as an approximation for new products that do not belong to, nor correspond to a specific category.

6   Conclusions

Considering that uncertainty is the main challenge of retail and logistics, numerous solutions have being developed to avoid, at every level, each of the issues that arise from it; the approach of this paper was to propose a novel method to calculate a Safety Stock equation to circumvent the variations associated with forecast inaccuracy. Nowadays, the use of Machine Learning models in the industry, is not as common as it should, it is known that sales and operations planning areas still retain the old ways in terms of their methodology, statistic models and calculations. The approached proposed in this paper has demonstrated to be really efficient and statistically accurate, which could lead to time and money savings by using Machine Learning models to predict expected behaviour of groups of products. The use and analysis of data for inventory management via Machine Learning could represent a great competitive advantage for any business, specially for the make-up production and retail industry where the usual aim is to work by exception due to the trouble that validating an extensive catalogue of products represents, adding the predictive engine to the equation not only on forecast but on supply parameters leads to a better and accurate production planning.

In the current work, the target was to predict accurately Safety Stock, but future works could include forecasting and lead time error predictions.

References

1. Yuri Vasconcelos de Almeida, T.C.W.: Artificial intelligence (fuzzy logic) for local safety stock forecasting in multinational companies. Gestão da Produção, Operações e Sistemas 14(4), 1–10 (2019)
2. Gupta, A., Maranas, C.D.: Managing demand uncertainty in supply chain planning. Comput. Chem. Eng. 27(8–9), 1219–1227 (2003)
3. Syntetos, A.A., Babai, Z., Boylan, J.E., Kolassa, S., Nikolopoulos, K.: Supply chain forecasting: theory, practice, their gap and the future. Eur. J. Oper. Res. 252(1), 1–26 (2016)

4. Bertsimas, D., Kallus, N., Hussain, A.: Inventory management in the era of big data. Prod. Oper. Manag. **25**(12), 2006–2009 (2016)
5. Li, H., Jiang, D.: New model and heuristics for safety stock placement in general acyclic supply chain networks. Comput. Oper. Res. **39**(7), 1333–19344 (2011)
6. Gonçalves, J.N., Carvalho, M.S., Cortez, P.: Operations research models and methods for safety stock determination: a review. Elsevier **7** (2020)
7. L'Oreal: L'oreal operations manual
8. Çolak, M., Hatipoğlu, T., Aydin Keskin, G., Fiğlali, A.: A safety stock model based on order change-to-delivery response time: a case study for automotive industry. Sigma J. Eng. Nat. Sci. **37**(3), 841–853 (2019)
9. Chapman, P., et al.: CRISP-DM 1.0. SPSS Inc. (2001)
10. Zhong, W., Zhang, L.: The prediction research of safety stock based on the combinatorial forecasting model. In: International Conference on Computational Science and Engineering (2015)

DBSCAN Parameter Selection Based on K-NN

Leonardo Delgado(✉) and Eduardo F. Morales

Instituto Nacional de Astrofísica, Óptica y Electrónica, Luis Enrique Erro 1,
Tonantzintla, Puebla, Mexico
{leonardodelgado,emorales}@inaoep.mx

Abstract. In this paper, we introduce a parameter selection for the algorithm DBSCAN based on the $K - neighborhood$. We change the parameters ϵ and min_points by a $K - neighborhood$ (named β), scale, and an α value. We use the scale parameter to balance the dataset. β is used to select $min\_points - \epsilon$ and α to reduce the value of min_points. We use homogeneity, completeness, and v-measure scores over datasets with balanced and unbalanced clusters to evaluate the performance. We compared our results against ACND and DBSCAN with the original parameter selection. Finally, we use our proposal to detect contour over 3D shapes. Our results show better performance in three of the eight datasets, and a better performance into border detection on 3D shapes.

Keywords: DBSCAN · ACND · Parameter selection · Clustering

1 Introduction

Clustering is a process that implies the organization of unlabeled data into similar groups called clusters. Each cluster have objects which are similar between them, and dissimilar to the objects in other clusters. For this process three things are needed: 1) a proximity measure, 2) a criterion function to evaluate a clustering and 3) an algorithm to compute clustering. There are different types of clustering methods, but the most common are partitioning, hierarchical, fuzzy, model-based and density-based methods. The first method is a division method which consist of creating k partitions and by iteratively relocating data points between clusters until the quality of the division is (locally) optimal [1,10,14]. The second method creates a hierarchical tree to decompose the given data [8,21]. The third category is the fuzzy methods in which each element has some degree of membership for each class (each point can belong to more than one cluster) [3]. The fourth is the model-based method where a model is assumed for each cluster and the clustering process consists of trying to find the most suitable data for each model. The density-based methods belong to the last category, which try to find clusters of arbitrary shape by looking for a dense region in the data space which is separated from others by a sparse regions [7,13,18]. In the last years, density-based methods has been widely used combined with

© Springer Nature Switzerland AG 2021
I. Batyrshin et al. (Eds.): MICAI 2021, LNAI 13067, pp. 187–198, 2021.
https://doi.org/10.1007/978-3-030-89817-5_14

the k-NN to improve the clustering performance [7,13,17,18]. One of the most used density-based method is DBSCAN which was proposed in 1996 [7]. The density is modeled in terms of a radius (ϵ) and a minimum number of points inside the radius (min_points), the points with higher density that contain at least min_points within the radius are named **core points**, the points with less that min_points but which are close enough to a core point by at least ϵ are named **border points**, finally the points that are neither core nor borders are named **outliers**. One of the main problems with DEBSCAN is the selection of its parameters, which are chosen in a heuristic way. Note that if the dataset has duplicate points, this can produce false core points. That can produce a wrong selection of the border and outlier points. In this paper, we present a heuristic for parameter selection in the DBSCAN algorithm. It is experimentally shown that with this heuristic, we get a better border definition. Our parameters can be selected in an intuitive way if the user has an *a priori* knowledge about the expected clustering results.

The rest of the paper is organized as follows. Section 2 provides a review of methods used to detect border, core, and outlier points based on the density approach. Section 3 introduces our heuristic for parameter selection based on k-NN. We present the experiments and results and a short discussion in Sect. 4 to show the effectiveness of our proposal. Finally, Sect. 5 gives our conclusions and directions for future research.

2   Related Work

The border algorithm proposed by Chenyi et al. [20] detects the boundary points using mainly the reverse of k nearest neighbor. They prove that the reverse k nearest neighbor decreases when the distance of the point to the center increases. ADACLUS is a clustering algorithm based on density, it can discover clusters of different shape and density and clustering boundaries, and has low computational complexity [13]. The BDDTS algorithm proposed by Baozhi et al. is based on the characteristics of the boundary point distribution, it does need to know the number of clusters, and can detect the boundary points [15]. The BPDG algorithm proposes by He Yu-Zhen et al. [11], uses a grid, where each cell is formed by the sum of densities of each point on the dataset. A convolution with 8 direction Prewitt's operators is applied to this 2D grid to detect boundary points. They improve the time complexity of the BORDER algorithm [20]. ACND algorithm (Adaptive Clustering algorithm based on k-NN and density) [18] proposed by Bing Shi et al. in 2018 is a clustering algorithm based on KD-trees and statistic metrics. The ACND algorithm does not require the user to specify the parameters which are determined during the process of clustering. This algorithm calculate a static measure of denseness for the whole sample distribution.

In almost all the algorithms previously mentioned, the user needs to define parameters to perform the clustering without a clear idea of how to select them. Some approaches, like ACND, propose a non-parametric algorithm, but the results are biased by global values that impact the clustering process. Our proposal introduce a parameter selection strategy in which we describe the impact that each of the parameters has on the clustering process.

Our algorithm is inspired by the ACND algorithm. Similarly to ACND, we use a KDTree to determine the parameters of DBSCAN. Our results show better performance on three out of eight datasets when compared against ACND and DBSCAN original parameter's selection.

3 DBSCAN Parameter Selection Based on K-Distance

Our proposed method uses three parameters, two to define ϵ and min_points ($K\text{-}neighborhood$ which will named β and α) and one to balance the dataset (scale). Scale is used to remove duplicated samples by multiplying each one by scale and round out. A variable with unique values is generated (pc_3). To calculate ϵ and min_points a KDTree[1] is generated with the unique values of the previous step. To get ϵ, the arithmetic mean of their closest distance to each p_i is evaluated, then the arithmetic mean over all p_i means is calculated. Equation 1 is used to get ϵ, where m represents the number of closest points to p_i, q_j represents each closest point, and n represents the total number of elements in the dataset. The min_points (see Eq. 2) are calculated as the arithmetic mean over the density of each p_i given ϵ minus α multiplied by the standard deviation over the density of each p_i given ϵ. Where the density is the number of points which are at least a distance ϵ from p_i. The entire algorithm is described in Algorithm 1 (see Fig. 1).

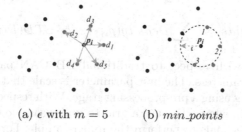

(a) ϵ with $m = 5$ (b) min_points

Fig. 1. a) Shows the neighborhood with five elements, where we use the arithmetic mean of the distances from p_i to d_i to calculate ϵ. Note that for each p_i we are going to have different arithmetic means. b) Shows how with an ϵ we have a different density for each p_i.

$$\epsilon = \frac{1}{n} \sum_{i=1}^{n} \frac{1}{m} \sum_{j=1}^{m} d(p_i, q_j) \tag{1}$$

---

[1] A KDTree is a space-partitioning data structure for organizing samples. The KDTree is used to search a K-neighborhood given a sample and search samples by a radius.

Algorithm 1. Parameter selection based on K-distance.

$DBSCAN\_L(dataset, \beta, scale, \alpha)$
return $cluster$
$pc\_2 \leftarrow dataset$
$pc\_2 \leftarrow pc\_2 * scale$
$pc\_2 \leftarrow round(pc\_2)$
$pc\_2 \leftarrow add\_index(pc\_2)$
$pc\_3 \leftarrow unique\_values(pc\_2)$
$tree\_pc \leftarrow KDTree(pc\_3)$
$dist \leftarrow []$
for $sample$ in $dataset$ **do**
    $dist.append(tree\_pc.query(sample, \beta))$
end for
$\epsilon \leftarrow ceil(mean(dist))$
$density \leftarrow []$
for $sample$ in $dataset$ **do**
    $density.append(tree\_pc.query\_radius(sample, \epsilon))$
end for
$min\_points \leftarrow floor(mean(density) - \alpha * std(density))$
$cls \leftarrow DBSCAN(\epsilon, min\_points, pc\_3)$
$labels \leftarrow cls.labels\_$
$labels[cls.core\_sample\_indices\_] \leftarrow -1$
$borders \leftarrow labels! = -1$
$pc\_3 \leftarrow add(pc\_3, borders)$
$pc\_2 \leftarrow merge(pc\_3, pc\_2)$
$dataset \leftarrow add\_index(dataset)$
$cluster \leftarrow merge(dataset, pc\_2, on = index)$

$$min\_points = \frac{1}{n} \sum_{i=1}^{n} Density(p_i, \epsilon) - \alpha \cdot STD(Density) \qquad (2)$$

Our algorithm, with respect to traditional DBSCAN parameter selection, introduces three parameters. The first parameter is scale that controls the influence of over-sampling using a pre-processing stage. With respect to the literature, this is similar to [11], where they use a grid to under-sample the data, however, we use significant decimals to perform the under-sample. For example, suppose that there is a normalized dataset with values between 0–1, then if we use a scale equal to ten, the values will be multiply by ten and rounded, conserving only the unique values. $[1, 2] \leftarrow [0.12, 0.13, 0.23, 0.25]$. The β parameter represents a neighborhood from which we obtain ϵ and part of min_points. This parameter captures the global density distribution in two ways: (i) First getting an ϵ value with the mean distance over all instances to its $k-neighbors$, and (ii) then using the ϵ value to calculate the min_points as the mean of the densities minus α by the standard deviation of them. Note that if all samples have a homogeneous distribution, the standard deviation will be zero, and min_points will be equal to the mean of the densities, and it will be equal to the β parameter. If the standard deviation is different from zero, we can adjust its influence to the min_points

through the α parameter. If we need to relax the cluster's criteria of a border, we can assign a value close to one, and if we want to be strict in the clusterization, we can assign an α close to zero. When compared with the original parameters' selection of DBSCAN, our approach allows us to adjust what we want to capture in the clustering. Compared with new proposals that are non-parametric, like ACND [18], they are limited to the developer's criteria to perform the clustering.

4   Experiments

In this section, the results are compared in two ways against DBSCAN (using or not the proposed heuristic to select parameters) and ACND. To compared the clustering performance we used the datasets: Spiral [6], Pathbased [6], Jain [12], Flame [9], Concentric [4], R15 [19], and D31 [19]. The ground truth clusters of these datasets are shown in Fig. 2. A second comparison is in terms of border definition over 3D shapes using ShapeNet [5].

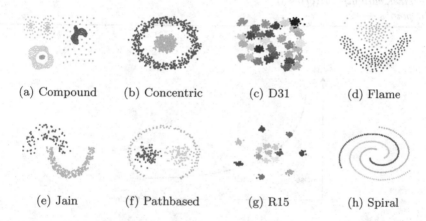

(a) Compound   (b) Concentric   (c) D31   (d) Flame

(e) Jain   (f) Pathbased   (g) R15   (h) Spiral

Fig. 2. The figure shows the data sets used in the comparison. Note that there are unbalanced datasets like (Concentric-Jain). Data sets that decrease in density as they move away from the center (Compound-Spiral), and datasets with a high number of groups (D31-R15).

4.1   Parameter Selection DBSCAN

As reference, the parameters of the original DBSCAN paper are used (see Algorithm 2). Let d be the distance of a point p to its k-th. nearest neighbor. Mapping each point to the distance from its k-th. nearest neighbor and sorting the dataset in descending order of their k-dist values, to produces a graph. The graph (sorted k-dist graph) of this procedure gives some hints concerning the density distribution in a dataset. The first point in the first "valley" of the sorted k-dist graph (see Fig. 3) will be ϵ, and the k value will be *min_points* (To get the valley, we first smooth the k-dist graph using a Gaussian filter and then we valuate the second derivative). Note that all points with a higher k-dist value (left of the threshold) are considered to be noise. All other points (right of the threshold) are assigned to some cluster.

Algorithm 2. Original DBSCAN parameter selection

$DBSCAN\_parameter\_selection(dataset, K)$
return ϵ, min_points
$tree\_pc \leftarrow KDTree(dataset)$
$dist \leftarrow []$
for $sample$ in $dataset$ **do**
    $dist.append(tree\_pc.query(sample, K))$
end for
$dist \leftarrow sort(dist)$
$smooth = gaussian\_filter1d(dist)$
$cross_0 = \text{diff}(\text{diff}(smooth))$
$\epsilon \leftarrow first\_valley\_point(cross_0)$
$min\_points \leftarrow K$

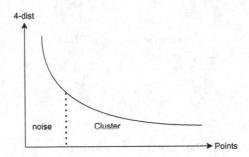

Fig. 3. The figure shows the k-dist graph. The dotted line shows the inflection point, which is where the ϵ value is selected. Note that this parameter selection does not consider the shape of the clusters, as it only uses a distribution to select the parameters.

4.2   ACND

The other compared method is ACND with is a DBSCAN variation with dynamic ϵ and min_points. To make a comparison against it we implemented the method in python. Our implementation can be found in: https://github.com/LeonardoDelgado/DBSCAN-Parameter-selection-based-on-K-NNfor-border-points.git.

4.3 Clustering

To evaluate the performance of the clustering we use Homogeneity, Completeness, and V-measure scores [16] which are defined as follows:

$$h = 1 - \frac{H(C|K)}{H(C)} \tag{3}$$

$$c = 1 - \frac{H(K|C)}{H(K)} \tag{4}$$

$$H(C|K) = -\sum_{c=1}^{|C|} \sum_{k=1}^{|K|} \frac{n_{c,k}}{n} log(\frac{n_{c,k}}{n_k}) \tag{5}$$

$$H(C) = -\sum_{c=1}^{|C|} \frac{n_c}{n} log(\frac{n_c}{n}) \tag{6}$$

where $H(C|K)$ is the conditional entropy of the classes given the cluster assignments and $H(C)$ is the entropy of the classes, with n equal to the total number of samples, n_c and n_k being the number of samples belonging to class c and cluster k respectively, and finally n_{ck} is the number of samples from class c assigned to cluster k. The conditional entropy of clusters given class $H(K|C)$ and the entropy of clusters $H(K)$ are defined in a symmetric manner. V-measure is defined as the harmonic mean of homogeneity and completeness:

$$v = 2\frac{h \cdot c}{h + c} \tag{7}$$

Table 1. Quantitative comparison, greater is better. For our proposal the values $scale = 100$, $\beta = 24$, and $\alpha = 0.2$ were used

Dataset	Homogeneity			Completeness			V-measure		
	ACND	Proposal	DBSCAN	ACND	Proposal	DBSCAN	ACND	Proposal	DBSCAN
Pathbased	0.15	0.78	0.74	0.26	0.78	0.57	0.19	0.78	0.65
R15	0.91	0.92	0.87	0.58	0.90	0.97	0.71	0.91	0.92
Compound	0.89	0.76	0.72	0.94	0.87	0.91	0.92	0.81	0.81
Concentric	0.97	0.95	1.00	0.14	0.95	0.94	0.24	0.95	0.97
D31	0.96	0.84	0.34	0.65	0.86	0.99	0.77	0.85	0.51
Jain	0.97	1.00	1.00	0.97	1.00	0.61	0.97	1.00	0.76
Spiral	1.00	0.54	0.97	1.00	0.46	0.85	1.00	0.50	0.90
Flame	0.01	0.91	0.01	0.18	0.86	0.18	0.02	0.88	0.02

A quantitative comparative is shown in Table 1 and a qualitative comparison is shown in Fig. 5. Our results were calculated with a $scale = 100$, $\beta = 24$, and $\alpha = 0.2$. The parameters were select in an empirical way. Quantitatively, we get better performance in Pathbased, Jain, and Flame. For Pathbased and Jain, the results obtained can be explained since one cluster is under-sampled as compared against the others, producing that the entire cluster is labeled as outliers. ACND classified in a correct way the JAIN dataset due to its dynamic $\epsilon>$. The original DBSCAN parameter selection has lower min_points that produce a misclassified for both dataset.

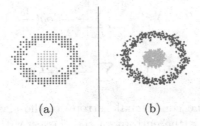

(a)   (b)

Fig. 4. a) Concentric dataset under-sampled and clustered, using a lower scale, as can be seen, the cluster is balanced as a result. b) After clustering, our algorithm assigned the label in the original scale.

The influence of our parameter over the results can be improved if we have knowledge about the dataset, such as over-sample, which means that there are several points very similar between them. We can use the parameter scale to deal with the problem. For example, for the concentric dataset, which has a cluster with higher density with respect to the others, we can reduce $scale = 10$ which reduces the influence of the cluster with higher density and allows recognizing the cluster with lower density (see Fig. 4). Another interest case is when the cluster is reducing its density on the border which is the case of Spiral. In that case, we can increase the value of α (see Fig. 6). Increasing the value of α decreases min_points in terms of the standard deviation.

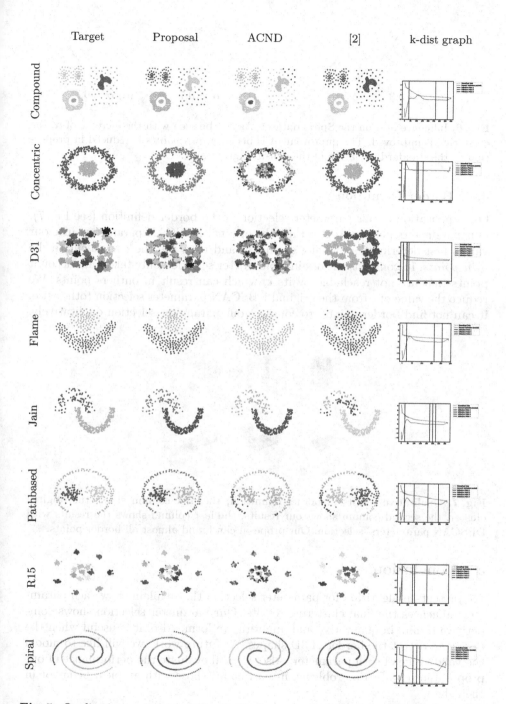

Fig. 5. Qualitative comparison. Qualitatively, our selection of parameters performs better with the exception of the spiral data set.

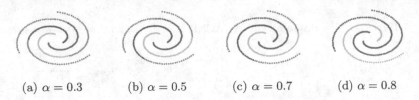

(a) $\alpha = 0.3$ (b) $\alpha = 0.5$ (c) $\alpha = 0.7$ (d) $\alpha = 0.8$

Fig. 6. Influence of α on the Spiral dataset. As can be seen with the increase of α, the clustering is improved. The improvement is because *min_points* is reduced in proportion to the standard deviation of the densities given an ϵ.

4.4   Border Definition

One application of our parameter selection is the border definition (see Fig. 7). Our parameters perform better than the original DBSCAN parameter since our values first determine a global ϵ given a β and then use the ϵ to calculate the *min_points*. In contrast, the original parameter selection only takes into account points that are not reachable by its ϵ, which can result in outliers points. We reduce the value of ϵ from the original DBSCAN parameter selection (otherwise, it can not find border points) to compare our parameter selection qualitatively.

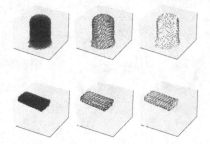

Fig. 7. Qualitative results of border definition, the first column shows the original cluster, the second column shows our results, the last column shows the results with DBSCAN parameters' selection. Our proposal can found almost all border points.

5   Conclusion

We present an algorithm for parameter selection that explains how each parameter influences the final clustering results. Our parameter selection shows competitive results in qualitative and quantitative form. Also, it is useful when the task is to detect the border of 3D shapes. As future work, we will like to model the valid range of each parameter. Also, we will evaluate the performance of our proposal into real case problems, in particular, those with atypical behavior in their data.

Acknowledgment. This work was supported by Project Grant CONACYT CB-2015-250938-Y. The first author is supported by scholarship No. CONACYT 707984.

References

1. Arthur, D., Vassilvitskii, S.: K-means++: the advantages of careful seeding. In: Proceedings of the Eighteenth Annual ACM-SIAM Symposium on Discrete Algorithms, SODA 2007, pp. 1027–1035. Society for Industrial and Applied Mathematics, USA (2007)
2. Atiya, A.F.: An unsupervised learning technique for artificial neural networks. Neural Netw. **3**(6), 707–711 (1990)
3. Bezdek, J.C., Ehrlich, R., Full, W.: FCM: the fuzzy c-means clustering algorithm. Comput. Geosci. **10**(2–3), 191–203 (1984). https://doi.org/10.1016/0098-3004(84)90020-7
4. Brownlee, J.: How to generate test datasets in python with scikit-learn, December 2018. https://machinelearningmastery.com/generate-test-datasets-python-scikit-learn/
5. Chang, A.X., et al.: ShapeNet: an information-rich 3D model repository. Technical report. arXiv:1512.03012 [cs.GR], Stanford University – Princeton University – Toyota Technological Institute at Chicago (2015)
6. Chang, H., Yeung, D.Y.: Robust path-based spectral clustering. Pattern Recogn. **41**(1), 191–203 (2008). https://doi.org/10.1016/j.patcog.2007.04.010
7. Ester, M., Kriegel, H.P., Sander, J., Xu, X.: A density-based algorithm for discovering clusters a density-based algorithm for discovering clusters in large spatial databases with noise. In: Proceedings of the Second International Conference on Knowledge Discovery and Data Mining, KDD 1996, pp. 226–231. AAAI Press (1996). http://dl.acm.org/citation.cfm?id=3001460.3001507
8. Fisher, D.H.: Knowledge acquisition via incremental conceptual clustering. Mach. Learn. **2**(2), 139–172 (1987)
9. Fu, L., Medico, E.: Flame, a novel fuzzy clustering method for the analysis of DNA microarray data. BMC Bioinform. **8**(1), 3 (2007). https://doi.org/10.1186/1471-2105-8-3
10. Hartigan, J.A., Wong, M.A.: Algorithm AS 136: a k-means clustering algorithm. Appl. Stat. **28**(1), 100 (1979). https://doi.org/10.2307/2346830
11. He, Y.Z., Wang, C.H., Qiu, B.Z.: Clustering boundary points detection algorithm based on gradient binarization. Appl. Mech. Mater. 263–266, 2358–2363 (2012). https://doi.org/10.4028/www.scientific.net/amm.263-266.2358
12. Jain, A.K., Law, M.H.C.: Data clustering: a user's dilemma. In: Pal, S.K., Bandyopadhyay, S., Biswas, S. (eds.) PReMI 2005. LNCS, vol. 3776, pp. 1–10. Springer, Heidelberg (2005). https://doi.org/10.1007/11590316_1
13. Nosovskiy, G.V., Liu, D., Sourina, O.: Automatic clustering and boundary detection algorithm based on adaptive influence function. Pattern Recogn. **41**(9), 2757–2776 (2008). https://doi.org/10.1016/j.patcog.2008.01.021
14. Pelleg, D., Moore, A.: X-means: extending k-means with efficient estimation of the number of clusters. In: Proceedings of the 17th International Conference on Machine Learning, pp. 727–734. Morgan Kaufmann (2000)
15. Qiu, B., Wang, S.: A boundary detection algorithm of clusters based on dual threshold segmentation. In: 2011 Seventh International Conference on Computational Intelligence and Security. IEEE, December 2011. https://doi.org/10.1109/cis.2011.276

16. Rosenberg, A., Hirschberg, J.: V-measure: a conditional entropy-based external cluster evaluation measure. In: Proceedings of the 2007 Joint Conference on Empirical Methods in Natural Language Processing and Computational Natural Language Learning (EMNLP-CoNLL), pp. 410–420. Association for Computational Linguistics, Prague, Czech Republic, June 2007. https://www.aclweb.org/anthology/D07-1043

17. Schubert, E., Sander, J., Ester, M., Kriegel, H.P., Xu, X.: DBscan revisited, revisited: why and how you should (still) use DBscan. ACM Trans. Database Syst. **42**(3) (2017). https://doi.org/10.1145/3068335. https://doi.org/10.1145/3068335

18. Shi, B., Han, L., Yan, H.: Adaptive clustering algorithm based on kNN and density. Pattern Recogn. Lett. **104**, 37–44 (2018). https://doi.org/10.1016/j.patrec.2018.01.020

19. Veenman, C., Reinders, M., Backer, E.: A maximum variance cluster algorithm. IEEE Trans. Pattern Anal. Mach. Intell. **24**(9), 1273–1280 (2002). https://doi.org/10.1109/tpami.2002.1033218

20. Xia, C., Hsu, W., Lee, M., Ooi, B.: BORDER: efficient computation of boundary points. IEEE Trans. Knowl. Data Eng. **18**(3), 289–303 (2006). https://doi.org/10.1109/tkde.2006.38

21. Zhang, T., Ramakrishnan, R., Livny, M.: BIRCH. ACM SIGMOD Rec. **25**(2), 103–114 (1996). https://doi.org/10.1145/235968.233324

Deep Learning Architectures Applied to Mosquito Count Regressions in US Datasets

Cuauhtemoc Daniel Suarez-Ramirez[1], Mario Alberto Duran-Vega[1],
Hector M. Sanchez C.[2], Miguel Gonzalez-Mendoza[1(✉)],
Leonardo Chang[1], and John M. Marshall[2]

[1] School of Engineering, Computer Science Department, Tecnologico de Monterrey,
Nuevo Leon, Mexico
{a01206503,a00755076}@exatec.tec.mx, {mgonza,lchang}@tec.mx
[2] School of Public Health, Epidemiology and Biostatistics Department,
University of California, Berkeley, Berkeley, USA
{sanchez.hmsc,john.marshall}@berkeley.edu

Abstract. Deep Learning has achieved great successes in various complex tasks such as image classification, detection and natural language processing. This work describes the process of designing and implementing seven deep learning approaches to perform regressions on mosquito populations from a specific region, given co-variables such as humidity, uv-index and precipitation intensity. The implemented approaches were: Recurrent Neural Networks (LSTM), an hybrid deep learning model, and a Variational Autoencoder (VAE) combined with a Multi-Layer Perceptron (MLP) which instead of using normal RGB images, uses satellite images of twelve channels from Copernicus Sentinel-2 mission. The experiments were executed on the Washington Mosquito Dataset, augmented with weather information. For this dataset, an MLP proved to achieve the best results.

Keywords: Deep Learning · Machine Learning · Computer Vision · Mosquito · LSTM · VAE · MLP

1 Introduction

Mosquito-borne diseases have been one of the main causes of mortality in humans for centuries [2,5,13]. These encompass the malaria parasite, and the Dengue (DENV), Yellow Fever (YFV), Chikungunya (CHIKV) and Zika (ZIKV) arboviruses [1,5,13]. According to the World Mosquito Program (WMP), more than 700 million infections occur each year, with at least 1 million of them resulting in death [14]. This is why WHO has declared some of these diseases as some of the main threats to mankind.

For the Arboviruses, the *Aedes* genus is the main cause of human-to-human transmission [13]; particularly, female individuals of *Ae. aegypti* species [1] are the main vectors of the disease.

© Springer Nature Switzerland AG 2021
I. Batyrshin et al. (Eds.): MICAI 2021, LNAI 13067, pp. 199–212, 2021.
https://doi.org/10.1007/978-3-030-89817-5_15

Quantifying the presence of the population of these mosquitoes is relevant to predict which zones will become foci of infection to take preemptive actions such as the deployment of insecticides or other population control techniques to prevent a fast paced increase in cases.

One of the efforts regarding monitoring mosquito populations, is the dataset created by Washington D.C. Health authorities [3]. This department has been trapping and testing mosquitoes for almost a decade, in response to the Zika outbreak in Latin America and the Caribbean. The Washington Mosquito Dataset (WMD) contains 2024 records from 28 sites and 36 traps collected across 8 wards from the District of Columbia. This data is collected and reported from April through October, which is usually the mosquito season in Washington D.C.

The task of predicting the mosquito populations sizes (regression) using Machine Learning (ML) or Deep Learning (DL) techniques, has been tackled with meager efforts. Sotomayor et al. [11] - inspired by Inception Net [12] - used a Multi-Layered Perceptron (MLP) to process tabular data (weather and trap related information), and a Convolutional Neural Network (CNN) to process Satellite information (images) in the CNN. The Hybrid Model (HM) combine later both outputs. The results were produced by using the Washington Mosquito Dataset (WMD) and they are shown on Table 1.

This work aims towards tackling the regression task of predicting the population size of the *Ae. aegypti* in Washington D.C., by combining diverse DL techniques which encompass the use of spatial and temporal information to create more robust architectures.

Table 1. Previous obtained results with Washington Mosquito Dataset

Metric	MLP	CNN	Hybrid
Mean error	14.58	75.05	6.86
SD	17.94	61.21	11.08
R^2	0.9816	0.091	0.9956

The remainder of this work is organized as follows Subsect. 2.1 presents a detailed analysis of multiple MLP implementations. Thereafter, the process of designing our custom dataset is described in Subsect. 2.2. Subsection 2.3 provides the implementation of a CNN using Google Maps imagery. Following, the testing of the Hybrid Model is detailed in Subsect. 2.4. The designing and implementation of the Sentinel Hybrid Model is described in Sect. 2.5. Subsequently, An Hybrid VAE Model description and Implementation is shown in Sect. 2.6. Then, a RNN approach is tested in Subsect. 2.7. Finally, analysis and conclusions are shown in Sect. 3.

Additionally, the datasets and the code for implementing the architectures here presented is located in https://github.com/CuauSuarez/Mosquito-Count-DL.

2 Methods

Sotomayor *et al.* implemented three DL approaches for the task of predicting mosquito populations over WMD [11]: a four-layer MLP; a pre-trained VGG19 network; and an HM which concatenates the output from the first and second model, to create an input for a final MLP. To improve upon these architectures, we implemented a total of seven architectures, four previously tested by Sotomayor *et al.* but slightly improved, and three new approaches. These methods are:

1. An MLP using just the categorical and numerical data from the weather reports (cleaned and pre-processed).
2. An MLP, with WMD augmented with weather information acquired from DarkSky API [7] (https://darksky.net) and World Weather Online (WWO) [9] (https://www.worldweatheronline.com), to improve our previous results with additional information related to the mosquito life-cycle.
3. A CNN using exclusively satellite imagery from Google Maps. This architecture was based on DarkNet-53 using pre-trained weights from YoloV3 [10].
4. An HM based on the work of Sotomayor *et al.* It concatenates the output of an MLP and a CNN, to create an input for the final MLP.
5. The previously presented HM, but instead of using images of three channels provided by Google Maps, it uses twelve-channel images from Copernicus Sentinel-2 mission. Which contains more information like humidity and vegetation.
6. A custom HM which concatenates a latent vector from the satellite image (obtained with a VAE), with the rest of the tabular data.
7. A LSTM, approach which fits naturally due to the temporal nature of the WMD plus weather. Since it is augmented with weather information of two weeks from the trap collection.

The process of developing these seven architectures was that of gradually increasing the complexity of it and including more data regarding time and space.

In the following sections, each of the architectures are further described.

2.1 Multi-Layer Perceptron (MLP)

The first approach we implemented was an MLP with WMD. To make data patterns easier for the model to learn. We pre-processed the data as follows:

1. We removed the columns *Lifestage*, *EggsCollected*, *LarvaeCollected*, *PupaCollected*, *Town*, *State*, and *County*. Since those columns contained the same data across the 2023 rows.
2. We removed duplicated data, such as *X*, *Y* and *Address*, which represent the location of the trap, while keeping *Latitude* and *Longitude* for that purpose.
3. We decomposed the dates into two numerical columns, *week* and *Year*; and we normalized all numerical columns.

We selected a K-Fold Cross-validation with a K of 5 for validating our model. To produce the best possible results by each model, we tried all the combinations of the next hyper-parameters: Learning rate (0.01, 0.001, 0.0001), Dropout (0, 0.1, 0.2, 0.3, 0.4, 0.5, 0.60), Hidden Units (10, 20, 30, 40, 50, 60, 70, 80, 90, 100), Hidden Layers (2, 3, 4).

Table 2. Results of each MLP with the corresponding number of layers in Washington D.C dataset.

MLP results				
Metric	Previous MLP	MLP-2	MLP-3	MLP-4
Mean error	14.58	8.82	8.79	8.78
R^2	0.9816	0.61	0.64	0.65
(Only Aedes) MLP results				
Metric	Previous MLP	MLP-2	MLP-3	MLP-4
Mean error	–	3.98	3.99	4.00
R^2	–	0.49	0.48	0.48

The best combination of hyper-parameters for all mosquito species were: RMSProp with learning rate of 0.001, 4 layers, with a hidden size of 70, and a dropout of 0.1. This resulted in a mean absolute error of **8.78**.

As shown in Table 2, our MLP architecture of 4 layers obtained a better median average than the MLP implemented by Sotomayor *et al.* However, we are not sure how R^2 was measured for the MLP presented by Sotomayor *et al.* Since a mean average of 14.58 is unlikely to lead to the $R^2 = 0.98$ they reported.

The WMD contain information of several species of mosquitoes, like *Culex pipens* or *Psorophora columbiae*. But *Aedes aegypti* is the main cause of human-to-human transmission of Arboviruses. We prepared a smaller version of the dataset, containing only samples of this species. We trained and tested our model with this dataset. The best mix of hyper-parameters were: RMSProp with learning rate of 0.01, 4 layers, with a hidden size of 60, and a dropout of 0.1. This led in a mean absolute error of **3.98** (Table 2).

The results obtained by our MLP, improved the results presented by the MLP of Sotomayor *et al.* However, one further improvement is to augment the WMD with weather information. Since weather information is related to the mosquito life-cycle and their population number.

2.2   Approach 2 (MLP + Weather Data)

Aedes aegypti has a life cycle that lasts around 1.5 weeks (with intense sunlight) or 3 weeks (in cold periods). It involves fours stages: egg, larva, pupa, and adult. Variables like temperature or food availability, affects the life cycle of the mosquito. In good environmental conditions, mosquitoes reach their adult form in around 10 days (Fig. 1) [11].

Fig. 1. Egg (2–7 days) larva (4 or more) pupa (2 days) emerging adult.

To include information about the mosquito life-cycle in our model, we built a WMD dataset augmented with weather information. To create this dataset, we requested information from the past 14 days since the trap collection date, and the GPS coordinates.

To build the aforementioned dataset, we used the DarkSky API [7]. This framework provides weather information according to a given latitude, longitude and date. Which is information available in WMD.

A previous approach proposed by Sotomayor *et al.* involved augmenting the WMD by requesting weather information from the day the trap was collected. However, since the life-cycle of *Aedes aegypti* depends of the weather conditions from the previous 10 to 14 days, we augmented WMD with weather information from the previous 14 days before the trap was collected. We trained our MLP with the WMD augmented with weather information (Table 3). The best combination of hyper-parameters for all mosquito species were: RMSProp with learning rate of 0.01, 3 layers, with a hidden size of 140, and a dropout of 0.3. This resulted in a mean absolute error of **9.06**.

Moreover, we repeated this experiment with a smaller dataset containing only *Aedes aegypti* (Table 3). The best mix of hyper-parameters were: RMSProp with learning rate of 0.01, 2 layers, with a hidden size of 120, and a dropout of 0.2. This led to a mean absolute error of **4.10**.

Results obtained by using WMD augmented with weather information were inferior when compared to the original WMD. This could be caused by the noise generated by the large amount of variables used to train our MLP. Therefore, we performed multiple feature selection approaches to improve our results (Pearson, Kendall, Spearman and PCA).

Table 3. Results of each MLP by number of layers, using WMD + Weather.

Metric	Previous MLP	MLP-2	MLP-3	MLP-4
Mean error	14.58	9.061	9.060	9.11
R^2	0.9816	0.64	0.60	0.67
(Only Aedes)				
Metric	Previous MLP	MLP-2	MLP-3	MLP-4
Mean error	–	4.10	4.15	4.16
R^2	–	0.54	0.55	0.50

We trained multiple MLP models selecting features from the aforementioned coefficients. However, even after performing feature selection, additional weather information from DarkSky did not improve the results. Thus, we used another weather information provider to confirm the validity of the information provided by DarkSky. For this additional test, we used World Weather Online (WWO) API. We downloaded all the weather information from 14 days before the trap was collected, and trained our MLP models. However, the result of our best model was a median error of **9.85**, which was inferior when compared with the results presented at Table 3. DarkSky API and WWO API return similar weather information, since we evaluated two providers of meteorological information. And despite we included attributes directly related to the mosquito life-cycle, our results did not improve. Thus, we think WMD is biased or noisy.

2.3   Approach 3: CNN

Mosquitoes like *Aedes aegypti*, use almost any kind of clean water container to lay eggs [4]. One of the findings of Sotomayor *et al.* [11], is that we can use satellite images of the trap's surroundings to extract spatial features and improve the mosquito count prediction (for example, water containers).

We implemented a CNN regression model, which takes as input the satellite image from the location of the trap (obtained from Google Maps, and scaled to 416×416). To process the aforementioned images, the selected architecture was a pre-trained Darknet-53 [10] followed by 3 dense layers. We performed a cross validation of 5 folds to validate our model and measure the accuracy (Table 4).

The best combination of hyper-parameters for all mosquito species were: RMSProp with learning rate of 0.001, 3 layers, weight decay of 0.004. This results in a mean absolute error of **13.11**. For *Aedes* the best combination were: RMSProp with learning rate of 0.001, 3 layers, with a hidden size of 120, and a weight decay of 0.004. This led to a mean absolute error of **5.22**.

According to the results, our model performed better than the CNN model presented by Sotomayor *et al.*

Table 4. Results of our pre-trained CNN

CNN		
Metric	Prev	Avg
Mean error	75.05	13.11
Aedes CNN		
Metric	Prev	Avg
Mean error	–	5.22

2.4 Approach 4: Hybrid Model (CNN + MLP)

The MLP we proposed in Subsect. 2.1 and the CNN model in Subsect. 2.3 improved the results of the same architectures implemented by Sotomayor *et al*. Following the idea of the hybrid architecture introduced in their work, we combined both architectures into one. By adding an additional NN at the end, which takes as input the output of the MLP and the CNN model. To evaluate this architecture, we used K-Fold Cross-validation of 5 folds, the results are shown in Table 5. The best mix of hyper-parameters for all mosquito species were: RMSProp with learning rate of 0.0001, 3 layers, weight decay of 0.004, and dropout of 0.3. This led in a mean absolute error of **10.85**. For *Aedes* the best combination were: RMSProp with learning rate of 0.0001, 3 layers, weight decay of 0.004, and dropout of 0.3. This resulted in a mean absolute error of **4.8**. For this hybrid architecture, results were inferior as the presented by Sotomayor *et al*.

Table 5. Results of our Hybrid Model compared to the previous Hybrid Model from Sotomayor *et al*. [11]

Hybrid model		
Metric	Prev	Avg
Mean error	6.86	10.85
Aedes hybrid model		
Metric	Prev	Avg
Mean error	–	4.8

2.5 Approach 5: Sentinel CNN + MLP

This approach was inspired by the hybrid model proposed by Sotomayor *et al*. However, our HM implementation showed in Sect. 2.4 was not able to improve or replicate the work of Sotomayor *et al*. Nevertheless, we implemented some improvements to the aforementioned model, en compare the results with the previous approaches.

The hybrid model presented by Sotomayor *et al.* improved the results of their predictor by using satellite images provided by Google Maps. However, despite their good results, we noted some possible improvements.

1. Google Maps API, does not provide images for given dates. To make a good prediction, we need to gather images which are closer to the day the trap collection date. To extract features like soil humidity and vegetation that might affect the mosquito count prediction.
2. *Aedes aegypti* can only flight around 1 km, thus, each image needs to cover that maximum space.
3. Google Maps API only offers images of 3 channels (RGB), but other APIs like Sentinel-02, offer images of 12 channels; which contain information about vegetation, soil humidity, and other features that could help the predictor to achieve better results.

Following these improvements, we compiled a dataset of Sentinel-02 images, one for each of the rows of WMD. Sentinel-02 images have 12 channels instead of 3, thus, we need to implement and train a Darknet-53 architecture capable of handling images of 12 channels. Unfortunately, our dataset is too small to train a Deep Neural Network such as Darknet-53. Therefore, we used a pre-trained Darknet-53 model trained for Yolov3. However, as Sentinel-2 images have 12 channels, we split each image of 12 channels into 3 images of 3 channels each, to make them compatible with the pre-trained Darknet-53. Each image contained information like vegetation and soil humidity.

We pass each of the three images generated by the Sentinel-02 through Darknet-53 to extract spatial features. Then, we concatenate the resulting 3 outputs with the result of the MLP network. Finally, we sent the resulting vector to a NN of 3 Dense Layers. The best combination of hyper-parameters for all mosquito species were: RMSProp with learning rate of 0.00001, 3 layers, weight decay of 0.004, and dropout of 0.3. This results in a mean absolute error of **11.10**. For *Aedes* the best combination were: RMSProp with learning rate of 0.00001, 3 layers, weight decay of 0.004, and dropout of 0.3. This resulted in a mean absolute error of **4.8** (Table 6).

Table 6. Results of our hybrid sentinel model

Hybrid sentinel model		
Metric	Prev	Avg
Mean error	6.86	11.10
Aedes hybrid sentinel model		
Metric	Prev	Avg
Mean error	–	4.8

2.6   Approach 6: VAE + MLP

Following the same line of thought, the next approach we tested was based on combining the spatial information (matrices processed with CNN) from satellite images plus the numerical information (vectors processed with MLP) from weather data. In this case, before directly using the images, we opted to encode them into latent vectors, which are lower dimensional representations of the data, as the most important data that is subtracted from the satellite images (Sentinel) relies on info like the water bodies, number of houses, potential nests, vegetation, etc., which does not require multiple dimensions to be represented (each of them). In other words, we transformed the matrix information into vectors.

Variational Autoencoder (VAE). Grouped inside the Generative Models, Autoencoders (AEs) is an architecture composed of an encoder part and a decoder one. The purpose of this architecture is to reduce the dimensionality of the input data (to save memory) and then being able to recreate it [8]. After the encoder section, data is represented by a low dimensionality vector in a "latent space" which, in concept, encodes the most important features of the input. Thus, it could be used in the same ways as dimensionality reduction architectures.

Traditional AEs present an issue in their latent space, it is not continuous. This has two main problems:

1. Trying to generate new similar images to the data-set from random vectors from the latent space won't always have good results.
2. Images that are similar do not necessarilty are closer in the latent space.

Variational Autoencoders (VAEs) are a modification to AEs to introduce a Divergence Measure (DM) to the latent space to assure that samples are distributed in a continuous matter; Usually, for the DM, the Kleiber Divergence (KD) is used, but this is perceived as restrictive and not preserving the variance across vectors. Thus, the Maximum Mean Discrepancy (MMD) was used as the error measure [6].

Note: Satellite data (from Google) was not used as it only gives images of the current time, not historical.

$$\mathrm{MMD}(p(z)\|\|q(z)) = \mathbb{E}_{p(z),p(z')}[k(z,z')] + \mathbb{E}_{q(z),q(z')}[k(z,z')] \\ - 2\mathbb{E}_{p(z),q(z')}[k(z,z')] \tag{1}$$

where $k(z,z')$ is a generic kernel; in this case, the radial basis kernel was used.

Based on this, we used the architecture shown in Fig. 2. Here, c is the number of channels of the original image, Sentinel-1 has 12 different channels and we tested using only the RGB ones (3 channels) or the 11 channels (only 11 channels are provided by the sentinel-02 web service). Moreover, the n_latent dimension was chosen to be 50 as it produced low reconstruction error.

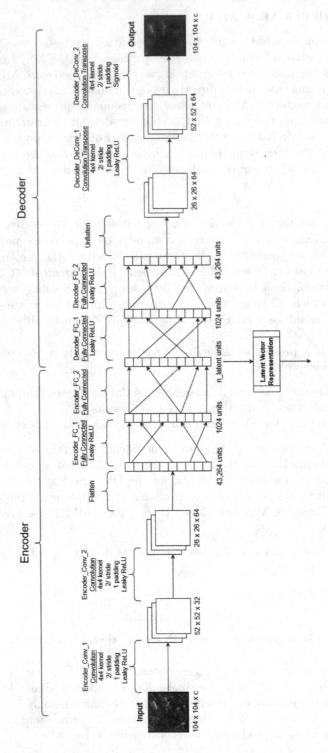

Fig. 2. VAE architecture for sentinel images.

Apart from the MMD, VAEs use the traditional loss function for comparing images (Negative Log Likelihood). The obtained errors where: **0.0001** when using just the RGB channels, and **0.0001** when using all the channels.

Adding the Numerical Data. Now that the images are processed, the rest of data was added following a mixed model approach. This means that the data from the tables is processed through a Neural Network (NN) having a lower dimension vector as an output, this is then concatenated to the Images information, and the resulting vector is then passed through another NN to obtain the final prediction.

The hyper-parameters that we chose to modify are: $e1$ as the number of the dimensions of the embeddings (for the categorical data), nX which are the different number of hidden layers marked on the architecture. Moreover, we also chose different optimizers (Adam and RMSProp), and learning rates (0.01, 0.001, 0.005, 0.0005, 0.0001). The final setup used the 5 most important parameters (according to the correlation coefficients), RMSProp as the optimizer (with learning rate 0f 0.0001), $e1$ was chosen for each particular categorical parameters, $n1 = 20$, $n2 = 10$, $n3 = 100$, $n4 = 50$. The error for this model was of **8.58** when measuring mean absolute error.

As an alternative, another architecture was used applying a NN also to the latent vector. After experimentation, the hyper parameters were chose as: RMSProp with learning rate of 0.0001, $n1 = 30$, $n2 = 10$, $n3 = 30$, $n4 = 50$, $n5 = 30$. This resulted in a mean absolute error of **8.89**.

In general, these both architectures are troublesome for choosing the correct hyper-parameters (and number of layers/units), it is usual to arrive to solutions where the same number is always the same number (highly non-convex topology of the solutions), and the results are plainly surpassed by a simpler NN where only the structured data is used. Thus, the efforts were redirected to a model that only uses this kind of data and considers the temporal nature of the information.

2.7 Approach 7: Recurrent Neural Network (LSTM)

As the weather information is obtained per day, it makes sense to use time-series-analysis techniques in these kind of data, considering the cycle of 14-days for the development of the mosquitoes. Then, we proposed a mixed model where the temporal data is processed through a LSTM+MLP, the numerical data is passed through a MLP, the categorical data is embedded and then passed through a MLP, and finally, either we take a weighted average of the 3 outputs (first approach), or concatenate the resulting vectors and pass it through a final MLP.

Weighted Average. For calculating the dimension of the LSTM, the following formula was used:

$$n1 = \lfloor \frac{batchsize}{a_h \times \#columns} + 1 \rfloor \tag{2}$$

The "TRAPDAYS" column was created by us to count the number of days that the trap was set. An improvement that can be done to this is to consider also the time of the day when the trap was set and recovered to include decimal values; is not the same to set the trap on the evening of day 1 and retiring on the morning of day 2, when compared to setting the trap on the morning of day 1 and retiring it on the evening of day 2.

Then, after trying several combinations of hyper-parameters, the chosen were: RMSProp with learning rate of 0.001, $a_h = 3$, $l1 = 2$, $n2 = 15$, $n3 = 15$, and a weighted average giving a value of 2 to the numerical data. This resulted in a mean absolute error of **4.80** which is much better than the models using images.

Mixed Model. This architecture was the most explored approach of the temporal ones. The best combination of hyper-parameters gave a mean absolute error of **4.36** which is considerably better than the weighted average approach.

This combination has a mean absolute error of **4.36** which is considerably better than the weighted average approach.

3   Analysis and Conclusions

Mosquitoes are one of the most dangerous creatures for the human beings due to the pathogens they transmit. The development of tools to predict mosquito populations, is of great interest to design control strategies that might help to reduce mosquito-related infections.

We implemented a total of 7 approaches for the prediction of mosquito populations. To present the results of each approach in a more readable manner, we have grouped them in Table 7. Surprisingly, the lowest error was obtained by applying a MLP in the WMD. However, we decided to analyze the dataset to check if the quality of the data was not a factor in getting poor results.

To have a better understating of the quality of WMD, we performed a correlation analysis, calculating Pearson, Spearman and Kendall coefficients. We found that the highest correlation (Spearman) with the dependent variable (Total of mosquitoes) was 0.29 and it corresponded to the "GENUS" variable; the rest of the variables had correlation coefficients lower than this (by 0.05 or more points). We concluded that WMD has poor data quality for this particular application, due to the low correlation coefficients that variables like temperature or precipitation, show for the total number of mosquitoes. Additionally, to corroborate the quality of the extracted weather information from DarkSky, we gathered weather information from WWO to compare output of the models. However, both datasets showed no improvement in the results. Thus, we think that the WMD is a noisy dataset or, at least, it is not fit for getting further insights from it.

Other possibility to consider is to test more methods to get data from images and different time-windows for the temporal architectures. As using these information increased the error of the model, implementing other techniques to get relevant attributes from this data could increase the value of the metrics.

One possibility could be to implement an image segmentation network to search the bodies of water in the images instead of processing the complete satellite image.

Table 7. Best results obtained for each of the used approaches

Approach	Mean absolute error
VAE + MLP	8.58
Pre-trained CNN	5.22
Sentinel CNN + MLP	4.8
Implemented hybrid model [11]	4.8
LSTM + Weighted average	4.80
LSTM + Mixed model	4.36
Simple MLP-3 + Weather data	4.10
Simple MLP-2	**3.98**

Although adding extra temporal weather information was not helpful for this dataset, using information directly related to the mosquito life-cycle might still work with other datasets since mosquitoes depend of rain, water containers, and temporal conditions like the temperature over time to develop. In the future, it would be worth to test these 7 approaches with other mosquito datasets besides WMD.

References

1. Arredondo-García, J., Méndez-Herrera, A., Medina-Cortina, H.: Arbovirus en latinoamérica. Acta pediátrica de México **37**(2), 111–131 (2016). http://dx.doi.org/10.18233/APM37No2pp111-131
2. Bhatt, S., et al.: The global distribution and burden of dengue. Nature **496**, 504–507 (2013). https://doi.org/10.1038/nature12060
3. D.C. Office of the Chief Technology Officer: Mosquito Washington Dataset (2019). https://opendata.dc.gov/datasets/DCGIS::mosquito-trap-sites/about. Accessed 9 Aug 2020
4. Christophers, S.R.: Aedes aegypti: the yellow fever mosquito. CUP Archive (1960)
5. Ferreira, A.G., Fairlie, S., Moreira, L.A.: Insect vectors endosymbionts as solutions against diseases. Curr. Opin. Insect Sci. **40**, 56–61 (2020). https://doi.org/10.1016/j.cois.2020.05.014. http://www.sciencedirect.com/science/article/pii/S221457452030078X
6. Gretton, A., Borgwardt, K.M., Rasch, M., Schökopf, B., Smola, A.J.: A kernel method for the two-sample problem. In: Advances in Neural Information Processing Systems, pp. 513–520 (2007)
7. Apple Inc: Dark Sky API (2020). https://darksky.net/. Accessed 9 Aug 2020
8. Kingma, D.P., Welling, M.: Auto-encoding variational bayes. arXiv preprint arXiv:1312.6114 (2014)

9. World Weather Online: World Weather Online (2020). https://www. worldweatheronline.com/. Accessed 11 Nov 2020
10. Redmon, J., Farhadi, A.: YOLOv3: An Incremental Improvement. arXiv preprint arXiv:1804.02767 (2018)
11. Sotomayor, L., Vallejo, E., Sánchez Castellanos, H.: Application of a hybrid neural network trained with satellite imagery and weather data to predict mosquito populations based on mosquito trap captures. Ph.D. thesis, Tecnologico de Monterrey, December 2019
12. Szegedy, C., et al.: Going deeper with convolutions. In: Computer Vision and Pattern Recognition (CVPR) (2015). https://doi.org/10.1109/CVPR.2015.7298594. http://arxiv.org/abs/1409.4842
13. Weaver, S.C., Charlier, C., Vasilakis, N., Lecuit, M.: Zika, chikungunya, and other emerging vector-borne viral diseases. Annu. Rev. Med. **69**(1), 395–408 (2018). https://doi.org/10.1146/annurev-med-050715-105122, pMID: 28846489
14. World Mosquito Program (WMP): What are mosquito-borne diseases? (2020). https://doi.org/10.1016/j.cppeds.2009.01.001. https://www.worldmosquitoprogram.org/en/learn/mosquito-borne-diseases. Accessed 9 Aug 2020

Causal Based Action Selection Policy for Reinforcement Learning

Ivan Feliciano-Avelino(ID), Arquímides Méndez-Molina$^{(\boxtimes)}$(ID),
Eduardo F. Morales(ID), and L. Enrique Sucar(ID)

Instituto Nacional de Astrofísica, Óptica y Electrónica Luis Enrique Erro # 1,
C.P. 72840, Tonantzintla, Puebla, México
{ivan.feliciano,arquimides.mendez,emorales,esucar}@inaoep.mx

Abstract. Reinforcement learning (RL) is the *de facto* learning by inter-action paradigm within machine learning. One of the intrinsic challenges of RL is the trade-off between exploration and exploitation. To solve this problem, in this paper, we propose to improve the reinforcement learning exploration process with an agent that can exploit causal relationships of the world. A causal graphical model is used to restrict the search space by reducing the actions that an agent can take through graph queries that check which variables are direct causes of the variables of interest. Our main contributions are a framework to represent causal information and an algorithm to guide the action selection process of a reinforcement learning agent, by querying the causal graph. We test our approach on discrete and continuous domains and show that using the causal struc-ture in the Q-learning action selection step, leads to higher jump-start reward and stability. Furthermore, it is also shown that a better perfor-mance is obtained even with partial and spurious relationships in the causal graphical model.

Keywords: Reinforcement learning · Causal graphical models · Action selection

1 Introduction

One of the goals of artificial intelligence (AI) is to create autonomous agents that learn through interaction with their environment [3]. One framework that emerges from that purpose is *reinforcement learning* (RL), which studies how an agent can learn to choose actions that maximize its future rewards through interactions with its environment [32]. RL algorithms have been shown to be effective in various domains such as video games [33], robotics [2], and medical care [15]. However, most of the current reinforcement learning systems suffer from some shortcomings, in particular, they do not take advantage of high-level processes to exploit patterns beyond the associative ones [13]. Among these high-level processes is causal reasoning [28].

Causal inference (CI) is a learning paradigm concerned at uncovering the cause-effect relationships between different variables [28]. CI addresses questions

© Springer Nature Switzerland AG 2021
I. Batyrshin et al. (Eds.): MICAI 2021, LNAI 13067, pp. 213–227, 2021.
https://doi.org/10.1007/978-3-030-89817-5_16

like: If I desire this outcome, what action do I need to take? So it can provide the information needed for an intelligent system to predict what may happen next so that it can plan better for the future. Learning causal relations in the real world is a challenging task for which many algorithms have been proposed according to different set of constraints. However, once the causal structure is known, it is possible to predict what would happen if some variables are intervened, and also, predict the outcomes of cases that are never observed before. This paper follows the manipulationist causality theory [36], in which the fundamental idea about causality can be described intuitively as: if A is a cause of X, then if A is manipulated, this will produce a change in X [7,35]; in particular, we use the manipulationist framework proposed by [28].

Reinforcement learning (RL) and causal inference (CI) have evolved independently and practically with no interaction between them. Nonetheless, recent work has focused on connecting both fields [9,12,18]. The goal of these works is to show how RL can be made more robust and general through causal mechanisms, known as *CausalRL* [20]. CausalRL attempts to mimic human behavior: learning causal effects from an agent interacting with the environment, and then optimizing its policy based on the learned causal relations.

In this work, we focus on one way of combining both fields, that is to use causal inference to improve reinforcement learning. In particular, we focus in the trade-off dilemma in RL between trying new actions (exploration) or selecting the best action based on previous experience (exploitation). Traditional exploration and exploitation strategies are undirected and do not explicitly chase interesting transitions. Using causal models is a promising way to cope with this problem. There are tasks in which an expert or even the algorithm itself can learn the latent causal model, such as: robotics, Animal AI [5], goal-directed tasks [27] and even some games [21].

We propose a method to guide action selection in RL tasks that have an underlying causal structure. The agent begins its search blindly, through trial-and-error interactions. However, the agent can, through interventions in a causal model, make queries of the type: *What if I do ...?* e.g., If I drop the passenger off here, will my goal be achieved? These interventions reduce the search space and allow the model to be used as an "oracle" to avoid performing actions that can lead to undesired states or to prefer actions that lead to a goal. We assume the presence of a causal model (partial or complete) for which both the structure and its parameters are known and no causal edge is pointing in the wrong direction. In the real world this may be a strong assumption, however we would like to point out that our main contribution is a framework to represent the causal information in RL settings and an algorithm to guide the action selection process of the agent, by querying the causal graph.

Experiments in different variants of the light switch scenario [27] show how an agent using a causal model achieves a higher reward in a shorter time compared to traditional RL. Even with an incomplete or partially incorrect causal model, it improves its performance.

2   Related Work

Exploitation and exploration strategies in RL are undirected and do not explicitly seek interesting transitions [23]. However, according to Hafner et al. [16], using prediction models seems a promising way to deal with this problem. There are several examples of using prior knowledge to guide an RL agent, as in [1,11,22,26,30]. However, causal models provide several advantages: (i) *Intervention*, evaluate changes in the effects given interventions in the causes; (ii) *Explanation*, to know why a certain sequence of decisions was chosen, and (iii) *Counterfactual*, to evaluate the potential impact of alternative actions.

The idea of using causal models to reduce the trial-and-error learning in RL is an area with little exploration but great promise. Commonly, the problem tackle by existing works is of the type *multi-armed bandit (MAB)*. Lattimore et al. [19] exploit causal information in the bandit problem and show how, through interventions, the rate at which actions with a higher reward are identified can be improved. In [4], the problem of unobserved confounders while trying to learn policies for RL models such as multi-armed bandits (MAB) is attacked. Without knowing the causal model, MAB algorithms can perform as badly as randomly taking an action at each time step. Specifically, the Causal Thompson Sampling algorithm is proposed to handle unobserved confounders in MAB problems. The reward distributions of the arms that are not preferred by the current policy can also be estimated through hypothetical interventions on the action (choice of arm). By doing this, it is possible to avoid confounding bias in estimating the causal effect of choosing an arm on the expected reward. To connect causality with RL, the authors view a strategy or a policy in RL as an intervention.

It is shown in [19] that adding causal information in a fixed budget decision problem[1] allows the decision maker to learn faster than if it does not considered causal information. Their work requires that the causal model is fully known to the decision maker, this requirement is relaxed later in [31] where the proposed system requires only that part of the causal model is known and allows interventions over the unknown part. In [19], a causal graphical model G is assumed to be known and a number of learning rounds T is fixed. In round $t \in [1, ..., T]$ the decision maker chooses $a_t = do(X_t = x_t)$ and observes a reward Y_t . After the T learning rounds, the decision maker is expected to choose an optimal action a^* that minimizes the expected regret, which is defined as $R_T = \mu^* - \mathbb{E}[\mu_{a^*}]$ where $\mu^* = \max \mathbb{E}[a]$. They show that the achieved regret is smaller than the regret obtained by non-causal algorithms.

In contrast to previous work, which focuses on non-model-based bandit algorithms, in [29] the authors propose a novel framework for learning causally correct partial models in model-based reinforcement learning settings. Those models does not fail to make predictions under a new policy, yet remain fast in high dimensional observation settings (e.g. images) because they do not need to fully model future observations.

---

[1] In this setting, each action is associated with a cost and the agent cannot spend more than a fixed budget allocated for all the task.

An interesting example focus on knowledge transfer in RL using causal inference tools [37]. Here, the problem is how to transfer knowledge across bandit agents in settings where causal effects cannot be identified by Pearl's do-calculus nor standard off-policy learning techniques. A new identification strategy is proposed that combines two steps: first, deriving bounds over the arm's distribution based on structural knowledge; second, incorporating these bounds in a novel bandit algorithm. Simulations show that their strategy is consistently more efficient than the current (non-causal) state-of-the-art methods.

Another problem in RL that is being attacked with causal elements is reward tampering [10]. This problem arises when an agent focuses on collecting small rewards and avoids the behavior that leads to the larger reward. For example, an agent who has to collect diamonds, but there are rocks that also give it rewards, may be inclined to only look for the rocks and not the diamonds. The authors use a causal influence diagram to attack this problem.

In contrast to previous work, we address the general case in which a decision problem can be represented as a Markov decision process (MDP); and propose a way to represent and use causal knowledge to accelerate the learning process in discrete and continuous spaces. Additionally, we experimentally show that even incomplete and partially incorrect causal models can improve performance, an important step towards simultaneously learning and using a causal model.

3   Proposed Method

Our work is focused on problems that can be posed as a goal-conditioned Markov decision processes. According to Nair et al. [27], this type of task has an underlying causal structure that describes the behavior of the environment. We define a goal-conditioned MDP as a tuple $(\mathcal{S}, \mathcal{A}, \mathcal{X}, \mathcal{D}, \mathcal{P}, \mathcal{G}, r, \gamma, \phi)$. The elements of the tuple are described as follows: $\mathcal{S}$ denotes the state space, $\mathcal{A}$ is the set of possible actions, $\mathcal{X}$ is the set of causal macro-variables[2] which describes the state of the environment at a high abstraction level (see [8]), $\mathcal{D}$ is a graph of the causal model ruling the agent's world, $\mathcal{P} : \mathcal{S} \times \mathcal{A} \times \mathcal{S} \to [0, 1]$ defines the probability transition function between states given an action, $\mathcal{G}$ is the goal space where its elements are vectors of variables on $\mathcal{X}$, $r : \mathcal{S} \times \mathcal{A} \times \mathcal{G} \to \mathrm{R}$ is the reward function where $r(s, a, g)$ yields the immediate reward conditioned on the goal $g \in \mathcal{G}$, γ is the discount factor, and $\phi : \mathcal{S} \to \mathcal{X}$ is a function which associates the state space to the macro-variables space. The goal of the RL system is to learn an optimal policy $\pi_g^* : \mathcal{S} \times \mathcal{G} \to \mathcal{A}$ such that it maximizes the expected total return $R = \sum_k^\infty \gamma^k r(s_k, a_k, g)$.

Often, an agent does not have enough information to define explicitly the transition dynamics of its environment, i.e., $\mathcal{P}$. However, there are tasks where some extra information can be given by a domain expert. This extra information can be encoded within the graph $\mathcal{D}$. An agent with a representation of the

---

[2] A high-level variable or macro variable is a function over a data structure, which in turn is defined from other variables [8]. These variables can be seen as a quantity that summarizes information about some aspect of the data structure.

underlying causal model of the environment, e.g., $\mathcal{D}$, is able to pursue actions that lead to desired states or to avoid choosing actions that take it into unwanted states. On the other hand, in many of the current problems in which RL-based solutions are applied, the information received by the agent is hard to model. For example, some Gym [6] environments are described by RGB images. Generating a causal model that represents all the information of the environment with which the agent interacts can be intractable. We propose to use a causal structure within goal-conditioned MDPs. Such causal structure describes the cause-effect relationships between actions and high-level variables that represent the observable states.

In this section, we describe our methodology to improve the performance of a classical reinforcement learning agent. Our aim is to decrease the agent's learning time by guiding its actions using the extra information given by the causal model that rules its environment.

3.1   Assumptions and Limitations

In this work, the following assumptions are considered:

1. It is assumed that a, possibly incomplete but partially correct, causal graphical model, $\mathcal{D}$, is known which relates state variables or meta-variables and actions with goals, sub-goals, and undesirable states.
2. In tasks where states are continuous, the RL agent also has access to ϕ. The latter function relates states to high-level variables.
3. It is assumed that there exist a mapping between a state's description and the variables used in the causal model. This mapping may be trivial, when the causal model is described by the same state variables used by the RL agent or can include a mapping involving for instance deep networks when states are represented by images.
4. $\mathcal{D}$ is considered to be a causal graph; that is, the Markov, minimality and faithfulness conditions (described in [17]) are assumed to be satisfied.

3.2   Action Selection

In the classical ϵ^3 greedy action selection policy, the agent randomly selects, based on the ϵ value, between exploring or exploiting its current knowledge. Therefore, the agent balances the exploitation and exploration trade-off. With ϵ probability, the agent selects a random action, and with $1 - \epsilon$ probability the agent takes the current best action. We extend the ϵ greedy strategy by adding an extra step, where the agent can also choose to query the causal model. In this work, this additional step is incorporated into the ϵ greedy policy, however, it is possible to use it in other action selection mechanisms.

In particular, we exploit the query: What action(s) leads me to meet the current goal? We can define a set $E = \{x \in \mathcal{X}$ such that x is not equal to its

---

[3] ϵ is a value between 0 and 1 that weights the relationship between exploration and exploitation.

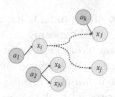

Fig. 1. Example of a causal graph $\mathcal{D}$. Dotted lines denote causal paths.

desired value given by g}, where its elements are those variables that have not reached the desired value. The desired value of a variable is indicated by the goal g. With this query, we want to know which variables in the set of actions affect the variables in E. What this means is that performing an action, i.e., assigning a value to an $a \in \mathcal{A}$, causes one or more variables in E to change their value, bringing the agent closer to its goal. Using E we obtain the benefit of narrowing the space of possible actions, i.e., we select those actions that bring us closer to the goal and at the same time we do not directly change those variables that have the desired value. By computing the set of predecessors of an effect $x \in \mathcal{X}$ we answer the previous query; so the agent prefers actions that are a parent of the variable of interest in the causal model, $\mathcal{D}$. Because the causal structure is in terms of high-level variables, ϕ, we need to map the observation s to a vector of macro variables $\mathbf{x} = [x_1, \ldots, x_N]$, where $x_i \in \mathcal{X}$.

A brief overview of the action selection policy is described as follows. With probability $1 - \epsilon$ the best action is chosen. For example, in the Q-learning algorithm the best action is given by that action which gives the largest value for the action value function, i.e., $\text{argmax}_a Q(s, a)$. Initially the probability of exploiting the best option is very low, i.e., ϵ starts with a value equal to or close to 1. If the best action is not chosen, then the causal structure is queried and if there is insufficient information in it, a random action is selected. In this work, the value of ϵ is linearly decremented throughout the agent's training.

Once the association between the state space and the high-level variables space is done, a list E of variables of interest is obtained, which will be consulted in the causal graph. The list E is obtained through a function f that calculates which variables have a different value between the goal g and the vector of macro variables $\mathbf{x}$. In some cases, the variables of interest may follow a causal order, which means that one goal depends on other subgoals. Therefore, the list E can be stored in a data structure such that its elements are ordered by a priority function. Thus, the agent choose first those actions that lead to the final goal. Figure 2 describes an example of how the list E is produced with respect to the causal graph in Fig. 1.

The next step corresponds to obtain an action $a \in \mathcal{A}$ for the agent to execute. For this, the agent makes a query to the causal structure. The query consists of

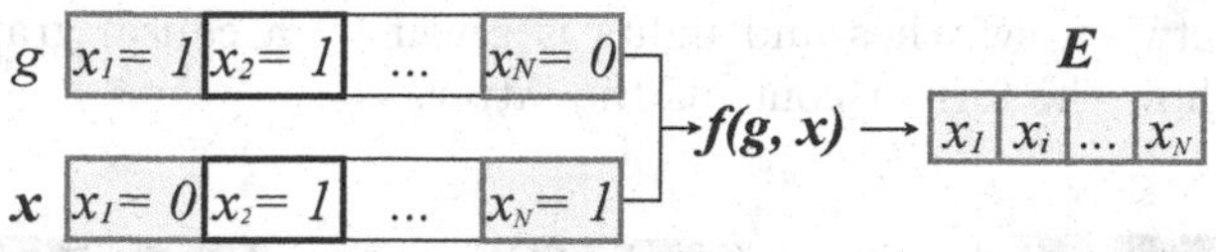

Fig. 2. E is computed finding those variables that have different values between $\mathbf{x}$ and g. x_i is stored in E if $|x_i - x_i'| = 1$ such that $x_i \in g$ and $x_i' \in \mathbf{x}$.

traversing the list of elements of E. If for an element $x \in E$ we found one action variable a as a predecessor, such action is performed[4].

4   Experimental Set Up

The experiments consist of integrating the causal graph to the classical ϵ greedy policy in the Q-learning algorithm. For this proof of concept, we assume that the action space $\mathcal{A}$ is discrete and the values of the actions are binary, $\{0, 1\}$. Furthermore, the elements of $\mathcal{X}$ are variables with binary values, $\{0, 1\}$. Even if the use of binary variables may seem as a simplified scenario, there are common use cases where we can interpret an action $a \in \mathcal{A}$ as performed or not. On the other hand, a binary variable $x \in \mathcal{X}$ can describe if a feature of interest is true or false, or if there is an object in an image or if an aim have been accomplished or if it is incomplete (e.g., in video games).

In the ϵ greedy policy, instead of keeping the value of ϵ fixed, we propose to start the learning by motivating the agent to use the causal model or explore. Then, we decrease ϵ to give more weight to exploitation (choose the best action according to the learned policy). Four algorithms are compared, where all of them follow the Q-learning approach. The main difference between them is the amount and quality of the information encoded in the causal graph. The following sections describe in detail the experiments performed and the results[5].

4.1   Environment

We conducted a series of experiments on the light switch control tasks testbed introduced by Nair et al. [27]. This is an episodic problem where in each episode an agent, which is provided with an initial lighting setting of the environment, aims to reach a previously specified configuration of the lights; i.e., which lights are on and which ones off. An example of an initial observation and the goal of the agent is shown in Fig. 3. Specifically, an agent has control over N light switches which control N lights. The agent performs an action (flipping a switch) and the environment feeds back with a new state of the lights and switches. The

---

[4] We consider problems where it is not necessary to perform several actions at the same time to affect x.

[5] The software developed is available at https://anonymous.4open.science/r/ cbdbb0ba-d371-4e0b-97b6-24613aff69ac/.

relationship between switches and lights is given by a causal graphical model which defines how the former controls the latter.

Fig. 3. A simple example of a trajectory of actions, from an initial lightning setting of the environment to the final lightning setting. A bird's eye view of the environment's state and the causal graph that rules the environment are shown. The blue and green vertices denote the switches and the lights, respectively. The bold blue node (on the left side) represents the manipulated variable to affect the darkest green node (on the right side). In the first case (at t_1), after manipulating the third blue node, the light is turned off in one room (represented by the fifth green node). In the second case (at t_2), after manipulating the second blue node the lights are turned on in another room (represented by the first green node). (Color figure online)

The testbed involves three different types of causal relationships between the switches and lights, shown in Fig. 4: *one to one*, where each switch controls only one light; *common cause*, where a single switch may control more than one light and each light is controlled by at most one switch; and *common effect*, where each switch maps to one light, but more than one switch can control the same light.

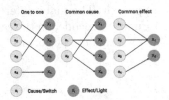

Fig. 4. Examples of the three types of latent causal structures in the light switch problem: one to one, common cause, and common effect.

4.2   Implementation and Compared Approaches

The proposed method is incorporated into the Q-learning algorithm [34]. The training process of the Q-learning algorithm remains identical, except for the guided action selection step. Four algorithms are compared, based on the Q-learning method, the differences lie in the quantity and quality of the additional information available:

- *Q-learning without additional information* (Q_1). It serves as a baseline. The algorithm, depending on the state space being worked on, is the classic Q-learning [34] or deep Q-learning with experience replay (DQN) [24], for discrete and continuous states, respectively.
- *Q-learning + complete causal structure* (Q_2). The agent has the complete and true causal structure of the environment $\mathcal{D}$.
- *Q-learning + partial causal structure* (Q_3). In this case, the agent has a subgraph $\mathcal{D}'$ of the graph $\mathcal{D}$.
- *Q-learning + incorrect causal structure* (Q_4). This algorithm queries a $\mathcal{D}''$ structure with spurious relationships and without some true links.

4.3   Evaluation Metric and Exploration Rate Decay

To measure the performance of the algorithms in each experiment, the average reward is evaluated over a series of simulations. Each simulation consists of running the learning algorithm for k episodes, in an environment with a fixed causal structure $\mathcal{D}$ and where the goal g is sought to be achieved. The average reward for the i-th. episode is given by $R^i = \frac{1}{H}\sum_{t=0}^{H} r(\mathbf{x}_t, g)$, where H corresponds to the size of the episode. The vector $\mathbf{R_i}$, of the i-th. experiment contains the average rewards for each episode, and is defined as $\mathbf{R_i} = (R^1, \ldots, R^k)$.

Thus, the comparison measure between algorithms is the average of the vectors $\mathbf{R_i}$, $i \in [1, M]$, obtained in M simulations. This measure, denoted as *average*, can be written as

$$average(\mathbf{R_1}, \ldots, \mathbf{R_M}) = \frac{1}{M}(\sum_i^M \mathbf{R_i^1}, \ldots, \sum_i^M \mathbf{R_i^k}), \tag{1}$$

where M is the number of simulations and $\mathbf{R_i^j}$ indicates the average reward obtained in the j-th. episode of the i-th. simulation.

The parameter ϵ decreases linearly, where at each action selection it decreases until a minimum value is reached. The ϵ update rule at time step t can be defined as $\epsilon = \max(\epsilon_{min}, \epsilon_{max} - \frac{|\epsilon_{max} - \epsilon_{min}|}{H \times k \times \delta} \times t)$, where $H = N$ and $0 < \delta \leq 1$, is a factor to control how fast the minimum ϵ value is reached, the closer to 0, the faster the exploration stage finishes.

5   Results

Three experiments are carried out. For the first two experiments, we directly provide the agent with the set of high-level variables, such that it works with a tabular version of the Q-learning algorithm. In the third experiment, we do not have the aforementioned set, so we only have access to $s \in \mathcal{S}$. Thus, for the latter case, we use the DQN [25] algorithm to deal with images as inputs.

The first experiment aims to measure the performance when modifying the causal structure $\mathcal{D}$ at different percentages to obtain $\mathcal{D}'$ and $\mathcal{D}''$. In the second experiment, we propose to change the decrease rate of ϵ to get faster or slower to the constant exploitation phase. In the last experiment, we test the algorithm when the high-level variables are not available as direct observations, therefore, it works on a continuous state space.

5.1 Modifying the Causal Graph

The goal of this experiment is to determine whether the information provided by an incomplete or partially incorrect model helps and does not negatively affect the performance of the RL algorithm. These subgraphs are generated from the $\mathcal{D}$, altering it to different levels. We modify $\mathcal{D}$ at three levels, to obtain the graphs $\mathcal{D}'$ and $\mathcal{D}''$. The percentage level of change is represented by the parameter p_{mod}. For each level, the $\mathcal{D}'$ subgraph is generated by removing a percentage p_{mod} of the edges of the $\mathcal{D}$ graph. To produce $\mathcal{D}''$, after removing edges in a similar way, for half of the missing connections, new ones, different from the initial ones, are created. We use three values to test the modification level: low, medium and high, with $p_{mod} = 25\%$, $p_{mod} = 50\%$, and $p_{mod} = 75\%$, respectively. The experiment is run on environments with the three possible types of structures: one to one, common cause, and common effect. The exploration rate is given by a $\delta = 0.5$, indicating that approximately halfway through the training, the probability of exploitation reaches its maximum value.

The results are summarized in Fig. 5. We can observe that in most cases where the algorithms use knowledge from the causal model ruling the agent's environment, there is a higher jumpstart and a faster converge than the Q-learning algorithm without additional information. In general, it can be seen that the algorithms Q_3 and Q_4 behave similarly. For the case where $p_{mod} = 25$, the algorithms with incomplete and incorrect information have a similar performance as the algorithm with the complete causal model, and the difference with the basic Q-learning is significant. Something similar occurs for $p_{mod} = 50$. For $p_{mod} = 75$, despite having modified the causal graph by a fairly high percentage, the small amount of information that remains and is correct, is enough to reach a higher reward much faster. This is most noticeable for common cause and common effect type structures. There is a strange behavior in tasks with a one to one type structure, which might be related to the exploration strategy.

5.2 Exploit or Keep Exploring

This section shows experiments to compare the performance of the algorithms with two different values for the factor controlling the decrement rate of ϵ. We control how fast we want to reach ϵ_{min} by varying δ. If the training time is divided into quarters, then the values of δ correspond to the quarter in which the minimum value for ϵ is reached. We can reach the ϵ_{min} value at different training times. To reach that value at the first, second, and third quarter of the training, we set δ to 0.25, 0.50, and 0.75, respectively. Since Sect. 5.1 includes the case of $\delta = 0.5$, we do not included it in this set of experiments. Our objective is to determine whether reducing or increasing causal graph queries throughout the learning process affects the performance of the algorithms. To get the graphs

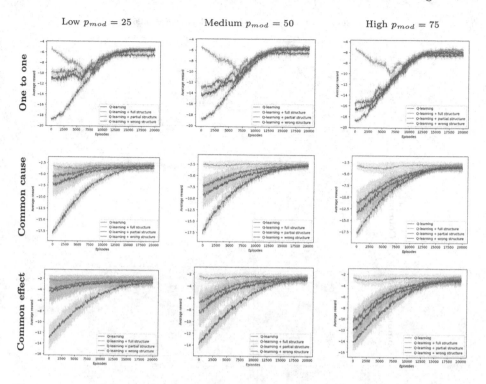

Fig. 5. Comparing the impact of a full/partial/incorrect causal model, over 10 simulations and $N = 9$. The plots show the average reward (and standard deviation) for each of the three types of structures and different values of p_{mod}, to compare the four algorithms. The vertical and horizontal axis correspond to the value of the evaluation metric and the episodes, respectively (best seen in color).

$\mathcal{D}'$ and $\mathcal{D}''$, the percentage of change is set to $p_{mod} = 25$. Figure 6 shows that the higher δ, the longer it takes to stabilize the learning algorithm without information to help it. This is expected, since it continues exploring for a longer time. In most cases, the Q_{2-4} algorithms are faster than Q_1, in the various environment configurations.

5.3   Using Visual Observations of the Environment

In this experiment the agent does not have access to the macro variables $\mathcal{X}$ directly; it receives images of the state of the environment as observations. The objective is to determine whether the causal model with variables in another space still retains the capacity of accelerating learning as in the discrete cases. The observations are images of 84×84 pixels in RGB color space, obtained from an eye's bird view of the environment (see Fig. 3). To associate the images to the high-level state space $\mathcal{X}$, the ϕ function is a convolutional neural network classifier [14]. The agent is provided with the classifier already trained.

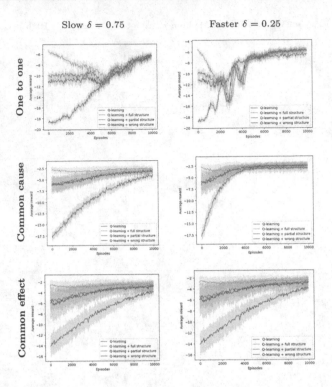

Fig. 6. Analyzing the impact of the exploration scheme over 10 simulations and $N = 9$. The plots show the average reward (and standard deviation) for each of the three types of structures and different values of δ, comparing the four algorithms. The vertical and horizontal axis in each plot correspond to the value of the evaluation metric and the episodes, respectively (best seen in color).

The outputs of the network represent the probability that the variable x_i takes the value 1, where $i \in \{1, \ldots, l\}$, $l = 9$. The causal graph alteration parameter is $p_{mod} = 25$. The rate of decrement of *epsilon* is controlled by the factor *delta* = 0.75. Since the observations are images, the DQN steady-state version of the Q-learning algorithm is used. The architecture and training hyperparameters are the same as those in the original paper [25]. The results depicted in Fig. 7 show that the algorithms using knowledge from the causal structures start with a higher reward (jumpstart) and stabilize faster than the DQN algorithm without additional information, in all three types of underlying structures.

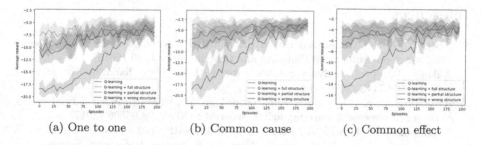

(a) One to one   (b) Common cause   (c) Common effect

Fig. 7. Results, over 10 simulations, for the continuous state space experiment. Each figure compares the performance of the four algorithms for the three types of structures. The vertical and horizontal axis indicate the value of the reward and the episodes number, respectively (best seen in color).

6   Conclusions

A methodology to guide the interaction step of an RL agent was presented, integrating a causal graph governing the agent's world, or at least a part of it, into the action selection policy. The method was tested in different experimental settings: modifying the causal network at different levels to have cases where only little or partially incorrect information is available, varying the rate at which the causal model is no longer consulted, and using images of the state of the environment as observations. Based on the results of the experiments, the following conclusions were reached:

- Incorporating causal knowledge into RL accelerates the learning process. This is relevant because the knowledge is expressed in terms of a causal model which can be used for explanations and reused in similar tasks.
- Providing an agent with a graph that preserves some true causal relationships continues to perform better than without guiding its choice of actions. This is important for learning at the same time the causal model.
- Reducing the probability of querying the causal graph at an early stage of training, i.e., giving more weight to exploitation than to exploration, does not affect the agent's performance. This could be because the little guided exploration experienced by the agent may have already biased its behavior.

The results indicate that the presented methodology is a suitable alternative for attacking interaction tasks, where the environment is governed by a causal model. Using the causal graph as a means of querying the information of the latent causal model can reduce the learning time with respect to a trial-and-error interaction with the environment. The proposed approach was implemented on commonly used reinforcement learning algorithms, Q-learning and DQN, however, it can be easily transfer to other algorithms. It is left as future work how to learn a causal model while using it to make decisions.

Acknowledgments. This work was partially supported by CONACYT, Project A1-S-43346 and scholarships 725976 (first author) and 754972 (second author) .

References

1. Abel, D., et al.: Goal-based action priors. In: Proceedings of the International Conference on Automated Planning and Scheduling, vol. 25 (2015)
2. Akkaya, I., et al.: Solving rubik's cube with a robot hand. arXiv preprint arXiv:1910.07113 (2019)
3. Arulkumaran, K., Deisenroth, M.P., Brundage, M., Bharath, A.A.: A brief survey of deep reinforcement learning. arXiv preprint arXiv:1708.05866 (2017)
4. Bareinboim, E., Forney, A., Pearl, J.: Bandits with unobserved confounders: A causal approach. In: Conference on Neural Information Processing Systems, pp. 1342–1350 (2015)
5. Beyret, B., Hernández-Orallo, J., Cheke, L., Halina, M., Shanahan, M., Crosby, M.: The animal-AI environment: training and testing animal-like artificial cognition (2019)
6. Brockman, G., et al.: OpenAI gym (2016)
7. Campbell, D.T., Cook, T.D.: Quasi-experimentation: Design & Analysis Issues for Field Settings. Rand McNally College Publishing Company, Chicago (1979)
8. Chalupka, K., Perona, P., Eberhardt, F.: Visual causal feature learning. In: Proceedings of the Thirty-First Conference on Uncertainty in Artificial Intelligence (UAI 2015), pp. 181–190. AUAI Press, Arlington, Virginia (2015)
9. Dasgupta, I., et al.: Causal reasoning from meta-reinforcement learning. CoRR abs/1901.08162 (2019). http://arxiv.org/abs/1901.08162
10. Everitt, T., Hutter, M.: Reward tampering problems and solutions in reinforcement learning: a causal influence diagram perspective. CoRR abs/1908.04734 (2019). http://arxiv.org/abs/1908.04734
11. Geibel, P.: Reinforcement learning with bounded risk. In: Proceedings of the Eighteenth International Conference on Machine Learning, pp. 162–169. Morgan Kaufmann (2001)
12. Gershman, S.J.: Reinforcement Learning and Causal Models. The Oxford handbook of causal reasoning, p. 295 (2017)
13. Gonzalez-Soto, M., Sucar, L.E., Escalante, H.J.: Playing against nature: causal discovery for decision making under uncertainty. CoRR abs/1807.01268 (2018). http://arxiv.org/abs/1807.01268
14. Goodfellow, I., Bengio, Y., Courville, A.: Deep Learning. MIT Press, Cambridge(2016)
15. Gottesman, O., et al.: Evaluating reinforcement learning algorithms in observational health settings (2018)
16. Hafner, D., Lillicrap, T.P., Ba, J., Norouzi, M.: Dream to control: learning behaviors by latent imagination. In: 8th International Conference on Learning Representations (ICLR). OpenReview.net (2020). https://openreview.net/forum?id=S1lOTC4tDS
17. Hitchcock, C.: Causal models. In: Zalta, E.N. (ed.) The Stanford Encyclopedia of Philosophy. Stanford University, Metaphysics Research Lab (2019)
18. Ho, S.: Causal learning versus reinforcement learning for knowledge learning and problem solving. In: Workshops of the The Thirty-First AAAI Conference on Artificial Intelligence. AAAI Workshops, vol. WS-17. AAAI Press (2017)
19. Lattimore, F., Lattimore, T., Reid, M.D.: Causal bandits: Learning good interventions via causal inference. In: Advances in Neural Information Processing Systems, pp. 1181–1189 (2016)

20. Lu, C., Schölkopf, B., Hernández-Lobato, J.M.: Deconfounding reinforcement learning in observational settings. CoRR abs/1812.10576 (2018), http://arxiv.org/abs/1812.10576
21. Madumal, P., Miller, T., Sonenberg, L., Vetere, F.: Explainable reinforcement learning through a causal lens. arXiv preprint arXiv:1905.10958 (2019)
22. Mazumder, S., et al.: Guided exploration in deep reinforcement learning (2019). https://openreview.net/forum?id=SJMeTo09YQ
23. McFarlane, R.: A Survey of Exploration Strategies in Reinforcement Learning. McGill University (2018). http://www.cs.mcgill.ca/cs526/roger.pdf
24. Mnih, V., et al.: Playing Atari with deep reinforcement learning. arXiv preprint arXiv:1312.5602 (2013)
25. Mnih, V., et al.: Human-level control through deep reinforcement learning. Nature **518**(7540), 529–533 (2015)
26. Nair, A., McGrew, B., Andrychowicz, M., Zaremba, W., Abbeel, P.: Overcoming exploration in reinforcement learning with demonstrations (2017)
27. Nair, S., Zhu, Y., Savarese, S., Fei-Fei, L.: Causal induction from visual observations for goal directed tasks. arXiv preprint arXiv:1910.01751 (2019)
28. Pearl, J.: Causality: Models, Reasoning, and Interference. Cambridge University Press, Cambridge (2009)
29. Rezende, D.J., et al.: Causally correct partial models for reinforcement learning. CoRR abs/2002.02836 (2020). https://arxiv.org/abs/2002.02836
30. Saunders, W., Sastry, G., Stuhlmüller, A., Evans, O.: Trial without error: towards safe reinforcement learning via human intervention. In: André, E., Koenig, S., Dastani, M., Sukthankar, G. (eds.) Proceedings of the 17th International Conference on Autonomous Agents and MultiAgent Systems (AAMAS), pp. 2067–2069. ACM (2018)
31. Sen, R., Shanmugam, K., Dimakis, A.G., Shakkottai, S.: Identifying best interventions through online importance sampling. In: Proceedings of the 34th International Conference on Machine Learning, vol. 70, pp. 3057–3066. JMLR. org (2017)
32. Sutton, R.S., Barto, A.G.: Reinforcement Learning: An Introduction. The MIT Press, Cambridge (2018)
33. Vinyals, O., et al.: Grandmaster level in Starcraft III using multi-agent reinforcement learning. Nature **575**, 350–354 (2019)
34. Watkins, C.J., Dayan, P.: Q-learning. Mach. Learn. **8**(3–4), 279–292 (1992)
35. Woodward, J.: Making Things Happen: A Theory of Causal Explanation. Oxford University Press, Oxford (2005)
36. Woodward, J.: Causation and manipulability. In: Zalta, E.N. (ed.) The Stanford Encyclopedia of Philosophy. Metaphysics Research Lab, Stanford University, winter 2016 edn. (2016)
37. Zhang, J., Bareinboim, E.: Transfer learning in multi-armed bandit: a causal approach. In: Proceedings of the 16th Conference on Autonomous Agents and MultiAgent Systems, pp. 1778–1780 (2017)

Performance Evaluation of Artificial Neural Networks Applied in the Classification of Emotions

Juan-José-Ignacio Lázaro-Lázaro[1], Eddy Sánchez-DelaCruz[1(✉)] [iD],
Cecilia-Irene Loeza-Mejía[1] [iD], Pilar Pozos-Parra[2],
and Luis-Alfonso Landero-Hernández[1]

[1] Departamento de Posgrado e Investigación, Tecnológico Nacional de México,
Campus Misantla, Veracruz, Mexico
[2] Universidad Autónoma de Baja California, Mexicali, Mexico

Abstract. Facial expressions are always manifested by people. For the human being, it is easy to recognize emotions. Technological advances applied to the recognition of expressions are growing rapidly and, in the same way, interest in research on this topic. However, at the computational level, it is a complicated task, some of the expressions of human beings are similar, for this reason, the computer can be confused at the moment of recognition. The use of machine learning models specifically artificial neural networks that have a good performance in emotions recognition is required to detect automatically feelings. This research shows the performance analysis of artificial neural networks applied to emotion datasets. The FER2013 and JAFFE datasets were used, a pre-processing of the data was carried out. For the classification, a comparison was made between artificial neural networks (Perceptron, VGG, and a Convolutional Neural Network). Optimal results were obtained in the detection of emotions.

Keywords: Artificial neural networks · Machine learning · Emotion recognition

1 Introduction

Optimal performance of neural networks in classifying images of human facial emotions is essential. The detection of emotions is a topic of interest and a challenge since there is little difference in some of the emotional states, which makes detection difficult. Technological advances applied to the recognition of expressions are growing rapidly and, in the same way, interest in research on this topic. The face shows changes in the eyes, nose, and mouth, these changes allow neural networks to perform the recognition of facial emotions in images.

Machine learning (ML) techniques have become powerful tools for finding patterns in data [5]. According to Beam and Kohane, ML is described as a program that learns to perform a task or make a decision automatically from data,

© Springer Nature Switzerland AG 2021
I. Batyrshin et al. (Eds.): MICAI 2021, LNAI 13067, pp. 228–238, 2021.
https://doi.org/10.1007/978-3-030-89817-5_17

rather than having explicitly programmed behavior [4]. Likewise, the amount of data is increasing, therefore, it is difficult for humans to analyze it, many decision-making problems are solved better by machines than by humans, in terms of accuracy and scalability [7]. Hence, ML has emerged as a method for developing practical software for computer vision, speech recognition, natural language processing, robot control, and other applications [8]. Experimental research in the field of computer science makes it possible to study problems related to society, in particular the recognition of human facial emotions. Traditional methods have poor performance in image recognition, that is, it is difficult for the computer to learn to recognize objects. The implementation of artificial neural networks allows recognizing objects faster and more accurately. The objective of this work is to analyze the performance of artificial neural networks applied in the classification of images of human facial emotions.

The rest of this work is organized as follows, Sect. 2 depicts representative works of the literature. Section 3 exposes the materials and methods. Section 4 illustrates the results and analysis of this work. Also, the results are compared with the state-of-the-art results. While in Sect. 5 the investigation is concluded and future works are shown.

2   Related Work

Siraj *et al.* conducted classification experiments of six types of primary emotions. Each emotion is represented by 82 variables, 21 represent the left eyebrows, 21 represent the right eyebrows, and 40 represent the lips. After capturing and extracting the images, image processing techniques were applied. The dataset was tested on a multilayer perceptron with the backpropagation algorithm and a learning rate of 0.1. The results show that a network with 10 hidden units obtained an accuracy of 97.50% [16].

Akakın *et al.* considered feature extraction schemes for the recognition of facial expressions. In one scheme, facial landmarks are tracked in successive video frames using an effective detector and tracker to extract reference paths. The characteristics were extracted using the Independent Component Analysis method. In the alternative scheme, the evolution of the expression of emotion on the face is captured by stacking normalized and aligned faces in a space-time face cube. The emotion descriptors were 3D Discrete Cosine Transformation characteristics of this prism. They used various classifier configurations and their performance is determined to detect the 6 basic emotions. The proposed method was evaluated with the Cohn-Kanade facial expression database and achieved a recognition performance of 95.34% [2].

Muhammad *et al.* proposed a facial expression recognition system to improve health care service since human expressions change with different health states. Applying a bandlet transformation to a facial image to extract sub-bands and then a central symmetric local binary pattern to each band block by block, the CS-LBP histograms of the blocks were concatenated to produce a feature vector. Then, they selected the most dominant feature that was fed into two classifiers: a Gaussian model and Support Vector Machine. A confidence score is produced when the classifiers by weight are merged, they are used to make decisions about the type of facial expression. The accuracy obtained in the experiment was 99.95% [10].

Abdul-Mageed *et al.* developed a deep learning model and achieved a new state-of-the-art in 24 types of emotions. They then divided the data into 80% training, 10% development, and 10% testing They considered a learning rate of 0.001, 3 dense layers, each with 1, 000 units, an attrition rate of 0.5, categorical cross-entropy for the loss function. For central modeling, they used closed recurrent neural networks, achieving an accuracy of 87.58% and 95.68%, modeling Robert Plutchik's 8 primary emotion dimensions [1].

Alizadeh and Fazel trained Convolutional Neural Network (CNN) models with different depths to classify gray-scale images of facial emotions into one of the seven categories of the FER2013 dataset that contains approximately 29,000 images. An accuracy of 65% was obtained in the validation set and 64% in the tests with a model of 4 convolutional layers, 2 fully connected layers, and a learning rate of 0.01 [3].

Fathallah *et al.* developed a CNN architecture adjusted with the Visual Geometry Group model to obtain better results. ConvNet contains 4 convolutional layers and 3 maximum grouping layers, followed by the fully connected layer. The input of the network is 165×165 and the output layer is Softmax with 6 expression classes. The model was tested with several mostly public facial image datasets. They achieved 99.33% accuracy with the CK + dataset, 87.65% with the MUG dataset, and 93.33% accuracy with the RAFD dataset [6].

Ribes developed a facial emotion recognition system in real-time using deep learning. Once the position of the face is identified, the Key Points are detected, then the emotion is classified and then an avatar is painted on top of the face. The datasets he used were JAFFE, CK, 10K, Yale, and OpenCV for face detection. He implemented different neural networks with different architectures and layers, along with a vector support machine. Using a CNN and the Jaffe dataset, a result of 73% and 71% was obtained with the Cohon-Kanade dataset [13].

Taghi Zadeh *et al.* performed classification experiments for the recognition of facial emotions using convolutional neural networks and Gabor filters for the extraction of characteristics. They applied two Gabor filters to the JAFFE dataset containing 213 images and used a network with 3 convolutional layers and 2 fully connected layers. They mentioned that the proposed features increase the speed and accuracy of network training. They achieved 91% and 82% accuracy without feature extraction [18].

Pranav *et al.* used a deep neural network model to classify 5 human facial emotions with a manually collected image dataset. The dataset consisted of 2040 images for training, 255 for validation, and 255 for testing. The network architecture consisted of 2 convolution layers and an abandonment layer after each convolution layer. The input image size was 32×32 and they used the Relu trigger function which makes the negative values zero and the positive ones stay the same. They gave the grouping layer a 2×2 size to reduce the image without losing information. The output layer had 5 units and the function used in the output layer was Softmax. They used the Adam optimizer, cross-entropy as loss function, 0.01 for the learning rate, and the network was trained with 4 epochs. The accuracy they achieved was 78.04% [11].

3 Materials and Methods

3.1 Data Acquisition

FER2013[1] dataset [12] and JAFFE[2] dataset [9] were selected. FER2013 [12] contains 35685 digital images of facial emotions (happiness, neutrality, sadness, anger, surprise, disgust, fear) in 48 × 48 pixel grayscale. The images belong to people of different ages, the most feasible images were manually selected for learning the model and the "Disgust" class was dropped because it has fewer images. FER2013 also contains a CSV file with the data such as grayscale pixels, classes to which each pixel belongs, and the last column indicates if it is for training or validation. A sample of the selected images is shown in Fig. 1. On the other hand, the JAFFE dataset [9] contains 213 images of 7 facial expressions (happiness, neutrality, sadness, anger, surprise, disgust, fear). The faces of the images have a front profile. The example of the images is shown Fig. 2. The datasets were divided into 70% for training and the rest for testing.

a) Angry   b) Fear   c) Happy   d) Neutral   e) Sad   f) Surprise

Fig. 1. Example of FER2013 dataset images.

a) Angry   b) Disgust   c) Fear   d) Happy   e) Neutral   f) Sad   g) Surprise

Fig. 2. Example of JAFFE dataset images.

3.2 Preprocessing

Some images are not conducive to classifying emotions, because some images contain letters, images without faces, images with emotions that do not belong

---

[1] Facial Expression Recognition 2013 https://datarepository.wolframcloud.com/resources/FER-2013.

[2] Japanese Female Facial Expression https://zenodo.org/record/3451524#.YP9_P-hKhPY.

to the class. Some of the images from the FER2013 dataset were manually discarded, examples of these images are shown in Fig. 3: a) images containing baby faces, b) images of animated faces, c) images without faces, d) images with letters, e) images of faces with a side profile. This dataset contains 7 types of human facial emotions, but 6 emotions were used (see Fig. 1) since the number of images in this class is less than the other classes.

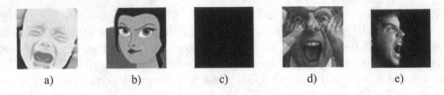

a)   b)   c)   d)   e)

Fig. 3. Examples of discarded images from the FER2013 dataset

The images in the JAFFE dataset contain images of 7 types of facial emotions (angry, disgusted, scared, happy, neutral, sad, and surprise). The Haar Cascade method [17] was applied, which allows identifying the face of a person in the image. After the face is identified, the part of the image where the face is found is extracted and in this way, part of the background of the images is discarded. Figure 4 shows two examples with images of the JAFFE dataset.

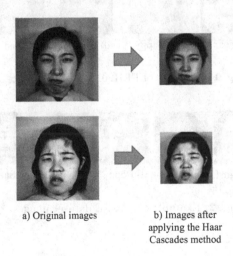

a) Original images   b) Images after
applying the Haar
Cascades method

Fig. 4. Haar Cascade method applied to JAFFE dataset

3.3   Artificial Neural Networks

The data classification procedure was carried out by implementing artificial neural networks. Three types of networks were chosen for the performance analysis in the classification of emotions. In the first emotion classification experiment, a simple perceptron was trained to classify facial emotions. In the second experiment, VGG architecture [15] was used. In addition, in the third experiment, a proposed CNN was implemented.

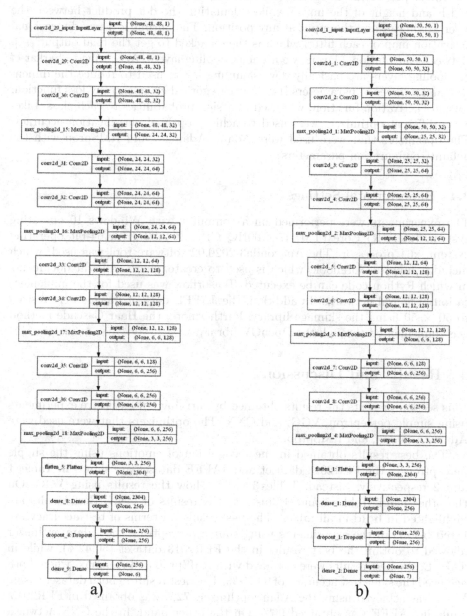

Fig. 5. CNN architecture: a) with FER2013 dataset & b) with JAFFE dataset.

Figure 5 shows the implemented architecture of the CNN, a) represents the architecture that was applied to classify the FER2013 dataset and b) to classify JAFFE dataset. The convolutional (Conv) layer is a set of filters. These filters help capture spatial features that are typically long and wide but cover the full depth of the image. In the forward step, the filter is slid across the width and height of the image while calculating the dot product between the filter attributes and the input at any position. The output is a two-dimensional activation map of each filter, which is then stacked to get the final output [14]. Between convolutional layers, we use max pooling layers. According to Sarkar *et al.* pooling layers are basically downsampling layers used to reduce the dimension and number of parameters [14]. The categorical cross-entropy loss function, softmax activation function, was used to assign probabilities to each class. Likewise, different optimizers were used to achieve optimal classification accuracy. The optimizers that were used were Adam, Adamax, and RMSprop with the default configuration parameters.

3.4   Experimental Settings

The experiments were performed on a computer with Windows 10 operating system, Intel (R) Core (TM) i7-7700HQ CPU @ 2.80 GHZ, 64-bit operating system, x64 processor. The Anaconda3-2020.02 software tool was used, which includes Jupyter Notebook, which is used to create, share and edit documents in which Python code can be executed. Tensorflow was used for the implementation of neural networks. In addition, the JAFFE dataset images were resized to 50 × 50 using the skimage library. Furthermore, the Haar Cascade method was implemented through the OpenCV library.

4   Results and Discussion

This section presents the results obtained by carrying out extensive experiments using simple perceptron, VGG, and CNN. The optimizers that were used were Adam, Adamax, and RMSprop.

The best results obtained in the recognition of emotions using the simple perceptron in the FER2013 dataset and JAFFE dataset are shown in Tables 1 and 2 respectively. Instead, Tables 3 and 4 show the results using VGG. On the other hand, Tables 5 and 6 illustrate the results of CNN. Best results are highlighted in **bold** in all tables. The best results in terms of test accuracy are listed below. In the experiments using simple perceptron, the Adam optimizer allowed to obtain the best results in the FER2013 dataset (50.42%), while in JAFFE, the best results were obtained with RMSprop and Adamax, where both achieved the same test accuracy of 93.75 %. The best result in both datasets using VGG was achieved using the Adam optimizer, 72% was obtained in FER2013, while in JAFFE was obtained 96%. On the other hand, in the CNN network, the best result in the FER2013 dataset was obtained by applying the RMSprop optimizer (73%), and the best result in the JAFFE dataset was obtained with the Adam optimizer (95%).

Experiments show overfitting using simple perceptron since there is a difference between the accuracy of training and tests. On the other hand, with the CNN and VGG networks, the results were more favorable, since a higher accuracy was obtained and the overfitting in the data is less. Therefore, we can say that a greater number of layers in artificial neural networks improves the performance of emotion recognition. In addition, the overfitting is decreased. In general, the best results were obtained with the Adam optimizer, which makes it suitable for implementation in multiclass classification problems such as emotion classification.

Table 7 shows the comparison of our best results in all experiments with those obtained in the state-of-the-art. When comparing our work with the literature, we can observe that the simple perceptron neural network was the one that

Table 1. Simple perceptron results in the FER2013 dataset

Optimizer	Training accuracy	Testing accuracy
Adam	**0.5974**	**0.5042**
Adamax	0.5967	0.4924
RMSprop	0.5258	0.4109

Table 2. Simple perceptron results in the JAFFE dataset

Optimizer	Training accuracy	Testing accuracy
Adam	**0.9933**	0.9062
Adamax	0.9866	**0.9375**
RMSprop	**0.9933**	**0.9375**

Table 3. VGG results in the FER2013 dataset

Optimizer	Training accuracy	Testing accuracy
Adam	0.7300	**0.7200**
Adamax	0.6600	0.6900
RMSprop	**0.7400**	0.5200

Table 4. VGG results in the JAFFE dataset

Optimizer	Training accuracy	Testing accuracy
Adam	**0.9200**	**0.9600**
Adamax	0.8200	0.8500
RMSprop	**0.9200**	0.9000

Table 5. CNN results in the FER2013 dataset

Optimizer	Training accuracy	Testing accuracy
Adam	0.7100	0.6800
Adamax	0.7300	0.7000
RMSprop	**0.7600**	**0.7300**

Table 6. CNN results in the JAFFE dataset

Optimizer	Training accuracy	Testing accuracy
Adam	**0.9600**	**0.9500**
Adamax	0.9100	0.9000
RMSprop	0.9500	0.8600

obtained the worst performance, whereas, with the CNN and VGG networks, competitive results were achieved with the literature. In our work, the accuracy in the FER2013 dataset exceeded that obtained by Alizadeh and Fazel [3]. In addition, in the JAFFE dataset, it was achieved surpassing the accuracy obtained by Ribes [13]. However, there are still challenges in performing the performance comparison due to private sets used by some authors in the past.

Table 7. State-of-the-art comparison of accuracy for classifying emotions

Reference	Neural network	Dataset	Accuracy (%)
Siraj *et al.* [16]	Multilayer perceptron	Own dataset	97.50
Alizadeh and Fazel [3]	CNN	FER2013	64.00
Fathallah *et al.* [6]	CNN	CK+	99.33
		MUG	87.65
		RAFD	93.33
Ribes [13]	CNN	JAFFE	73.00
		CK	71.00
Pranav *et al.* [11]	CNN	Own dataset	78.04
Our work	Simple perceptron	Fer2013	50.42
		JAFFE	93.75
Our work	VGG	FER2013	72.00
		JAFFE	96.00
Our work	CNN	FER2013	73.00
		JAFFE	95.00

5   Conclusion and Future Directions

The results obtained in the classification of emotions with the multiclass datasets allowed us to analyze the performance of artificial neural networks: simple perceptron, VGG, and a proposed CNN. In this research, better results were obtained with convolutional networks and VGG networks, little overfitting was presented in the data, however, they have a higher computational cost. Data preprocessing and feature selection algorithm allowed higher performance in human facial emotion classification experiments. In general, the best results were obtained with the Adam optimizer, which can be an indicator that its implementation is feasible in multiclass problems. However, one of the challenges of this study was the use of own datasets in previous works, which makes comparison difficult. For the continuation of this research, the following is proposed:

- Use other databases with a different number of classes.
- Implement other algorithms for the selection of characteristics or image segmentation.
- Classify facial emotions with the use of other algorithms.
- Develop an expert system based on the research carried out.

References

1. Abdul-Mageed, M., Ungar, L.: Emonet: fine-grained emotion detection with gated recurrent neural networks. In: Proceedings of the 55th Annual Meeting of the Association for Computational Linguistics (volume 1: Long papers), pp. 718–728 (2017)
2. Akakın, H.Ç., Sankur, B.: Spatiotemporal features for effective facial expression recognition. In: Kutulakos, K.N. (ed.) ECCV 2010. LNCS, vol. 6553, pp. 207–218. Springer, Heidelberg (2012). https://doi.org/10.1007/978-3-642-35749-7_16
3. Alizadeh, S., Fazel, A.: Convolutional neural networks for facial expression recognition. arxiv 2017. arXiv preprint arXiv:1704.06756 (2017)
4. Beam, A.L., Kohane, I.S.: Big data and machine learning in health care. JAMA 319(13), 1317–1318 (2018)
5. Biamonte, J., Wittek, P., Pancotti, N., Rebentrost, P., Wiebe, N., Lloyd, S.: Quantum machine learning. Nature 549(7671), 195–202 (2017)
6. Fathallah, A., Abdi, L., Douik, A.: Facial expression recognition via deep learning. In: 2017 IEEE/ACS 14th International Conference on Computer Systems and Applications (AICCSA), pp. 745–750. IEEE (2017)
7. Hernández Ávila, R.: Módulo de clasificación de imágenes para el FRAMEWORK JCLAL. B.S. thesis, Universidad de Holguín, Facultad de Informática-Matemática, Departamento de ... (2018)
8. Jordan, M.I., Mitchell, T.M.: Machine learning: trends, perspectives, and prospects. Science 349(6245), 255–260 (2015)
9. Lyons, M.J., Kamachi, M., Gyoba, J.: Coding facial expressions with gabor wavelets (ivc special issue). arXiv preprint arXiv:2009.05938 (2020)
10. Muhammad, G., Alsulaiman, M., Amin, S.U., Ghoneim, A., Alhamid, M.F.: A facial-expression monitoring system for improved healthcare in smart cities. IEEE Access 5, 10871–10881 (2017)

11. Pranav, E., Kamal, S., Chandran, C.S., Supriya, M.: Facial emotion recognition using deep convolutional neural network. In: 2020 6th International Conference on Advanced Computing and Communication Systems (ICACCS), pp. 317–320. IEEE (2020)
12. Research, W.: Fer-2013 (2018)
13. Ribes Gil, H.: Desarrollo de un sistema de reconocimiento de emociones faciales en tiempo real. Bachelor's thesis, Universitat Autònoma de Barcelona (2017)
14. Sarkar, D., Bali, R., Sharma, T.: Pract. Mach. Learn. Python. Apress, A Problem-Solvers Guide To Building Real-World Intelligent Systems. Berkely (2018)
15. Simonyan, K., Zisserman, A.: Very deep convolutional networks for large-scale image recognition. arXiv preprint arXiv:1409.1556 (2014)
16. Siraj, F., Yusoff, N., Kee, L.C.: Emotion classification using neural network. In: 2006 International Conference on Computing & Informatics, pp. 1–7. IEEE (2006)
17. Viola, P., Jones, M.: Rapid object detection using a boosted cascade of simple features. In: Proceedings of the 2001 IEEE Computer Society Conference on Computer Vision and Pattern Recognition (CVPR 2001), vol. 1, pp. I-I. IEEE (2001)
18. Zadeh, M.M.T., Imani, M., Majidi, B.: Fast facial emotion recognition using convolutional neural networks and Gabor filters. In: 2019 5th Conference on Knowledge Based Engineering and Innovation (KBEI), pp. 577–581. IEEE (2019)

Machine Learning Algorithms Based on the Classification of Motor Imagination Signals Acquired with an Electroencephalogram

Paula Rodriguez[✉], Alberto Ochoa Zezzatti, and José Mejía

Doctorado en Tecnología, Universidad Autónoma de Ciudad Juárez, Ciudad Juárez, Mexico
al206578@alumnos.uacj.mx

Abstract. Recent studies of brain-computer interface (BCI) have focused on the use of machine learning algorithms for the classification of brain signals. These algorithms find patterns in brain waves to distinguish between one class and another to turn them into control commands. To discuss the efficiency of classification algorithms in BCI for the classification of electroencephalogram (EEG) signals with motor images (MI), in this study twelve machine learning (ML) classifiers are applied and analyzed. The algorithms used are: 1) Convolutional network-Long short term memory (CNN-LSTM), 2) Convolutional network-gate recurrent unit (CNN-GRU), 3) Convolutional-bidirectional long short term memory (CNN-BiLSTM), 4) convolutional-bidirectional gated recurrent unit (CNN-BiGRU), 5) Random Forest, 6) Decision tree (DT), 7) Multilayer Perceptron (MLP), 8) Gaussian Naive Bayes, 9) Support Vector Machine (SVM), 10) Logistic Regression, 11) AdaBoost, 12) K-nearest neighbor (KNN). As classification tests, four mental tasks were registered, which are, the imagination of the movement of the left foot, the imagination of the movement of the left hand, state of relaxation, and mathematical activity, these mental tasks were obtained by a portable electroencephalogram (EEG) device. In the tests carried out it was found that the highest rate was 97% and the low rate was 22%.

Keywords: Brain-computer interface · Machine learning · Electroencephalogram · Motor imagination

1 Introduction

BCIs are communication systems between individuals and devices. Such communication is carried out by capturing the signals emitted by the brain during the process of executing mental tasks.

The signals are acquired, processed, and classified, generally converted into control commands either for neurorehabilitation or recreation [1]. Some studies have focused on the development of systems that have made it possible to control, for example, wheelchairs [2]; spellers [3], robotic arms [4], home automation [5], and rehabilitation [6].

I. Batyrshin et al. (Eds.): MICAI 2021, LNAI 13067, pp. 239–249, 2021.
https://doi.org/10.1007/978-3-030-89817-5_18

Signal acquisition methods are classified as invasive and non-invasive. Among the non-invasive methods, the most widely used method is the electroencephalogram (EEG). This is mainly due to portability and cost [7]. EEG signals are recorded from the cerebral cortex as temporal and spatial waves. The bioelectric potentials observed in the skin are caused by the flow of electrical currents generated by the nervous tissue [8]. The EEG signals are classified primarily based on their amplitude, morphology, and frequency [9]. It is worth mentioning that in the EEG some frequency bands are labeled as delta (0.5 Hz–4 Hz), theta (4 Hz–7 Hz), alpha (8 Hz–12 Hz), and beta (13 Hz–30 Hz) [10]. It should be noted that generating EEG signals requires mental tasks. One of the most used techniques is motor imagination (MI). This process consists mainly of imagining the movement of some part of the body [11].

In this research, it is proposed to classify signals acquired with a portable EEG device and generated from four MI tasks. For this, twelve algorithms of machine learning (ML) will be analyzed. The contribution of this work is mainly the comparison of the accuracy of twelve classifiers, to generate antecedents in the use of classifiers applied to the same database. Hence, this research contributes to future investigations of IM signal classification.

The rest of the paper is organized as follows; the second section has a brief description of the algorithms used for the classification of EEG signals. The third section describes the methodology used for the acquisition, processing, and classification of brain signals. In the fourth section, we present and analyze the experimental results on the performance of the models used. Finally, the fifth section shows the observations of the results obtained.

2 Background

Recently, to increase the accuracy of the machine learning algorithms, the studies carried out with EEG have focused on different methods of generation of patterns in brain signals as well as on the processing and classification, some of the algorithms most used in recent years according to the literature are PCA, SVM, LDA, NB, Logistic Regression, KNN, LSTM, CNN, GRU, Hybrid algorithms, Fourier transform, Wavelet transforms, Spatial and temporal filters [12]. Below we present the definitions of the algorithms used in this research to facilitate the understanding of the models.

2.1 CNN

Convolutional Neural Networks (CNN) is mainly applied in signal and image processing. Initially, this network performs a convolution of the input signal with an internal kernel, this operation is used for local detection of the characteristics of the signal. Generally, a convolution layer is composed of several kernels. The function that is used as activation is usually the ReLU function, whose expression is given by:

$$f(x) = maximum(0, x) \tag{1}$$

In addition, the max-pooling layer is responsible for grouping a particular area to reduce the sampling of the characteristics obtained from the convolutional layers, this to reduce the amount of calculation and the degree of overfitting of the network, in addition to increasing performance of the CNN model.

2.2 LSTM

Long short-term memory (LSTM) is a type of recursive neural network that is intended mainly to process sequential data, this is done through gates. The LSTM uses the input signal and converts it into four variables, which pass through three gates and a hyperbolic tangent. The output of the network depends on the data input, and in turn on the weights and parameters of the above output during model training.

2.3 GRU

Gated recurrent unit (GRU) is used for temporary data, it uses two entry gates of update and restarts. The update gate helps the model decide how much past information should be passed on to future iterations, while the reset gate determines what past information should be forgotten.

2.4 Bidirectional RNN

The bidirectional recurrent neural networks (BRNN) are models that are used in sequential data where the data of stopped iterations have inference in future inputs, also this process is carried out in reverse where future iterations infer in past interactions, all this to label a current data. this type of network is used in both LSTM and GRU.

2.5 DT

Decision trees (DT) is a method used in a classification where the model creates a prediction of the value of a variable where it learns different simple decision rules about the characteristics of the data, this is done by approximating a sinus which is an approximation gradual.

2.6 Random Forest

Random forest is a classifier that fits multiple decision tree classifiers in various subsamples of the data set, the results of the classifiers are averaged to improve accuracy and reduce overfitting.

2.7 MLP

Multilayer perceptron (MLP) belongs to neural networks, and consists of the input layer, a hidden layer, and an output layer, where it receives the input data, use backpropagation in its hidden layers, and finally classifies the data. MLPs tend to approximate any continuous function and solve problems that are not linearly separable.

2.8 Gaussian Naïve Bayes

Gaussian Naive Bayes follows a Gaussian distribution, it is also based on the Naive Bayes theorem, which is used to create its model which is defined by

$$P(y|X) = P(X|y).P(X)P(y) \tag{2}$$

Where, $X = x1, x2, x3,..., xn$, are a list of independent predictors, y is the class label and $P(y \mid X)$ is the probability of label y given the predictors X.

2.9 SVM

A support vector machine (SVM) is a supervised machine learning model which has two main advantages: higher speed and better performance than neural networks. The operating principle of SVM is to take the data and generate a hyperplane that divides the classes, the hyperplane generates a decision area for classification.

2.10 LDA and Logistic Regression

Linear discriminant analysis (LDA) is responsible for projecting the input data and separating it between classes, maximizing the separation of classes and decreasing the separation between the data. This serves to prepare the data entry for logistic regression. For its part, logistic regression is a discriminative model that only requires estimating the weights of the input data to calculate the posterior probability of each class.

2.11 AdaBoost

AdaBoost is an adaptive algorithm that performs classifications through simpler and weaker deductions, which are in states -1 and 1, this algorithm could adapt iteratively by performing a summation and finally generating a final output.

2.12 KNN

K-nearest neighbor (KNN), is characterized by being a parametric algorithm that is based on instances, for its operation, it calculates the distance between the data, finds the closest data, and finally labels the data, there are various methods to calculate the distance, the most used is the Euclidean distance, which is given by the expression:

$$Deuc(xi, xj) = \sqrt{\sum_{r=1}^{p} (x_{ri} - x_{rj})^2} \tag{3}$$

where p is the attribute vector and $xi = (x1, x2, x3, \ldots, xpi)$ belong to x.

3 Methodology

In this section, the materials and methods used for the acquisition, extraction of characteristics, and classification are described.

For the acquisition of signals, the Muse device was used, which is a non-invasive EEG headband. The device detects signals through four channels, which are located according to the international system in AF7, AF8, TP9, and TP10.

For the recording of the signals, a subject participated who generated signals with a duration of thirty seconds, of which 50 recordings of the imagination of the movement of the left foot, 50 recordings of the imagination of the movement of the left hand, 50 recordings in state of relaxation, and 50 recordings of mathematical activity.

A total of 200 recordings were obtained, data augmentation was applied to obtain a total of 1000 recordings. Each recording consisted of frames of 3000 samples with four channels. From this, the input shape of the data is (1000,3000,4).

Feature extraction was done using the Python Time series feature extraction library (TSFEL) package. TSFEL can extract features in the statistical, temporal, and spectral domains. In this experiment, the ML classic algorithms used were implemented using Sklearn [13], which is one of the most common machine learning libraries in python.

Finally, for the classification task, twelve different classifiers were used, which are: CNN-LSTM, CNN-GRU, CNN-BiLSTM, CNN-BiGRU, Random Forest, Decision Tree, MLP, Gaussian NB, SVM, LDA LR, AdaBoost, KNN. These were chosen because they are the most common algorithms used in classification and to compare results between them to be able to discern in future classification works of EEG signals. In Fig. 1, it is shown a block diagram of this process.

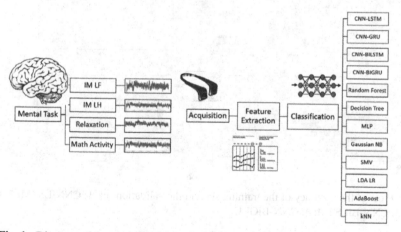

Fig. 1. Diagram of the process of the acquisition and classification of EEG signals.

4 Results

For the evaluation of the obtained results, the accuracy graph metrics were used in the case of CNN-LSTM, CNN-GRU, CNN-BiLSTM, CNN-BiGRU. Confusion matrix for each of the classifiers, ROC curves for the representation of the Random forest, Decision tree, CNN-LSTM, CNN-LSTM, CNN-GRU, CNN-BiLSTM, CNN-BiGRU classifiers. In addition, general graphs of accuracy, recall, and precision are shown.

Figure 2 shows the accuracy graphs provided in the training of neural networks. In the case of graph 1, which is the graph of the CNN-LSTM classifier, it is observed that the model was trained with 100 epochs, in addition, that the accuracy of the validation set was growing along with the accuracy of the training set, however, the validation set obtained a higher number of relapses and recoveries.

In the case of graph 2, which belongs to the training and validation of the CNN-GRU network, it is observed that the accuracy of the validation set increased according to the accuracy of the training set, however, at time 80 the validation set had a decline decreasing its value to 0.5, from which it recovered instantly.

Graph 3 belongs to the CNN-BiLSTM network, in this case, it is observed that both the accuracy of the validation set, and the accuracy of the training set go hand in hand, both of which suddenly decreased their accuracy in epochs 50 and 80, gradually recovering.

In graph 4, which belongs to the CNN-BiGRU network, it is observed that the accuracy of both the validation set and the training set start at 0.78, however, they fall from time 25 to 0.7 and 0.6 respectively and have another relapse at time 58, in the case of this network only 60 epochs were used in training.

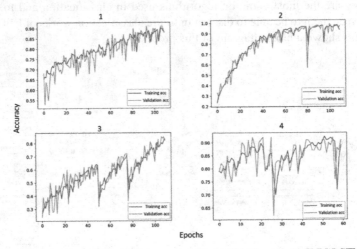

Fig. 2. Plots of the accuracy of the training set and the validation set. 1: CNN-LSTM; 2: CNN-GRU; 3: CNN-BiLSTM; 4: CNN-BiGRU.

The calculation of the normalized multiclass confusion matrix, which is a metric that helps us to visualize the performance of each classifier. In our case, it shows us the classifications of the four mental tasks, which are the imagination of the movement of the left foot, the imagination of the movement of the left hand, a state of relaxation, and mathematical activity.

For the calculation of the general accuracy of the classifier, which is shown in Eq. 4, it is necessary to calculate the precision that identifies the number of tasks classified correctly, where the best result tends to be 1 and is defined by Eq. 5.

$$Accuracy = (TP + TN)/(TP + FP + TN + FN) \tag{4}$$

$$Precision = TP/TP + FP \tag{5}$$

In addition, it is required to calculate the recall, this identifies the number of mental tasks classified correctly among the total number of tasks performed, where the best result tends to be 1 and is represented by the expression (6).

$$Recall = TP/TP + FN \tag{6}$$

Where TP is the true positive, TN is the true positive, FP is the false positive, and FN is the false negative of all classes in the confusion matrix. Figure 3 shows the normalized confusion matrix (CM) for each of the classes, in which it is observed that the columns represent the real classes, while the predicted classes are shown in the rows.

For the CM DT it is observed that the model discriminated the imagination class of the movement of the left hand as zero, that is, this class was confused with the imagination of the movement of the left foot. The CM Gaussian NB shows that the model best classified the imagination class of the movement of the left hand with 0.68. The CM AdaBoost shows us that the highest-ranked class was the left-hand movement imagination with 0.73. The CM Random forest shows a good classification in each of its classes, with the imagination of the movement of the left foot being the best classified with 1.0 which is the maximum value. The CM Gaussian NB shows an accuracy of 1.0 in the class of mathematical activity; however, the other classes are very low being the state of relaxation and the imagination of the movement of the left hand zero. The CM LDA RL shows that the best-classified class was that of mathematical activity with 0.94. The CM KNN shows that the class best classified by the model is that of mathematical activity with 0.78. The SVM CM shows that the best-ranked class was the imagination of the movement of the left foot with 1.0, however, the algorithm confused most of its classes with the imagination of the movement of the left foot.

On the other hand, within the neural network models, we find the CM of CNN-BiGRU, which shows us a high classification in all its classes, the state of relaxation being the highest with 0.96. The CM LSTM shows us that the class that obtained the highest heat was the one with the mathematical activity of 0.94. The CM GRU obtained in the classes of imagination of the movement of the left foot and the mathematical activity the value 1.0. And finally, in the CM BiLSTM, we observed that the maximum value obtained was in the relaxation state class with the value 0.96.

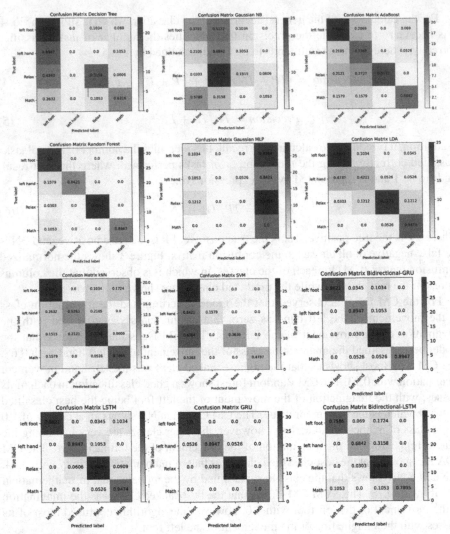

Fig. 3. Normalized confusion matrix for each algorithm for the classes MI left foot, MI left hand, state of relaxation, mathematical activity.

Figure 4 shows the graph of the accuracy of each algorithm, the algorithms that achieved the highest accuracy were CNN-GRU with an accuracy of 97%, followed by Randon forest with 93% accuracy, CNN-BiGRU with 91% accuracy. While the algorithms with the lowest accuracy were SVM with 53%, Naive Bayes with 31%, Decision Tree with 29%, and MLP with 22%. The kNN and AdaBoost algorithms obtained 65%, LDA LR obtained 75%, CNN-LSTM 82%, and finally CNN-BiLSTM 88%.

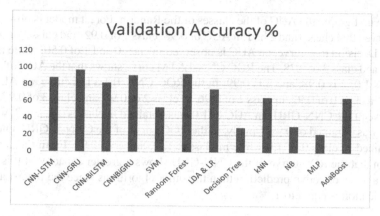

Fig. 4. Comparison of classification accuracy of algorithms.

Figure 5 shows a graph where the precision of each class is compared for each of the algorithms, in this, it is observed that the precision of each class depends directly on the model.

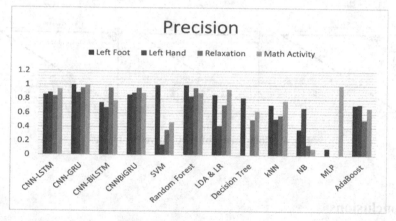

Fig. 5. Comparison of the precision of each class by the classifier.

Finally, an analysis of the ROC curves of the models is provided. The ROC curves show the performance of the model where the TPs are compared with the FPs. From the representation of the ROC curves, the area under the curve (AUC) is calculated, which shows the probability that a data is classified as positive, the area under the curve should be as close to 1 as possible. Figure 6 shows the graphs of the ROC curves of Random Forest, Decision Tree, CNN-LSTM, CNN-GRU, CNN-BiLSTM, CNN-BiGRU. For this, class 0 is considered as the imagination of the movement of the left foot, class 1 as the imagination of the movement of the left hand, class 2 as the state of relaxation, and class 3 as the mathematical activity.

In the first graph, the AUC of the classes of the Random Forest model is observed, in it is observed that class 0 and 2 had an AUC of 0.98, class 1 of 0.92, and class 3 of 0.95. In the ROC Decision tree curve, an AUC is observed in class 0 and 1 of 0.50, while in class 2 of 0.72 and class 3 of 0.78. The ROC CNN-LSTM curve shows that the AUC of classes 0 and 1 is 1.0, 2 is 0.98, and 3 is 0.99. In the ROC CNN-BiLSTM curve, an AUC was obtained in class 0 of 0.97, in class 1 of 0.96, in class 2 of 0.98, and class 3 of 0.97. In the ROC curve of the CNN-GRU, an AUC of 1.0 was obtained in classes 0.2, and 3, while in class 3 it was obtained of 0.99. Finally, in the ROC curve of the CNN-BiGRU, and AUC was obtained in class 0 of 1.0 while in classes 1, 2, and 3 of 0.99. From these curves, we obtain that the ROC curve of the CNN-GRU shows a greater capacity of this model to make a more effective prediction than the other algorithms since the macro-average of classification is equal to 1.

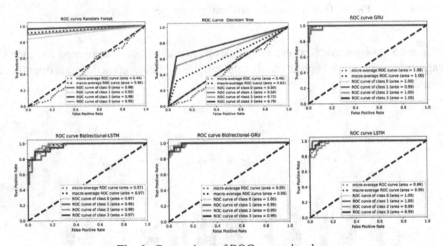

Fig. 6. Comparison of ROC curves by class.

5   Conclusions

In this article, the classification of EEG signals acquired by motor imagination with a portable four-channel acquisition device was proposed. For this, twelve algorithms were used, from which results that vary between 22% and 97% accuracy were obtained.

Therefore, according to the results obtained, the convolutional network in conjunction with a GRU network is the most appropriate to classify space-time EEG signals generated with IM, since both in the confusion matrix and in the ROC curve it obtained the best result. However, the combination of the TSFEL library together with the Random Forest classifier, generated promising results, this being the one that gave the second-best result with an accuracy of 93% in the classification of the four classes.

For future work, it is proposed to classify a greater amount of data obtained through motor imagination. For this, new recordings of the signals will be made. An EEG device that has eight acquisition channels will be used. Databases will be created using signals from 3 different people. The acquired signals will be classified using the CNN-GRU and Random Forest algorithms. The signals will be converted into control commands to control a robotic arm in real-time. This experiment is proposed as future work for an application within Industry 4.0.

References

1. Mudgal, S.K., Sharma, S.K., Chaturvedi, J., Sharma, A.: Brain-computer interface advancement in neurosciences: applications and issues. Interdiscip. Neurosurg. **20**, 100694 (2020)
2. Zgallai, W., et al.: Deep learning AI application to an EEG driven BCI smart wheelchair. In: 2019 Advances in Science and Engineering Technology International Conferences (ASET), pp. 1–5. IEEE (2019 March)
3. Lin, Z., Zhang, C., Zeng, Y., Tong, L., Yan, B.: A novel P300 BCI speller based on the triple RSVP paradigm. Sci. Rep. **8**(1), 1–9 (2018)
4. Schwarz, A., Höller, M.K., Pereira, J., Ofner, P., Müller-Putz, G.R.: Decoding hand movements from human EEG to control a robotic arm in a simulation environment. J. Neural Eng. **17**(3), 036010 (2020)
5. Yang, D., Nguyen, T.-H., Chung, W.-Y.: A bipolar-channel hybrid brain-computer interface system for home automation control utilizing steady-state visually evoked potential and eyeblink signals. Sensors **20**(19), 5474 (2020)
6. Xu, J., Liu, T.L., Wu, Z., Wu, Z., Li, Y., Nürnberger, A.: Neurorehabilitation system in virtual reality with low-cost BCI devices. In: 2020 IEEE International Conference on Human-Machine Systems (ICHMS), pp. 1–3. IEEE (2020, September)
7. Ramadan, R.A., Vasilakos, A.V.: Brain computer interface: control signals review. Neurocomputing **223**, 26–44 (2017)
8. Rocchi, L., et al.: Disentangling EEG responses to TMS due to cortical and peripheral activations. Brain Stimul. **14**(1), 4–18 (2021)
9. Akin, M.: Comparison of wavelet transform and FFT methods in the analysis of EEG signals. J. Med. Syst. **26**(3), 241–247 (2002)
10. Antoniou, E., et al.: EEG-based eye movement recognition using the brain-computer interface and random forests. Sensors **21**(7), 2339 (2021)
11. Duan, L., et al.: Zero-shot learning for EEG classification in motor imagery-based BCI system. IEEE Trans. Neural Syst. Rehabil. Eng. **28**(11), 2411–2419 (2020)
12. Rashid, M., et al.: Status, challenges, and possible solutions of EEG-based brain-computer interface: a comprehensive review. Front. Neurorobot. **14**, 25 (2020)
13. https://scikit-learn.org/stable/

Image Processing and Pattern Recognition

Touchless Fingerphoto Extraction Based on Deep Learning and Image Processing Algorithms; A Preview

Marlene Elizabeth López-Jiménez$^{(\boxtimes)}$ ⓘ, Víctor Rubén Virgilio-González ⓘ, Raúl Aguilar-Figueroa ⓘ, and Carlos Daniel Virgilio-González ⓘ

Biometria Aplicada SA de CV, Innovation Department, CDMX, Mexico City, Mexico
{mlopez,vvirgilio,raguilar,cvirgilio}@biometriaaplicada.com
https://biometriaaplicada.com/sitio/

Abstract. In recent years, touchless fingerprint recognition systems have become a reliable alternative to the conventional touch fingerprint recognition system. Furthermore, with the current health crisis caused by the emergence of SARS-COV 2, the implementation of technologies that allow us to avoid direct contact with readers and devices arises as an urgent need. This article shows a system for fingerprint segmentation, filtering, and enhancement by fingerphoto technology. The dataset was acquired from smartphones on uncontrolled conditions. The proposed fingerprint recognition scheme provides an efficient preview of an automated identification system that can be extended to numerous security or administration applications. Skin model segmentation presents an accuracy of 95% over other solutions for background removal in the state of the art. For fingerphoto extraction, results were evaluated with the NIST Finger Image Quality $NFIQ$ of the National Institute of Standards and Technology $NIST$.

Keywords: Fingerphoto · Touchless · Biometrics · Digital onboarding · Deep learning · U-Net

1 Introduction

For many years, the commercial industry had not faced as big a challenge as SARS-COV 2 [20]. With social distancing as a basic rule and the need to ensure business continuity, companies of all sizes and sectors have migrated their products and 100 % digital ecosystem services [21]. Common and highly standardized methods such as bank and public institutions onboarding onboarding, which implies that the client physically moves to the institution and performs long face-to-face procedures, were replaced by digital onboarding systems [1]. One step in digital onboarding consists of authenticating the identity of the client from the capture of identification photographs, face and, in more strict cases, fingerprints [2]. Fingerphoto technology has been appearing in the academic area for years, but in recent months it has become very popular in the biometric

© Springer Nature Switzerland AG 2021
I. Batyrshin et al. (Eds.): MICAI 2021, LNAI 13067, pp. 253–264, 2021.
https://doi.org/10.1007/978-3-030-89817-5_19

and digital identity area. In this situation, it becomes highly necessary to have remote fingerprint extraction systems and at low cost (compared to the traditional method that requires the purchase of specialized sensors), these systems must interact in uncontrolled environments, facing lighting, color, background texture as well as perspective and rotation changes in the fingerphoto capture [3,7,10]. In this paper, we propose a preview of fingerphoto extraction system with mobile cameras for capture. For now, the system is made up of background removal fingerprint segmentation stages, which are the most affected stages by the method of capture. According to what is reported in the state of the art, this is the first approach to fingerphoto background removal using deep learning, with U-Net as a proposal for finger-skin semantic segmentation. Final images are compatible and interoperable with such touch-based fingerprint systems. The results reported can provide a preview of a potential solution to biometric digital onboarding.

2   Related Work

Most of the current fingerprint recognition systems perform biometric acquisitions using touch-based devices, such as optical and solid-state sensors [9]. In Fig. 1 is shown the main components of these traditional systems, which include image acquisition, enhancement, feature extraction, and matching [15], sometimes enhancement filters are not necessary because the acquisitions images result so clearly.

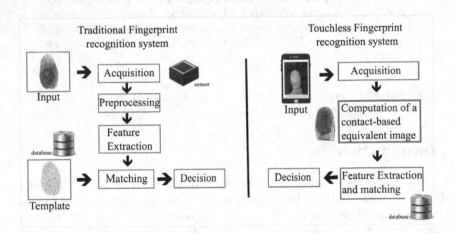

Fig. 1. Traditional fingerprint recognition system

Touchless recognition systems are generally based on images captured by CCD cameras. These images are vastly different from those obtained using touch-based acquisition sensors [4,9]. Most fingerprint recognition systems in the literature can be characterized by three primary steps: acquisition, computation of a touch-equivalent fingerprint image, and feature extraction and matching [4]. This section presents reported techniques for acquiring, segment and enhance touchless fingerprint examples based on one single capture (two-dimensional samples).

2.1 Image Acquisition

Fingerprint sensors are designed for biometric use. On the other hand, the main challenge of any fingerphoto system is that smartphone cameras are not designed for fingerprint collection [3,4,7]. Derawi et al. [11] shows the acquisition of fingerprints using two different smartphones, Nokia and HTC. They established a series of parameters for the acquisition protocol like the led flash on, ISO speed, and auto-focus tools provided by the smartphone in use, to ensure better performance of the proposal. However, they show the obstacles presented in any touchless fingerprint recognition system, such as image blurring, illuminance saturation, and perspective distortions. The recognition of fingerphoto technology with a low-resolution camera, in a fixed position and under laboratory conditions, was tested in [13], they proposed a continuous shooting mode that was used for the camera to capture multiple photos of test subjects at once in one session and select the best images. Mueller et al. [14] names his work 'Biometric Identity with low-cost equipment', where they carry out a series of experiments, with attachments and observations, to a web camera that must be a configuration to the sharpness and focus of the web camera with which to want to work to avoid reflections produced by the flash. In [8,12], the challenges to be solved in touchless fingerprint detection systems are described, such as segmentation in uncontrolled environments and the extraction of robust characteristics to the distortions associated with the absence of a support surface (the feature extraction is not part of this investigation). However, distortion corrections will possibly be a technique to employ for future work.

2.2 Background Removal

For background subtraction, available state of the art schemes rely on capturing a single finger with a controlled background [10,11,13,14]. Stein et al. [10] shows a traditional process based on image processing algorithms using the red channel of the input image and set a fixed value for thresholding to find the skin zone. In [5,10,13] present an adaptive skin thresholding algorithm to find the skin finger regions; alongside that connected component technique is used to isolate the suggested skin zone. All these proposes need a controlled background and illumination conditions. Raghavendra et al. [7] performs the background segmentation using color information by applying the Mean Shift Segmentation Algorithm in uncontrolled background, achieving a 98% of correct segmentation. Finally, [3] presents a multi-finger identification system based on HSV color space and analyzes the color probabilities frequency, helped by otsu segmentation and grab cut filter, to find the fingerprint object. In comparison [3,7] design a system without background and illumination conditions controlled, they develop a mobile app to decrease these extrinsic variations.

2.3 Fingerprint Enhancement

Another potential step in the fingerprint recognition process is image enhancement [4,9]. In traditional systems, global approaches are usually adopted during the enhancement of latent fingerprints or for the computation of synthetic fingerprint images [4]. A series of algorithms have been applied in diferent fingerphoto proposes like, geometric normalization, passband filters [10], histogram equalization [6], PCA [5], and adaptative histogram equalization [3], these algorithms has not been evaluated isolated, they have been evaluated by the FAR and ERR results by identity probes. This step is used just in case the system will be compatible with traditional fingerprint systems.

3   Methods and Materials

In this section, it is explained the foundations that constitute this research. The proposal is our company's property technology. However, we provide an overview of the main components of the system.

3.1 Fingerphoto Acquisition

It has been shown that fingerphoto acquisition is the first and most critical component of a system [4,13]. The null access to a validated and standardized database for the capture of touchless fingerprints (for commercial purposes) in uncontrolled environments, requires the generation of its database that is adapted to the specific purposes of this research. It was established a series of rules to capture the fingerphotos guided by state of the art recommendations [3,4,7,10], as illustrated in Table 1.

Table 1. Protocol's steps recommended to fingerphoto acquisition

Minimum image size/resolution	640 * 480
Flash	Mode on
Color space	RGB
Distance between fingerprints and smartphone	10–13 cm
Auto-focus	Mode on

For each individual several 10 images were required, one for each finger, and it was needed to use their smartphones. By the pandemic situation, the individuals were not in the same location, and each set of images had uncontrolled capture conditions, finally, the images were sent via email.

3.2   Background Removal

Background removal is a necessary step for fingerphoto naturally. There are documented lots of algorithms based on image processing and machine learning that could be used for this process. We can assume that just the finger is the skin zone presented in the image, the rest is the background. This research proposes to use of deep learning for the challenge involved in capturing fingerphotos for uncontrolled backgrounds.

Deep Learning. The algorithms that operate under the deep learning paradigm, provide the possibility of extracting characteristics directly from the raw input data, without the need to apply preprocessing techniques, in such a way that, through the approach of learning representations and through architectures composed of multiple layers that determine the depth of the model. Deep learning methods automatically learn successive layers of increasingly sophisticated and meaningful representations of the raw input data [8].

U-Net for Semantic Segmentation. U-Net is a CNN architecture that was designed by Ronneberger et al. [17] to be applied in related biomedical problems with the semantic segmentation of cells and neuronal structures. The goal of semantic image segmentation is to label each pixel of an image with a corresponding class of what is being represented. For fingerphotos segmentation case, the pixels that represent the finger skin and contains the fingerprint can be assigned to the finger class, any other pixel contained in the image will be assigned to the background class.

Dice Coefficient Metric. Intuitively, a successful prediction is one which maximizes the overlap between the predicted and true objects. There are different metrics for this goal, in this proposal Dice coefficient was used:

$$Dice\ coefficient = \frac{\|A \cap B\|}{\|A\| + \|B\|} \tag{1}$$

Here, A and B are two segmentation masks for a given class (but the formulas are general, that is, you could calculate this for anything, e.g. a circle and a square),$\|A\|$ is the norm of A (for images, the area in pixels), and $\cap$ is the intersection.

3.3   Fingerphoto Segmentation and Enhancement

The first step for fingerprint segmentation and enhancement established in state of the art is to convert the RGB input images in grayscale fingerphoto. A bandpass filter is applied to denoise the original image and a thresholding algorithm to highlight ridges and valleys of the fingerphoto. To read more about the Wahab filter, we can consult [18].

3.4   Equivalent Touch-Based Image

Samples captured by Smartphone cameras (touchless sensors) cannot be directly used by recognition methods designed for touch-based fingerprint images [4]. Touchless images must also be normalized to a fixed resolution. This process helps to obtain a touch-equivalent fingerprint. Touchless fingerprint now can be used by matching techniques based on minutiae features. The standard fingerprint scale is 500 dpi resolution (calibration output images captured by contact sensors). The final normalization task is registration of the minor axis of the image to the standard measure of 9/10 of the height of the final touch-equivalent image. Fingerprint images are stores in grayscale and PNG format.

3.5   NIST Fingerprint Image Quality

The performance of biometric systems is dependent on the quality of the acquired input samples [19]. The NIST Fingerprint Image Quality (NFIQ) algorithm is a standard method to assess fingerprint image quality. NFIQ algorithm was designed to predict the performance of minutiae matches. NFIQ is an expression of quality based on utility that reflects the predicted positive or negative contribution of an individual sample to the overall performance of a biometric system. NFIQ has an interval of $[1, 5] \epsilon Z$, being 1 the best value and 5 the worst, NFIQ = 1,2,3 are good quality acquisition and NFIQ = 4,5 bad objects. In addition, NIST is active in the ISO/IEC JTC1 SC37 standardization activities on biometric quality and sample conformance (ISO/IEC 29794).

4   Experiments and Results

To evaluate the proposed method for fingerphoto extraction, we perform two experiments. The first experiment consisted of background removal by the U-net skin model training and evaluation. The second experiment shows a test of the quality of the data set using the NFIQ metric.

Figure 2 shows the general process done by the method proposed, with two main phases to deployment. The input image is cropped by hand to have the

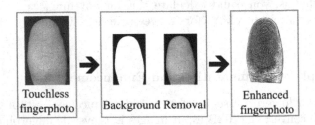

Fig. 2. Touchless fingerprint extraction method

initial ROI (finger, no finger regions). Then U-net CNN architecture is used to perform the semantic segmentation task. U-net takes the cropped image on a binary format like target and processes the background removal. U-net generates an output hoped in binary format. Then the binary mask and the input image are multiplied pixel by pixel, so the result images represent a reduced area that contains the fingerprint. The enhancement process has a minimal quantity of information to take care of, so this process details are described in Sect. 3.3.

4.1 Fingerphoto Dataset

By the pandemic situation, the fingerphoto capture was done by remote acquisition. There are fingerphotos from Mexican population located in different states. There was recollected an amount of 478 fingerphotos with a confidentiality consent established. Each participant make use of their smartphone camera and follow the recommendations mentioned in Sect. 3.1.

4.2 U-Net for Background Removal

Once the data was collected, the next step was to manually select the bounding box of the area of interest (the fingerprint), generating the expected binary segmentation map for each image. After constructing the ground truth, was the data augmentation technique performed using the Python ImageAug library, generating synthetic images with random variations of rotation and brightness. Images were re-scaled to a resolution of 512×512 pixels. Subsequently, three datasets were generated as follows: 70% train, 10% validation, and 20% test. Adam optimizer was established, with a learning rate of $1e^{-4}$ and as loss function $Binary\ cross-entropy$. To generate the performance history of the model, the metrics $accuracy$ and $Dice\ coefficient$ were used to provide accuracy in the categorization of each pixel and information about the overlap of the expected and predicted zones, respectively. The was established 500 epoch for training and a patience $= 100$. Google Colaboratory was used to train the convolutional network and take advantage of free Google GPU per session. The best performance of the segmentation model, concerning the validation set, was obtained in epoch 321 with an validation accuracy of **98.76%** with a dice coefficient of 0.9799.

A comparison was done for evaluating the viability of the proposed method versus other proposals reported in state of the art [5,6,10]. Figure 4 shows the results of two different input images with uncontrolled backgrounds. Figure 4 corresponding to the input image in RGB space, next d) Ground truth for finger segmentation is presented.

In terms of visualization c) Stein et al. [10], Terawi et al. [5], Sankaran et al. [6], methods were replicated. All those approaches are based on traditional image processing methods and are focused on color attributes that present inconsistency in the uncontrolled acquisition process. In comparison, the CNN method proposed Fig. 4d) is capable of distinguishing shapes and textures that improve finger segmentation for background removal tasks.

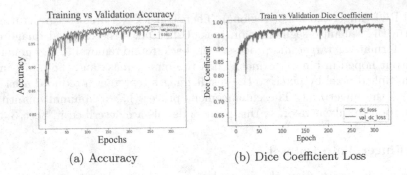

(a) Accuracy   (b) Dice Coefficient Loss

Fig. 3. Training metrics for U-net semantic segmentation

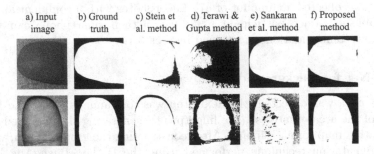

Fig. 4. Comparison between proposed method versus state of the art proposals

Then, an evaluation for each method was done by the Dice coefficient metric in the test set. Table 2 reports the evaluation results.

The table is divided into four columns: research replicated, the segmentation method, average accuracy calculated, and the dice coefficient average observed in the test set proposed. In comparison, the method proposed achieves an *accuracy* equals to 94.40% a *dice coefficient* = 0.9592 over other methods reviewed. These results represent more stable process on uncontrollable image acquisition and a much-improved reconstruction of the finger.

4.3 Fingerphoto Extraction

Neuro Verifinger and NFIQ were used to evaluate the quality and compatibility of touchless images. The short-term objective is to have a system compatible and interoperable with commercial technology, and the long-term goal is to be compatible with legacy databases. In traditional systems, NFIQ is a primary step used to check the quality of fingerprints. Despite the NFIQ metric is not compatible with touchless finger images, we decide to evaluate the dataset to know how the data is distributed. The first part was dedicated to evaluating the segmented images, just applying adaptive skin thresholding, an example is shown in Fig. 5d).

Table 2. Comparative table for finger semantic segmentation

Research	Segmentation method	Accuracy	Dice coefficient value
Stein et al.	Red-channel thresholding	81.73%	0.8608
Sankaran et al.	Adaptive skin color thresholding on CMYK color space and connected component scheme	80.36%	0.8041
Terawi and Gupta	Adaptive skin thresholding (OTSU method), morphological operations and a connected component scheme	73.44%	0.7436
Proposed method	**U-net for semantic segmentation**	**94.49%**	**0.9592**

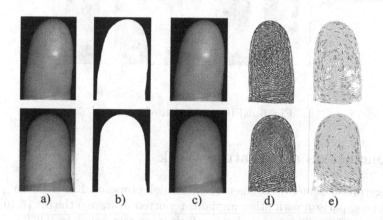

Fig. 5. Fingerphoto processing method and evaluation

Figure 5 shows the different processing steps that the input image suffers in the system. a) is the input RGB image and b) is the binary mask predicted by U-Net skin model, c) is the multiplication of a) × b) to do the background removal, then d) is the resulting image to the first valleys binarization, Neuro Verifinger is used to preview the d) minutiae.

The dataset evaluation obtained an NFIQ 1–3 score of 47% and an NFIQ score of 4–5 of 53% (Being NFIQ = 1 the highest quality score and NFIQ = 5 the worst). The next step was to apply the image enhancement process. This step is an important module established for touchless fingerprints. The graphic bar shows the NFIQ predominance in Fig. 6. NFIQ = 1–3 score of 46% and an NFIQ = 4–5 score of 54% were obtained. The results were very similar for both probes. Nevertheless, images are different if we evaluate by a visual inspection and the enhancement images have a less false minutiae presence. Images extracted are compatible with commercial biometric software, reading it and finding minutaes.

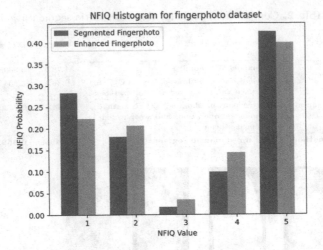

Fig. 6. NFIQ evaluation of dataset

5   Conclusions and Feature Work

The fingerphoto semantic segmentation model proposed has a better performance in comparison with other methods reported in state o the art [7,10], with an accuracy of 94.49% and a dice coefficient metric equal to 0.9592. By the test evaluation, the proposed model achieves a better performance in comparison with other approaches reported in the state of the art with a test accuracy lower than 90%. This model can improve the accuracy with more images and different types of data augmentation. Background removal is a critical step in fingerphoto extraction due to lots of background colors, shapes and brightness can be featured. U-net models can learn from the attributes, shapes, and texture of the skin color, unlike traditional methods that usually take information from a single characteristic, which are generally the skin color attributes.

The dataset evaluation shows that almost half of individuals have bad quality for enrollment purposes, next to that, trying to increase the quality a fingerprint enhancement was applied, and NFIQ quality does not improve a lot but fingerprint reconstruction using Neuro Verifinger was better. Most of the images in the dataset have not good quality, caused by remote uncontrolled acquisition, these examples does not affect to background removal stage but figerphoto enhancement is highly affected.

This research is a preview. For future work, it is necessary to continue working on data acquisition protocols to improve the background removal and achieve a better NFIQ distribution in the data recollected. It is fundamental to be compatible with commercial software and legacy databases, for that finger detection, geometric distortions, and perspective changes need to be solved. Finally, to be aware of NIST actualizations for touchless biometrics.

References

1. FINTRAIL: COVID-19 and the rush to Digital Onboarding (2020). https://www.fintrail.co.uk/news/2020/10/28/covid-19-and-the-rush-to-digitalonboarding,. Accessed 13 Apr 2021
2. Electronic Identification: Digital Onboarding: definition, characteristics and how it works (2021). https://www.electronicid.eu/es/blog/post/digital-onboardingprocess-financial-sector/en, Accessed 13 Apr 2021
3. Carney, L.A., et al.: A multi-finger touchless fingerprinting system: mobile fingerphoto and legacy database interoperability. In: Proceedings of the 2017 4th International Conference on Biomedical and Bioinformatics Engineering, pp. 139–147, November 2017
4. Labati, R.D., Piuri, V., Scotti, F.: Touchless Fingerprint Biometrics. CRC Press, Boca Raton (2015)
5. Tiwari, K., Gupta, P.: A touch-less fingerphoto recognition system for mobile handheld devices. In: 2015 International Conference on Biometrics (ICB), pp. 151–156, May 2015
6. Sankaran, A., Malhotra, A., Mittal, A., Vatsa, M., Singh, R.: On smartphone camera based fingerphoto authentication. In: 2015 IEEE 7th International Conference on Biometrics Theory, Applications and Systems (BTAS), pp. 1–7, September 2015
7. Raghavendra, R., Busch, C., Yang, B.: Scaling-robust fingerprint verification with smartphone camera in real-life scenarios. In: 2013 IEEE Sixth International Conference on Biometrics: Theory, Applications and Systems (BTAS), pp. 1–8. IEEE, September 2013
8. Valdés González, F.M.: Reconocimiento de huellas dactilares usando la cámara de un dispositivo móvil (2015)
9. Labati, R.D., Genovese, A., Piuri, V., Scotti, F.: Touchless fingerprint biometrics: a survey on 2D and 3D technologies. J. Internet Technol. **15**(3), 325–332 (2014)
10. Stein, C., Nickel, C., Busch, C.: Fingerphoto recognition with smartphone cameras. In: 2012 BIOSIG-Proceedings of the International Conference of Biometrics Special Interest Group (BIOSIG), pp. 1–12. IEEE, September 2012
11. Derawi, M.O., Yang, B., Busch, C.: Fingerprint recognition with embedded cameras on mobile phones. In: International Conference on Security and Privacy in Mobile Information and Communication Systems, pp. 136–147. Springer, Berlin, May 2011
12. Morales Moreno, A.: Estrategias para la identificación de personas mediante biometría de la mano sin contacto (2011)
13. Hiew, B.Y., Teoh, A.B.J., Yin, O.S.: A secure digital camera based fingerprint verification system. J. Vis. Commun. Image Represent. **21**(3), 219–231 (2010)
14. Mueller, R., Sanchez-Reillo, R.: An approach to biometric identity management using low cost equipment. In: 2009 Fifth International Conference on Intelligent Information Hiding and Multimedia Signal Processing, pp. 1096–1100. IEEE, September 2009
15. Maltoni, D., Maio, D., Jain, A.K., Prabhakar, S.: Handbook of Fingerprint Recognition. Springer Science & Business Media, London (2009)
16. Manders, C., et al.: Robust hand tracking using a skin tone and depth joint probability model. In: 2008 8th IEEE International Conference on Automatic Face & Gesture Recognition, pp. 1–6, September 2008
17. Ronneberger, O., Fischer, P., Brox, T.: U-net: convolutional networks for biomedical image segmentation. In: International Conference on Medical Image Computing and Computer-assisted Intervention, pp. 234–241, October 2015

18. Wahab, A., Chin, S.H., Tan, E.C.: Novel approach to automated fingerprint recognition. IEE Proc. Vis. Image Sig. Proces. **145**(3), 160–166 (1998)
19. Tabassi, E.: NIST Fngerprint Image Quality (NFIQ) Compliance Test. US Department of Commerce, National Institute of Standards and Technology (2005)
20. El cambio de paradigma en la industria de retail y centros comerciales postCovid-19 (2020). https://cincodias.elpais.com/cincodias/2020/06/22/legal/1592854555_991564.html. Accessed 13 Apr 2021
21. The digital-led recovery from COVID-19: Five questions for CEOs (2020). https://www.mckinsey.com/business-functions/mckinsey-digital/our-insights/the-digital-led-recovery-from-covid-19-five-questions-for-ceos. Accessed 13 Apr 2021

Real Time Distraction Detection by Facial Attributes Recognition

Andrés Alberto López Esquivel[1], Miguel Gonzalez-Mendoza[1]([⊠]),
Leonardo Chang[1], and Antonio Marin-Hernandez[2]

[1] Tecnologico de Monterrey, Monterrey, Mexico
mgonza@tec.mx
[2] Universidad Veracruzana, Veracruz, Mexico

Abstract. The deficit of attention on any critical activity has been a
principal source of accidents leading to injuries and fatalities. Therefore
the fast detection of it has to be a priority in order to achieve the safe
completion of any task and also to ensure the display the maximum
capabilities of the user when achieving the respective activity. While
multiple methods has been developed, a new trend of non-intrusive vision
based methodologies has been strongly picked by both the research and
industrial communities as one with the most potential effectiveness and
usability on real life scenarios.

In this paper, a new attention deficit detection system is presented.
Low-weight Machine Learning algorithms will allow the use in remote
applications and a variety of goal devices to avoid accidents caused by the
lack of attention in complex activities. This article describes its impact,
its functioning and previous work. In addition, the system is broken
down into its most basic components and its results in various evaluation
stages. Finally, its results in semi-real environments are presented and
possible applications in real life are discussed.

Keywords: Distraction detection · Machine learning · Video
processing · Facial attributes

1 Introduction

Attention deficit has been a huge problematic for users when an specific task
must get done, multiple environments rely on human's attention to reach an
specific goal and sometimes high values assets like a human life can be affected
by the lack of attention on a complex task. According to the World Health
Organization, Road traffic accidents are the 8th leading cause of death globally
with more than 1.2 million people killed in 2010 and for individuals aged between
fifteen and twenty-nine, road traffic accidents even represent the 1st leading cause
of death worldwide [14].

In an industrial setting some workplace distractions and interruptions are
unavoidable but others – if not properly controlled or regulated—could lead to

© Springer Nature Switzerland AG 2021
I. Batyrshin et al. (Eds.): MICAI 2021, LNAI 13067, pp. 265–276, 2021.
https://doi.org/10.1007/978-3-030-89817-5_20

injuries, lost productivity, and a decrease in worker morale. A study by Screen Education, Digital Distraction & Workplace Safety [2], reported that 26% of reported accidents had occurred in their workplace because someone was distracted causing in 75% of them property damage and injury or death in 58%.

What we can conclude from this statistics is that the attention deficit problematic involves multiple environments and can lead to disastrous consequences, therefore a solution is needed to reduce the risk of attention deficit on complex and high-focus tasks. To achieve this a new low-weight real time distraction detection algorithm is presented, utilizing video processing and machine learning algorithms using a simple camera installed on the workplace of a person, can be effective to detect the presence of attention deficit after a specific training of the model.

1.1   Previous Work

Currently, there are already multiple investigations that have tried to solve this problem using the latest technology innovations, One of the most complex ones are the ones that constantly monitors the driver brain signals to identify dangerous patterns like multitasking activity or secondary tasks on the brain that can distract the on-road attention of the driver [13].

On the other hand, there is a non-intrusive option that uses eye gaze detection using cameras and IR illuminators, but they opt for a more heuristics option to generate the algorithm and therefore the detection [12]. Even if both options generate correct results, it is believed that the integration of Machine Learning Techniques can boost our reliability to identify dangerous situations.

Osama worked on a Machine Learning Hierarchical Classification to identify if the driver was on a secondary task utilizing multiple data obtained from the car like speed, throttle position acceleration among others also obtained a high accuracy around 99% proving that this type of techniques can be applied successfully, in that case, their goal was to just identify which situation was and they didn't use that identification to generate any other output [8].

De Naurois attacked the problematic in a similar way than this paper is proposing but their work still involves intrusive systems into their Artificial Neural Network, this research had also a very high accuracy around 97% and utilized a similar idea of the controlled environment test bench [6].

R. Naqvi published his work on gaze detection using Deep Learning techniques on drivers and using a NIR Camera Sensor got a higher accuracy and overall better results, the paper also states that the system can also work on normal cameras with some limits, this information can be very useful and can be directly transcendent into this research [5].

Zhang worked on developing a system that recognizes the style of driving that the driver is having, even though his work is oriented in efficiency for electric cars, his methodology can be analyzed to apply its driving recognition classifier into distracted driving recognition [10].

De Castro's research was based on driver distractions, mainly focused on facial movements. He uses a camera to make an algorithm and this classifies the position of driver's face, visual, and mouth movements as input to determinate across the software if the conductor is capable to drive [4].

Summarizing all this related work, it is possible to find that there is a trend regarding the implementation of new technologies into cars to monitor the performance of the driver, also, it is important to find its relevance on multiple areas and that multiple objectives can be attacked by implementing systems like the ones instated above. After the analysis, it can be observed that all this literature can help develop new systems that can outwork past research on specific topics and environments and on the other hand, that there is a gap in the implementation of these systems allowing this proposal to generate a bigger impact.

2   Method and Data

This system only peripheral is a camera positioned in front of the person to analyze, and it will be connected to a computer that track their facial attributes and do the classification using those attributes to detect if it is distracted or not. This computer may be a personal computer (PC), laptop or can be a microcontroller embedded in a device (like cars, HMI, HMU, etc.) since the focus of this research is its low weight and real time application capabilities.

In the inside this system has 4 main modules, these modules are:

- Image acquisition
- Face detection
- Facial attributes recognition
- Classification of attention

These module's inputs and outputs are used through the whole process of detection and varies over time so a flexible system was created to adapt for multiple types of task and workplaces.

2.1   Image Acquisition

For the image acquisition module a normal camera is used (Logitech c270) but the module wouldn't have major changes for different cameras. This module will extract the pixels and reduce its resolution to 1280×720 pixels, reducing the resolution of the image can improve performance but accuracy may be impacted so 1280×720 is a good balance between both of them, but depending on the hardware that is running, resolution may change to maintain performance.

2.2   Face Detection

After the image is acquired, MediaPipe's sub-millisecond Neural Face Detection module named BlazeFace [1] is applied to start to identify a bounding box around the face, this is used to facilitate the work done by the facial attributes recognition module to work on a limited space instead of searching on the whole image, thus accelerating the process.

2.3   Facial Attributes Recognition

The following module, the Facial Attribute Recognition module uses the bounding box generated to extract variables of the face that will help the model to predict if the user is being focused to its task or if it is distracted from it. This module uses also the FaceMesh module on MediaPipe, which is based on the work of kartynnik, Artsiom, Igrishchenko and Grundman [3] which maps 468 3D estimated points using a ultra real time residual neural network architecture to the face and generates an X, Y, Z coordinates on the image to keep track of all the features for further classification. The 468 mapped points are distributed as shown in Fig. 1.

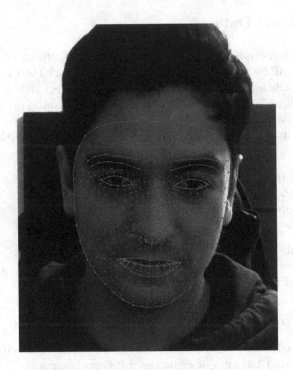

Fig. 1. Facial attributes detection

2.4   Classification

This research focus on 5 main variables which are:

- Gaze on center.
- Eyelid closure (PERCLOS).
- Head Pitch Estimation.
- Head Yaw Estimation.
- Head Roll Estimation.

Fig. 2. Data visualization dashboard

Gaze on Goal (Centered). This variable is obtained by selecting the vertices from the 468 point mesh that surround the eyes and creating a contour that will be used to extract the pixel values of the RGB image, the extracted eyes from the image will then be passed trough a segmentation stage where a simple binary segmentation will be performed to extract the iris from the rest of the eye.

Finally given the position of the iris respect to the region extracted from the points, we can measmain variables whichure the deviation from the center and therefore calculate if this deviation enters between two center limits which were previously calculated from video and data analysis of attention at driving scenarios.

After analysis and testing the center limits were established at 0.45 and 0.72 which is the middle of the data +− 1.5 standard deviations of the position of the iris when performing attention on complex activities.

This value is then converted to a Boolean variable taking the current numerical value and evaluating if the value is in a window between 0.45 and 0.72, this value will give us a relative centered view and therefore help us analyze if the user is being focus to its target.

To finally convert that value to the desired attribute we will store the last 80–90 frames of data which correspond to 3000 ms on a moving window strategy as seen in the figure below and take the sum of all the elements, this will help the model analyze a time frame of this variable instead of just fix moments, when false positives or false negatives could be detected, by doing this we can analyze the time behaviour of the variable and use it for classification. A simplified version of this Moving Window System is shown in Fig. 3.

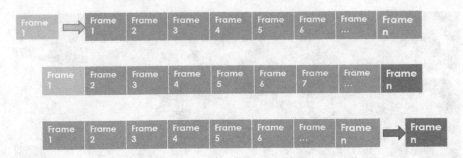

Fig. 3. Moving window system

Eyelid Closure (PERCLOS). Eyelid closure also called PERCLOS, give us important information about drowsiness of the user, to do so, we use a relative similar approach as the one described above.

First we use the points assigned by the recognition module and use it as Soukupová and Cech (2006) [11], to take the Eye Aspect Ratio (EAR) to analyze the current state of the eye, After collecting this information, this is then converted to a Boolean variable evaluating if the numerical value (width of eye divided by the height) is more than 5.0, this corresponds to the mean of the data +1.5 standard deviations of the relation of width and height of the eye of a person performing attention on complex activities like driving. This Boolean variable will be stored on the same moving window strategy with 80–90 of its last readings to analyze its time behaviour and therefore be useful for the classification task.

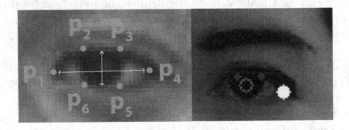

Fig. 4. EAR presented by Soukupov and Cech and its application in this research.

Head Angles (Pitch, Yaw, Roll). A Head pose estimation algorithm was implemented to give valuable variables for the classification task, using the set base model points that correspond to a face looking straight front, we can use those points as a reference and calculate the rotation of those points respect to the current points to calculate the Euler angles and use them for classification.

Given the set base points, the current model points and the intrinsic properties matrix describing the properties of the camera given by Eq. (3).

$$\begin{bmatrix} x \\ y \\ 1 \end{bmatrix} = s \begin{bmatrix} f_x & 0 & c_x \\ 0 & f_y & c_y \\ 0 & 0 & 1 \end{bmatrix} \begin{bmatrix} X \\ Y \\ Z \end{bmatrix} \tag{1}$$

We can obtain the rotation and translation vector from the OpenCv function "SolvePnP" to estimate the pose of the object in a 3D world from a 2D image. After obtaining the rotation vector, we can obtain the rotation matrix using the Rodrigues Formula which converts a rotation vector to rotation matrix or viceversa.

The Rodrigues Formula is given by:

$$R = \cos(\theta)I + (1 - \cos(\theta))rr^T + \sin(\theta) \begin{bmatrix} 0 & -r_z & r_y \\ r_z & 0 & -r_x \\ -r_y & r_x & 0 \end{bmatrix} \tag{2}$$

To obtain a rotation matrix with the form of Eq. (2):

$$\begin{bmatrix} X \\ Y \\ z \end{bmatrix} = \begin{bmatrix} r_{00} & r_{01} & r_{02} & t_x \\ r_{10} & r_{11} & 12 & t_y \\ r_{20} & r_{21} & r_{22} & t_z \end{bmatrix} \begin{bmatrix} U \\ V \\ W \\ 1 \end{bmatrix} \tag{3}$$

After finding the rotation matrix, we can obtain the Euler angles by getting the factor as a product of three rotation matrices with rotation orders of:

$$R_z R_y R_x \tag{4}$$

After obtaining the angles we store the Pitch, Yaw and Roll angles in a set of 80–90 values or on a period of 3000 ms, the average of that set is calculated, creating a moving average smoothing that gives stability to the readings.

Dataset Acquisition. The data used to train and test the model was extracted from the work of Ortega et al. [7] with the Driver Monitoring Dataset (DMD) which was devised to address different scenarios in which driver monitoring must be essential in the context of automated vehicles.

This dataset contains video of 37 participants and around 40 h of driving video in different driving scenarios, from driving simulators to real driving on a specialized car with multiple cameras, where only data from the front camera was used.

The data that was given to the model was an extract of around 40 min of video which was previously and carefully labeled in small clips that later were used to feed and train the model. Another subset of the dataset of around 45 min of video was used to test the model in multiple types of evaluations declared below.

A set of distinct video scenarios from the Driver Monitoring Dataset (DMD) is shown in Fig. 5.

Fig. 5. Upper section, RGB image while paying attention, middle section, IR image of an attention scenario, lower section, IR image of a no attention scenario.

By doing this, a variety of information of different driving situations can be gathered, which was manually selected and classified to obtain the database used to train the classifier model, being 0 an entry classified as being focus on the task and 1 as an entry classified as distracted.

Classification Model. Different classification models were implemented and measured utilizing different metrics and strategies. A regular test/train partition inside the same database was first used, after that a Cross-Validation strategy with 5 folds was implemented and finally a real time test was used, testing real time video from the DMD declared above.

The models compared in this paper are extracted from the Python Module SciKit Learn [9] and are listed below.

- K-Nearest-Neighbor (KNN)
- Linear Discriminant Analysis (LDA)
- Support Vector Machines (SVM)
- Naive Bayes (NB)
- Random Forest (RF)

3 Results

3.1 Classic Evaluation

The first evaluation phase in which the database was separated in a train/test partition with a 70/30 ratio outputted the results shown below.

Table 1. First evaluation phase results.

Classifier	Accuracy	Recall	Precision
KNN	0.9768	0.9619	0.9696
LDA	0.9799	0.9599	1.0
SVM	0.9399	0.9656	0.9393
Naive Bayes	0.9869	1.0	0.9122
Random Forest	0.9758	0.9934	1.0

3.2 5-Fold Cross Validation Evaluation

For the second evaluation phase 5-Fold Cross Validation was performed by partitioning the database in 5 evenly distributed folds, using 4 folds for training and the remaining one for testing the models, changing the fold used for testing at each iteration. The second evaluation phase using 5-fold Cross-validation outputs the results shown below:

Table 2. Second phase evaluation results.

Classifier	Iter1	Iter2	Iter3	Iter4	Iter5
KNN	0.9987	1.0	1.0	0.9486	1.0
LDA	1.0	1.0	1.0	0.9486	1.0
SVM	0.9987	1.0	1.0	0.9499	0.8908
Naive Bayes	0.9448	1.0	1.0	1.0	0.8973
Random Forest	1.0	1.0	1.0	0.9448	0.9537

3.3   Real Time Environment Evaluation

It is important to consider that in both past evaluations, the evaluations methods works with already collected data and also a thing to consider is that the time required to generate the data is not a factor that can alter its evaluation. In the third evaluation a more realistic environment was created in the driving simulator to evaluate the performance of the models, for this, the same test was made using testers on the driving simulator with different traffic environments and distractions generated sporadically.

By doing this, it is possible to evaluate the effectiveness of each model, for this a score was given to each model using a range from 0 to 100, being 100 the perfect score for a model with the best classifying score and the shortest time to output a prediction measured by its impact on the frames per second of the system.

The third and final evaluation phase output the next results:

Table 3. Third phase evaluation results.

Classifier	Evaluation	Observation
KNN	75	High execution time
LDA	85	Good overall classification and execution time
SVM	95	Good overall classification and faster execution time
Naive Bayes	80	Easily affected by disturbances
Random Forest	90	Medium to low execution time

4   Discussion

After analyzing the results of the three evaluations, it can be concluded that, even though the results were relatively close on the first two evaluations, the third evaluation was the one with the biggest significant differences among the models, giving the models with the best overall results being Linear Discriminant Analysis (LDA) and Support Vector Machine (SVM) from the SkLearn module on Python. It is important to denote that with the presented work, a group of linear and low weight classifiers can perform at high standards on real life situations and even outperform more robust models in both efficiency and also performance where in the case of real time classification is a key factor.

Both models were tested with different users and different architectures, because for the last evaluation the impact on performance of a real time test would not be as significant for a PC with high computational power, to properly test this the system was ported to a single board computer, in this case a Raspberry Pi 3 was used. By doing this, it was possible to not just measure the performance impact of the models, but also the overall performance of the system when running on limited computational resources.

4.1 Future Work

This System could be benefited with additional specific modules depending on the use context and environment, those modules could be more monitored cameras on key positions or other type of data that is not image related, for example in the case of vehicle distraction, the implementation of car-related data that is accessible from the internal network of the vehicle, or internal data managed by the embedded devices on industrial machinery can significantly increase the precision of the system.

Another module that can increase the capabilities of the system is the integration of an active learning algorithm, in which the system uses the classification done by the model on real time to increase the database on a feedback loop when high confidence on the prediction is obtained, by doing this the model can adjust to the user in a more personalized way and also would increase the flexibility of the system to adjust to different users and use-cases.

5 Conclusions

In this article an attention deficit detection system is presented, this system main objective is the prevention of risk-prone situations caused by the lack of attention on multiple environments, contexts and users. The flexibility of this system is vital for its efficacy and therefore the system should be capable to work autonomously and be able to complement itself with various systems already developed or new systems to be developed.

A system of this matter generate a high variety of use cases apart from those mentioned in this article such as industrial jobs and the automotive industry, this system could be implemented in various contexts such as the aeronautical industry, monitoring pilots, in the transportation industry, monitoring their workers and even in the medical industry with monitoring systems for patients and medical personnel in the operating room, etc.

Any monitoring system that can prevent any risk prone situation and therefore help save human lives or economic resources should be a priority for further new generation technological developments. The application of new technologies pushes the human development in many aspects of the ordinary life and a low amount of efforts are being shown in the aspects and applications presented in this article.

References

1. Bazarevsky, V., Kartynnik, Y., Vakunov, A., Raveendran, K., Grundmann, M.: Blazeface: Sub-millisecond neural face detection on mobile GPUs (2019)
2. Education, S.: Digital distraction in the workplace. In: A National Survey of Full-Time U.S. Employees, Screen Education, Cincinnat (2019)
3. Kartynnik, Y., Ablavatski, A., Grishchenko, I., Grundmann, M.: Real-time facial surface geometry from monocular video on mobile GPUs (2019)
4. Michael Jay, C., et al.: Distraction detection through facial attributes of transport network vehicle service drivers. In: Proceedings of the 2018 International Conference on Information Hiding and Image Processing, p. 112 (2018)
5. Naqvi, R.A., Arsalan, M., Batchuluun, G., Yoon, H.S., Park, K.R.: Deep learning-based gaze detection system for automobile drivers using a NIR camera sensor. Sensors (14248220) **18**(2), 456 (2018)
6. de Naurois, C.J., Bourdin, C., Stratulat, A., Diaz, E., Vercher, J.L.: Detection and prediction of driver drowsiness using artificial neural network models. Acc. Anal. Prevent. **126**, 95–104 (2019). https://doi.org/10.1016/j.aap.2017.11.038, http://www.sciencedirect.com/science/article/pii/S0001457517304347
7. Ortega, J.D., et al.: A large-scale multi-modal driver monitoring dataset for attention and alertness analysis. In: Proceedings of the European Conference on Computer Vision (ECCV) Workshops (2020, Accepted)
8. Osman, O.A., Hajij, M., Karbalaieali, S., Ishak, S.: A hierarchical machine learning classification approach for secondary task identification from observed driving behavior data. Acc. Anal. Prev. **123**, 274–281 (2019). https://doi.org/10.1016/j.aap.2018.12.005, http://www.sciencedirect.com/science/article/pii/S000145751831114X
9. Pedregosa, F., et al.: Scikit-learn: machine learning in python. J. Mach. Learn. Res. **12**, 2825–2830 (2011)
10. Qingyong, Z., Zhenfei, L., Yaru, W., Weiping, L.: Driving pattern recognition of hybrid electric vehicles based on multi-hierarchical fuzzy comprehensive evaluation. Jord. J. Mech. Ind. Eng. **14**(1), 157–163 (2020)
11. Soukupová, T., Cech, J.: Real-time eye blink detection using facial landmarks. In: 21st Computer Vision Winter Workshop (2016)
12. Vicente, F., Huang, Z., Xiong, X., De la Torre, F., Zhang, W., Levi, D.: Driver gaze tracking and eyes off the road detection system. IEEE Trans. Intell. Trans. Syst. **16**(4), 2014–2027 (2015)
13. Wang, Y.K., Chen, S.A., Lin, C.T.: An EEP-based brain-computer interface for dual task driving detection. Neurocomputing **129**, 85–93 (2014). https://doi.org/10.1016/j.neucom.2012.10.041, http://www.sciencedirect.com/science/article/pii/S0925231213009806
14. WHO: Global Status Report on Road Safety 2018. World Health Organization, Geneva (2018)

Urban Perception: Can We Understand Why a Street Is Safe?

Felipe Moreno-Vera[1](✉) ⓘ, Bahram Lavi[2](✉) ⓘ, and Jorge Poco[2](✉) ⓘ

[1] Universidad Católica San Pablo, Arequipa, Peru
felipe.moreno@ucsp.edu.pe
[2] Fundação Getulio Vargas, Rio de Janeiro, Brazil
{bahram.lavi,jorge.poco}@fgv.br

Abstract. The importance of urban perception computing is relatively growing in machine learning, particularly in related areas to Urban Planning and Urban Computing. This field of study focuses on developing systems to analyze and map discriminant characteristics that might directly impact the city's perception. In other words, it seeks to identify and extract discriminant components to define the behavior of a city's perception. This work will perform a street-level analysis to understand safety perception based on the "visual components". As our result, we present our experimental evaluation regarding the influence and impact of those visual components on the safety criteria and further discuss how to properly choose confidence on safe or unsafe measures concerning the perceptional scores on the city street levels analysis.

Keywords: Urban perception · Urban computing · Interpretability · LIME · Computer vision · Perception computing · Deep learning · Street-level imagery · Visual processing · Street view · Cityscape · ADE20K · Place pulse · Perception learning · Segmentation

1 Introduction

"Cities are designed to shape and influence the lives of their inhabitants" [14]. Various studies have shown that the visual appearance of cities plays a key role in human perception that could cause variant reactions (e.g., abnormality) in the city's environments, such as "The image of the city" [16]. A notable example is the Broken Window Theory [40] which delivers that visual signs of environmental disruption, such as broken windows, abandoned cars, trash, and graffiti, can induce social outcomes like an increase in crime levels. This theory has greatly influenced on public policy makers that lead to aggressive police tactics to control the manifestations of social and physical disorders. For example, in social experiments and studies on the perceived quality of life in the streets of New York, [11] reports the high correlation between graffiti or garbage presence and dangerous places. On the other hand, clean places present high correlation with safety places. Similar results were reported by [16,31,38] concluding that

© Springer Nature Switzerland AG 2021
I. Batyrshin et al. (Eds.): MICAI 2021, LNAI 13067, pp. 277–288, 2021.
https://doi.org/10.1007/978-3-030-89817-5_21

in places where "the rules are violated", none of the rules will be fulfilled in that place negatively influenced by the environment (e.g. graffiti, garbage). In addition, other studies have shown that the visual aspect of environments of a city affects the psychological state of its inhabitants [10,14]; Other studies show that the impact of green areas in urban cities has a positives relation to safety perception [13,30,39].

In this study, we present a methodology to analyze the influence of objects on a street image by taking into account their corresponding perceptual scores. Furthermore, we investigate machine learning techniques to alleviate the relationship between urban visual components and their perceptions score.

This paper is organized as follows: Sect. 2 explains the related works; Sect. 3 introduces our methodology in perception score analysis; Sect. 4 presents our experiments and discussion about our achieved results by providing some signs over the limitations on this research field; and finally, Sect. 5 concludes our work.

2   Related Works

Previous works have difficulty explaining the direct relation between the visual appearance of a city and its corresponding non-visual attributes. Therefore, these works focused on finding the relation between the data from police records and census statistics (e.g., robbery rate, house prices, population density, graffiti existence) with the the visual appearance of a city area. In the following, we will highlight and discuss them in details.

2.1   Urban Perception

Some studies have addressed urban perception analysis by examining different methods in computation and extracting knowledge from various resources (visual and non-visual components) – aiming to seek a proper correlation among them. The work proposed in [7] attempts to address the key question on the appearance of the Paris, "What makes Paris look like Paris?". The work was developed to compare, differentiate, and correlate the visual representation between 12 cities. Similarly, the work in [25] addressed another proposal as "What Makes London Look Beautiful, Quiet, and Happy?", which explores nearly to 700,000 street images through an online web survey. In [3], the work studied the correlation between non-visual-attributes from the city along with its visual appearance using some datasets containing the data from crimes statistics, robbery rate, house pricing rate, population density, graffiti presence, and a perception survey.

In addition, MIT Media Lab releases the PlacePulse dataset [28] which is composed of images over different streets from many capital cities like New York, Boston, Linz, and Salzburg. They also provided the associated perceptual scores for each of the images.

This work was born from the attempt to relate people's perception of a street through an online survey. This dataset conducted new studies for the problems like urban mapping [23] which performs as a classification/regression task to compare the performance of features extractors like Gist, SIFT+Fisher Vectors, and DeCAF [8]. In [20], a StreetScore approach proposed to compare a set of low-level features such as GIST, Geometric Probability Map, Text on Histograms, Color Histograms, Geometric Color Histograms , HOG 2x2, Dense SIFT, LBP , Sparse SIFT histograms, and SSIM features extractors doing a similar research on urban perception analysis. Following their methodology, a similar study was performed over the city of Bogotá, Colombia [1].

In summary, these works have difficulty in extracting information about the natural image because they use traditional image representations including Hog+Color descriptor, Locality-Sensitive Hashing, Gist, HOG+color [3], SIFT Fisher Vectors, DeCAF features [23], geometric classification map, color Histograms, HOG2x2, and Dense SIFT [20]. Besides, to train those features non-linear methods are used like SVM [4], Linear Regression [23], SVR [20], RankingSVM [24], Multi Task Learning [17], Transfer Learning based models and pre-trained networks in [1,9,12,13,18,21,41].

2.2 Model Interpretation

Model interpretation methods allow us to get insights and understand the behavior of the learning model in its training phase. In the line with those methods, there are several works whose purpose is to understand and explain predictions. Previous works such as LIME [26], SHAP [15], and Anchor [27] explain a model based on their local and global level feature components. Other approach based on gradient attribution methods used to generate feature maps of an input to provide a visual idea about the explanation like Saliency Maps [35], Gradient [34], Integrated Gradients [37], DeepLIFT [33], Grad-CAM [32], Guided Back Propagation [36], Guided gradCAM[32], and SmoothGrad [2]. These methods can ease and assist us in explaining simple/complex models aiming to identify the dependence of variables and determine whether one of them can be isolated or not; to ascertain which one has a better representation for prediction depending on the input sample.

In this work, our primary goal is to understand the behavior of urban perception. First, we extract the objects segmented from Place Pulse images denominating our visual components. Then, we train those components, considering them as a feature vector using the standard classifiers: SVC model with RBF and Linear kernels, Logistic Regression, and Ridge Classifier. Finally, we aim to understand the impact of the visual component representations by adopting the LIME (black-box) method to analyze the behavior of the predictions.

3   Methodology

In this section, we describe our methodology in urban perception analysis. Our main goal is to analyze the key components that mainly affect a security perception, such as safety. To this end, we first explain our utilized urban datasets, mainly they contain the perceptional scores for the safety criteria. Then we explain a possible solution for learning the key components to understand the importance of their presence in the images, which they have assigned as safe or unsafe.

3.1 Datasets and Data Pre-processing

PlacePulse has two versions. The first one is Placepulse 1.0 which composed by a set of street views images and provides their corresponding perceptual scores [28]. At the end of 2013, Place Pulse 1.0 was organized with a total of 73,806 comparisons of 4,109 images from 4 cities: New York City (including Manhattan and parts of Queens, Brooklyn and The Bronx), Boston (including parts of Cambridge), Linz and Salzburg of two countries (US and Austria) and three types of comparisons: *safe, wealth,* y *unique.* This dataset has been pre-processed for quick use, containing information on the position of each image (latitude and longitude), perception score for each category, an image identifier and the city to which said image belongs.

The second dataset is PlacePulse 2.0 [9] that contains a set of comparisons between image pairs, and include the latitude and longitude points for each. In addition, each comparison has the respective winner (or draw). In 2016, Place Pulse 2.0 already contained around 1.22 million comparisons of 111,390 images of 56 cities in 28 countries across the 5 continents and six types of comparisons: *safe, wealth, depress, beautiful, boring,* and *lively.* This dataset contain 8 columns: image ID (left and right), latitude and longitude (of each image), the result of the comparison, and the respective evaluated category (Table 1).

Table 1. Data summary about Place Pulse 1.0 and their respective category mean.

Place Pulse 1.0				
City	# images	*safe mean*	*wealth mean*	*unique mean*
Linz	650	4.85	5.01	4.83
Boston	1237	4.93	4.97	4.76
New York	1705	4.47	4.31	4.46
Salzburg	544	4.75	4.89	5.04
Total	4136			

We perform the method proposed by [29] to pre-process all comparisons in the dataset: for each compared image i with other images j many times in different categories, we define the intensity of perception of any image i as the percentage of times that the image was selected. Besides, the intensity of j affects i intensity. Due to this, we define the positive rate W_i (1) and the negative rate L_i (2) of an image i corresponding to a specific category:

$$W_i = \frac{w_i}{w_i + d_i + l_i} \tag{1}$$

$$L_i = \frac{l_i}{w_i + d_i + l_i} \tag{2}$$

Where, w_i is the number of wins, l_i number of loses, and d_i draws; From the Eqs. 1 and 2 we can calculate the perceptual score associated for each an image i called Q-score with notation $q_{i,k}$ in a category k:

$$q_{i,k} = \frac{10}{3}(W_{i,k} + \frac{1}{n_{i,k}^w}(\sum_{j_1} W_{j_1,k}) - \frac{1}{n_{i,k}^l}(\sum_{j_2} L_{j_2,k}) + 1) \tag{3}$$

The Eq. 3 is the perceptual score of the image i to be ranked, where j is an image compared to i, n_i^w is equal to the total number of images i beat and n_i^l is equal to the total number of images to which i lost. Besides, j_1 is the set of images that loses against the image i and j_2 is the set of images that wins against the image i. Finally, Q is normalized to fit the range 0 to 10; this scale is a standard measurement whereby one can evaluate the perceptions [22]. In this score, 10 represents the highest possible score for a given question. For example, if an image receives a calculated score of 0 for the question "Which place looks safer?" indicating that specific image is perceived as the least safe image in the dataset.

Table 2. Statistics obtained after process all comparisons from Place Pulse, containing information about images per city in each continent and the mean score for each requested category.

Place Pulse 2.0		
Continent	# cities	# images
America	22	50,028
Europe	22	38,747
Asia	7	11,417
Oceania	2	6,097
Africa	3	5,101
Total	56	111,390

(a)

Place Pulse 2.0		
Category	# comparisons	mean
Safety	368,926	5.188
Lively	267,292	5.085
Beautiful	175,361	4.920
Wealthy	152,241	4.890
Depressing	132,467	4.816
Boring	127,362	4.810
Total	1,223,649	

(b)

3.2 Visual Components Extraction

In this work, we will focus on Boston city for our experiments. We use two segmentation Network: (i) PSPNet [42] and (ii) DeepLabV3+ [5]. We define as "visual components" the object pixel presence extracted from (i) and (ii) per image of the city. As we show in Fig. 1, our main idea is to transform the percentage of objects present in each image. As we know, different images could present different % pixel ratio segmented (see Fig. 1(a), (b), (c), and (d)), we perform a Standardization of all features obtained by segmentation. We perform both extractions to compare these methods using the PSPNet as the baseline.

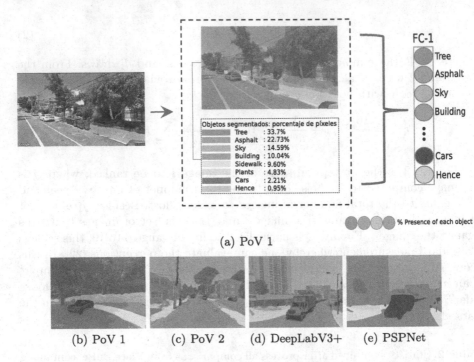

(a) PoV 1

(b) PoV 1   (c) PoV 2   (d) DeepLabV3+   (e) PSPNet

Fig. 1. (a) Input image and output features (based on the object pixel ratio segmented). (a) y (b): Different Point of View of the images, each object presence will depends of the image evaluated. (c) y (d): Different pixel ratios extracted by DeepLabV3+ and PSP-Net. As you can see in (d) DeepLabV3+ detects the street light, but failed to detect by PSPNet.

Both object segmentation extractors were trained in ADE20K [43], using networks like ResNet101 and Xception as backbone, respectively. We prefer to use the ADE20K dataset pre-trained weights instead of Pascal, COCO, or CityScapes [6], due to the number of classes between the datasets. ADE20K present 150 classes and a hierarchical tree of indoor-outdoor classes, CityScapes provides 50 classes (most of them contained in ADE20K), COCO 91 classes, and Pascal-VOC 20 classes.

Next, after extract our features, we train them using a Support Vector Classifier with 10 KFold cross-validations varying our regularization parameter l_2. To perform our classification task to predict is a street image is safe or not safe. In order to classify, we need to select subsets from each city dataset. To do this, we define a parameter called δ with a value between 0,05–0,5. This delta will create a subset using the binary labels $y_{i,k} \in \{1, -1\}$ for both training and testing as:

$$y_{i,k} = \begin{cases} 1 & \text{if } (q_{i,k}) \text{ in the top } \delta\% \\ -1 & \text{if } (q_{i,k}) \text{ in the bottom } \delta\% \end{cases} \tag{4}$$

Since we know from previous works results [19, 23, 28], we focus our study on the worst case reported, corresponding to $\delta = 0.5$ using all labels divided into safe and not safe.

4 Experiments and Discussions

This work presents a methodology to learn and explain which features have more impact on the prediction of safe or not safe categories. We perform our experiments in Boston city (1327 images) from Place Pulse 2.0 dataset. We focus only in the safety perception due to the larger number of image compared in that category per city (see Table 2(b)). We extract our visual components using both methods as mentioned above (PSP-Net, and DeepLabV3+). In order to visualize our distribution of visual components, we perform a mini-process in our pixel ratio extracted for each feature $X_{i,k}$:

$$X_{i,k} = \begin{cases} 1 & \text{if object } (k) \text{ is present in image } X_i \\ 0 & \text{if object } (k) \text{ is not present in image } X_i \end{cases} \tag{5}$$

Then, sum by column and divided by the total number of images, we got an object distribution presence in the whole Boston city for each method (see Fig. 2). Then, we train both visual components extracted by both methods (PSP-Net and DeepLabV3), we can see that both features yielded a poor performance on the dataset (see Table 3).

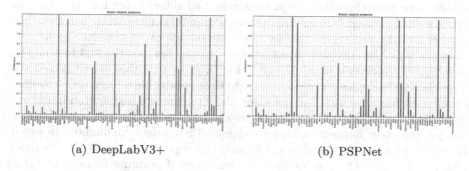

(a) DeepLabV3+   (b) PSPNet

Fig. 2. (a) DeepLabV3+ object distribution presence. (b) PSPNet Object distribution presence. As we can see, there are objects detected by only DeepLabV3 method.

Model Explanation

In this work, we want to understand why our street images are predicted as "safe" or "not safe". To do this, we use the black-box model explainer LIME: Local Interpretable Model-agnostic technique. LIME explains a black-box model by simulating local candidates close to the original prediction. By using these prediction outcomes, LIME generates a random distribution set of possible predictions based on L_2 distance called "local fidelity" taken as a reference to the original prediction.

First, based on the results of training using different models like Logistic Regression, Ridge Classifier, SVC with kernel Linear and RBF presented in the Table 3. Then, we choose the Logistic Regression model to analyze the feature importance and influence. We started analyzing the object presence in both subsets divided by safe and not safe categories.

Table 3. We report Test classification for $\delta = 0,5$ (Worst case) for each feature extractor method. We note that DeepLab and PSPNet have a poor performance in this task, but DeepLab present a better learning process.

Features	Metric	RBF-SVC	Linear SVC	Logistic regression	Ridge
PSPNet	*AUC*	0.46465	0.47036	0.465	0.48551
	ACC	0.44516	0.48065	0.47097	0.48387
	F1	0.42282	0.5752	0.50602	0.47712
DeepLabV3	*AUC*	**0.55553**	0.51255	0.51895	**0.56066**
	ACC	0.5129	0.50323	**0.52258**	0.51935
	F1	0.47018	**0.59043**	0.53459	0.52396

Second, we analyze the Feature Presence in whole images and the respective subsets corresponding to safety and not safety. At this step, we divided in two sub-subsets: miss-classified and correct classified subsets (see Fig. 3). Then, we perform the following Feature Importance, Permutation Importance in the whole dataset and the subsets divided by safe and not safe categories. In Fig. 3(a), we show the global and divided object presence, we can see that the first four objects in safety are the reverse of not safety. In Fig. 3(b), we note that object presence in miss-classified samples correspond more to the opposite class (e.g., miss-classified as safe that has presented like not safe), these regions are highlighted in red and green: red correspond to not safe class, which is present in miss-classified safe images, the same happens with miss-classified not safety highlighted in green. In Fig. 3(c) we present the object influence, this report obtained by the methods mentioned above. The results show the Influence of permute features or visual components and the relevance in the predictions, as the component 'tree'. The feature of 'tree' component has higher permutation importance in not safe class.

Our last step is to use the model explainer LIME to determine the object importance of predictions. This step shows the main results of LIME. The most highlighted ones are tree, sky, building, and road which are the most descriminant features to the predictions method in comparison with cars and sidewalk. This can be explained due to the high frequently appearance of these objects in all the images. Besides, objects like grass, plants, earth, and fence are very related to a particular category. In safety class, we have grass and plants, while in not safe class, it observes more with earth and fence components.

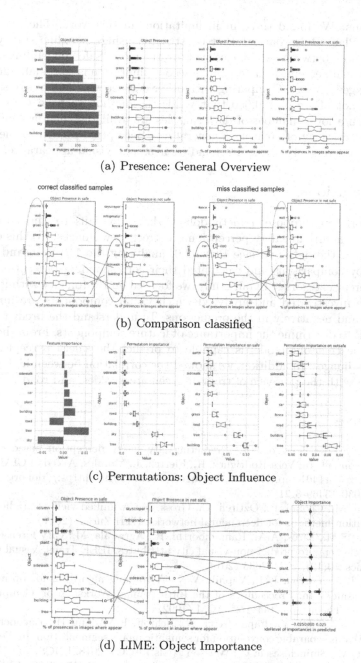

(a) Presence: General Overview

(b) Comparison classified

(c) Permutations: Object Influence

(d) LIME: Object Importance

Fig. 3. (a) General object presence in the whole dataset divided by safe and not safe categories. (b) sub-subsets of correct/miss classified. (c) Object Influence after permutations on features. (d) Object importance calculated by LIME, red lines mean the match between presence in safety/not safety and the importance in predictions.

Limitations: We found three main limitations in this work. The first one is about the Place Pulse dataset that constructed using an online survey. Each volunteer chose between two images that are the most "safe" depending on their biased personal perception criteria. The second limitation is the small number of sample images per city. Comparing with other dataset with millions of samples, in total is not above of 100,000 that yeilds the method to have week performance when a lower number of data samples are available. The last limitation is the impracticality of creating a general city perceptual predictor due to the significant difference between cities and their unique visual appearance.

5   Conclusions

In this work, we propose a methodology that allows us to understand the behavior of the urban safety perception on street view images. To this end, we pre-processed the dataset Place Pulse 2.0, analyzing the 110 thousand images obtained by comparisons and calculated their corresponding perception scores in six different categories. In our study, we focused on Boston city with its safety scores. We investigated and analyzed which visual components are impacting positively and negatively in the predictions. To understand the predictions, we used LIME to determine the importance of feature components. From the result, we conclude that our model is capable to predict the safety perception from street view images. In addition, we showed the correlation between higher safety perception with the presence of trees or green areas, skies, and roads.

References

1. Acosta, S.F., Camargo, J.E.: Predicting city safety perception based on visual image content. In: Vera-Rodriguez, R., Fierrez, J., Morales, A. (eds.) CIARP 2018. LNCS, vol. 11401, pp. 177–185. Springer, Cham (2019). https://doi.org/10.1007/978-3-030-13469-3_21
2. Ancona, M., Ceolini, E., Öztireli, C., Gross, M.: A unified view of gradient-based attribution methods for deep neural networks. ETH Zurich (2017)
3. Arietta, S.M., Efros, A.A., Ramamoorthi, R., Agrawala, M.: City forensics: using visual elements to predict non-visual city attributes. IEEE Trans. Visual Comput. Graphics **20**(12), 2624–2633 (2014)
4. Boser, B.E., Guyon, I.M., Vapnik, V.N.: A training algorithm for optimal margin classifiers. In: Proceedings of the Fifth Annual Workshop on Computational Learning Theory, pp. 144–152. ACM (1992)
5. Chen, L.-C., Zhu, Y., Papandreou, G., Schroff, F., Adam, H.: Encoder-decoder with Atrous separable convolution for semantic image segmentation. In: Ferrari, V., Hebert, M., Sminchisescu, C., Weiss, Y. (eds.) ECCV 2018. LNCS, vol. 11211, pp. 833–851. Springer, Cham (2018). https://doi.org/10.1007/978-3-030-01234-2_49
6. Cordts, M., et al.: The cityscapes dataset for semantic urban scene understanding. In: Proceedings of the IEEE Conference on Computer Vision and Pattern Recognition, pp. 3213–3223 (2016)
7. Doersch, C., Singh, S., Gupta, A., Sivic, J., Efros, A.: What makes paris look like paris? (2012)

8. Donahue, J., et al.: DeCAF: a deep convolutional activation feature for generic visual recognition. In: International Conference on Machine Learning, pp. 647–655 (2014)

9. Dubey, A., Naik, N., Parikh, D., Raskar, R., Hidalgo, C.A.: Deep learning the city: quantifying urban perception at A global scale. CoRR (2016)

10. Kaplan, R., Kaplan, S.: The Experience of Nature: A Psychological Perspective. Cambridge University Press, Cambridge (1989)

11. Keizer, K., Lindenberg, S., Steg, L.: The spreading of disorder. Science **322**(5908), 1681–1685 (2008)

12. León-Vera, L., Moreno-Vera, F.: Car monitoring system in apartments' garages by small autonomous car using deep learning. In: Lossio-Ventura, J.A., Muñante, D., Alatrista-Salas, H. (eds.) SIMBig 2018. CCIS, vol. 898, pp. 174–181. Springer, Cham (2019). https://doi.org/10.1007/978-3-030-11680-4_18

13. Li, X., Zhang, C., Li, W., Ricard, R., Meng, Q., Zhang, W.: Assessing street-level urban greenery using google street view and a modified green view index. Urban For. Urban Greening **14**(3), 675–685 (2015)

14. Lindal, P.J., Hartig, T.: Architectural variation, building height, and the restorative quality of urban residential streetscapes. J. Environ. Psychol. **33**, 26–36 (2013)

15. Lundberg, S.M., Lee, S.I.: A unified approach to interpreting model predictions. In: Guyon, I., et al. (eds.) Advances in Neural Information Processing Systems, vol. 30, pp. 4765–4774. Curran Associates, Inc. (2017)

16. Lynch, K.: Reconsidering the image of the city. In: Rodwin, L., Hollister, R.M. (eds.) Cities of the Mind, pp. 151–161. Springer (1984). https://doi.org/10.1007/978-1-4757-9697-1_9

17. Min, W., Mei, S., Liu, L., Wang, Y., Jiang, S.: Multi-task deep relative attribute learning for visual urban perception. IEEE Trans. Image Process. **29**, 657–669 (2019)

18. Moreno-Vera, F.: Performing deep recurrent double Q-learning for Atari games. In: 2019 IEEE Latin American Conference on Computational Intelligence (LA-CCI), pp. 1–4 (2019). https://doi.org/10.1109/LA-CCI47412.2019.9036763

19. Moreno-Vera, F.: Understanding safety based on urban perception. In: Huang, D.-S., Jo, K.-H., Li, J., Gribova, V., Hussain, A. (eds.) ICIC 2021. LNCS, vol. 12837, pp. 54–64. Springer, Cham (2021). https://doi.org/10.1007/978-3-030-84529-2_5

20. Naik, N., Philipoom, J., Raskar, R., Hidalgo, C.: StreetScore: predicting the perceived safety of one million streetscapes. In: 2014 IEEE Conference on Computer Vision and Pattern Recognition Workshops (2014)

21. Naik, N., Raskar, R., Hidalgo, C.A.: Cities are physical too: using computer vision to measure the quality and impact of urban appearance. Am. Econ. Rev. **106**(5), 128–32 (2016)

22. Nasar, J.L.: The evaluative image of the city (1998)

23. Ordonez, V., Berg, T.L.: Learning high-level judgments of urban perception. In: Fleet, D., Pajdla, T., Schiele, B., Tuytelaars, T. (eds.) ECCV 2014. LNCS, vol. 8694, pp. 494–510. Springer, Cham (2014). https://doi.org/10.1007/978-3-319-10599-4_32

24. Porzi, L., Rota Bulò, S., Lepri, B., Ricci, E.: Predicting and understanding urban perception with convolutional neural networks (10 2015)

25. Quercia, D., O'Hare, N.K., Cramer, H.: Aesthetic capital: what makes London look beautiful, quiet, and happy? In: Proceedings of the 17th ACM Conference on Computer Supported Cooperative Work & Social Computing (2014)

26. Ribeiro, M.T., Singh, S., Guestrin, C.: Why should i trust you?: Explaining the predictions of any classifier. In: Proceedings of the 22nd ACM SIGKDD International Conference on Knowledge Discovery and Data Mining, pp. 1135–1144. ACM (2016)

27. Ribeiro, M.T., Singh, S., Guestrin, C.: Anchors: high-precision model-agnostic explanations. In: Thirty-Second AAAI Conference on Artificial Intelligence (2018)

28. Salesses, M.P.: Place pulse: measuring the collaborative image of the city. Ph.D. thesis, Massachusetts Institute of Technology (2012)

29. Salesses, P., Schechtner, K., Hidalgo, C.A.: The collaborative image of the city: mapping the inequality of urban perception. PLoS ONE **8**, e68400 (2013)

30. Sampson, R.J., Morenoff, J.D., Gannon-Rowley, T.: Assessing "neighborhood effects": social processes and new directions in research. Ann. Rev. Sociol. **28**(1), 443–478 (2002)

31. Schroeder, H.W., Anderson, L.M.: Perception of personal safety in urban recreation sites. J. Leisure Res. **16**(2), 178–194 (1984)

32. Selvaraju, R.R., Cogswell, M., Das, A., Vedantam, R., Parikh, D., Batra, D.: GRAD-CAM: visual explanations from deep networks via gradient-based localization. In: Proceedings of the IEEE International Conference on Computer Vision, pp. 618–626 (2017)

33. Shrikumar, A., Greenside, P., Kundaje, A.: Learning important features through propagating activation differences. arXiv preprint arXiv:1704.02685 (2017)

34. Shrikumar, A., Greenside, P., Shcherbina, A., Kundaje, A.: Not just a black box: learning important features through propagating activation differences (2016)

35. Simonyan, K., Vedaldi, A., Zisserman, A.: Deep inside convolutional networks: visualising image classification models and saliency maps (2013)

36. Springenberg, J.T., Dosovitskiy, A., Brox, T., Riedmiller, M.: Striving for simplicity: the all convolutional net. arXiv preprint arXiv:1412.6806 (2014)

37. Sundararajan, M., Taly, A., Yan, Q.: Axiomatic attribution for deep networks (2017)

38. Tokuda, E.K., Cesar Jr., R.M., Silva, C.T.: Quantifying the presence of graffiti in urban environments. CoRR abs/1904.04336 (2019). http://arxiv.org/abs/1904.04336

39. Ulrich, R.S.: Visual landscapes and psychological well-being. Landscape Res. **4**(1), 17–23 (1979)

40. Wilson, J.Q., Kelling, G.L.: Atlantic monthly. Broken windows **249**(3), 29–38 (1982)

41. Zhang, F., et al.: Measuring human perceptions of a large-scale urban region using machine learning. Landscape Urban Plann. **180**, 148–160 (2018)

42. Zhao, H., Shi, J., Qi, X., Wang, X., Jia, J.: Pyramid scene parsing network. In: Proceedings of the IEEE Conference on Computer Vision and Pattern Recognition, pp. 2881–2890 (2017)

43. Zhou, B., et al.: Semantic understanding of scenes through the ADE20K dataset. Int. J. Comput. Vis. **127**, 302–321 (2018)

Continual Learning for Multi-camera Relocalisation

Aldrich A. Cabrera-Ponce[1]([⊠]) [iD], Manuel Martin-Ortiz[1] [iD],
and J. Martinez-Carranza[2,3] [iD]

[1] Faculty of Computer Science, Benemérita Universidad Autónoma de Puebla
(BUAP), 72592 Puebla, Mexico
aldrichcabrera@inaoep.mx

[2] Department of Computational Science, Instituto Nacional de Astrofísica,
Óptica y Electrónica (INAOE), 72840 Puebla, Mexico
carranza@inaoep.mx

[3] Department of Computer Science, University of Bristol, Bristol BS8 1UB, UK

Abstract. Visual relocalisation is a well-known problem in the robotics community, where chromatic images are used to recognise a place that is being re-visited or re-observed again. Due to the success of deep neural networks in several computer vision tasks, convolutional neural networks have been proposed to address the visual relocalisation problem as well. However, these solutions follow the conventional off-line training in order to generate a model that can be used to regress a camera's pose w.r.t to an input image. In this work, we present a methodology based on continual learning to address the visual relocalisation problem aiming at performing on-line model training, seeking to generate a model that is updated continuously to learn new acquired images associated with GPS coordinates. Moreover, we apply this methodology to the multi-camera case, where 8 images are acquired from a multi-rig camera, seeking to improve the localisation accuracy, this is, by using a multi-camera, we obtain a set of images observing different viewpoints of the scene for a given GPS position. Therefore, by using a voting scheme, our on-line learned model is capable of performing visual relocalisation with an accuracy of 0.78, performing at 50 fps.

Keywords: Relocalisation · Multi-camera · Continual learning · GPS

1 Introduction

Relocalisation is one of the main problems within robotics and industry that requires knowledge of the environment to perform tasks such as tracking, detection, and pose estimation. This issue mainly relates to navigation tasks, requiring additional information to recover and recognise a place using LiDAR sensors, GPS devices, and onboard cameras. However, the growth of technology has made available tools to develop suitable methods and strategies to tackle this problem. Therefore, vision algorithms have been used to localise a system using RGB

I. Batyrshin et al. (Eds.): MICAI 2021, LNAI 13067, pp. 289–302, 2021.
https://doi.org/10.1007/978-3-030-89817-5_22

images only, obtaining its camera position and orientation. These methods are widely used in visual-based pose estimation [9], VO (VO), and Simultaneous Localisation and Mapping (SLAM) [19].

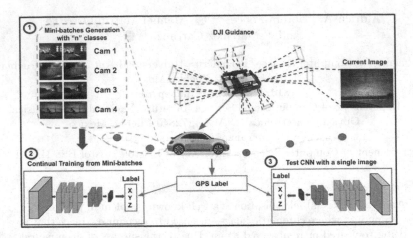

Fig. 1. Continual learning approach for multi-camera relocalisation. 1) Mini-batches generation during the navigation; 2) CNN continual training from the previously created mini-batches; 3) The available model is used to classify a current single image to give a discrete label that can be mapped to a GPS mean coordinate.

Nonetheless, deep learning has been popular in the research community in recent years due to the feasibility of working with images and processing information using fewer computational costs. Besides, it does not need to work with camera calibration and other parameters in contrast to visual methods. On the one hand, there are Convolutional Neural Networks (CNN) in combination with visual odometry and visual slam to obtain the localisation in real scenarios and the position of the camera [20]. On the other hand, novel architectures have been built for pose estimation from a single image [15] using a dataset and regression layers. However, increasing the number of views of a scenario leads to a better interpretation of the area, improving the result and the accuracy of localisation [1,9,27].

Therefore, relocalisation can be carried out using CNN architectures and a single camera to estimate camera poses and even GPS coordinates. Thus, a geolocalisation system builds upon visual relocalisation could be applied to localise a UAV while navigating outdoor areas [4]. On the other hand, if more than one camera is used (multi-camera) in combination with a CNN architecture, it is expected that pose estimation could be improved since several images offer redundancy in the CNN-based inference step [6]. These works prove that deep learning is suitable to face the relocalisation task despite the different architectures and models available. However, the main limitation is the relationship between training time and the number of examples in the dataset since a CNN network needs multiple instances to learn to perform the task. As a result, the training time increases proportionally to the amount of data, forcing us to wait for the model to become available once training has finished.

Motivated by the above, in order to carry out a relocalisation task for a system with an onboard camera, we propose a multi-camera approach with a continual learning method from multi-view images and a CNN architecture (Fig. 1). This CNN architecture is a MobileNet based on the ImageNet with an online training strategy proposed in [21], designed to create a model that can be carried out on devices without relying on big hardware. Our work aims to train a CNN continuously with information coming into the system in mini-batches, voiding the generation of a large dataset to create a learning model. The mini-batches are saved with image information and a mean GPS coordinate that represent sections along the trajectory. In this manner, the system consists of three-part: 1) the mini-batches generation with examples of the eight views from the multi-camera sensor in a real-world scenario; 2) the continual learning of the CNN while new data arrives at the ground station; 3) the evaluation of the CNN once that we have the learning model. We also carried out experiments in two ways: i) using all the images captured with the multi-camera for the testing trajectory; ii) with a voting scheme where all the images acquired by the multi-camera in a time step contribute with a vote about the corresponding location. Then the majority of votes will determine the final location to be issued as the prediction. For our experiments, we report results in terms of accuracy and processing time.

In order to present our work, this paper is organised as follows. Section 2 provides a literature study of the deep learning approaches using a multi-camera for the relocalisation task. Section 3 describes the methodology process to generate the dataset from a multi-camera sensor and associate them with GPS coordinate and the continual learning framework. Section 4 describes the experimental design and the recover GPS coordinates using a voting method for all the sequences. Finally, conclusions and future work are outlined in Sect. 5.

2   Related Work

In recent years, several works have employed multiple methods for a relocalisation task using monocular cameras and vision algorithms due to the interest of estimating the pose of a system during a mission in real-world scenarios [1,12]. Thus, these works start from mapping [16], VO (VO) [14], and Simultaneous Localisation and Mapping (SLAM) [19,23] methods to obtain the pose of a camera in an environment. For example, some of them have been carried out to get the position in indoor and outdoor scenarios for detection, tracking, and navigation [28]. Therefore, monocular cameras are the most commonly used sensors for tasks whose information can be manipulated to obtain a suitable result.

The study of using a multi-camera system has led to new methods in which multi-view information is exploited. Thus, the works presented in [26,27] created a multi SLAM system with multi-cameras for pose estimation using RGB-D camera parameters. The authors in [13,17] use the RANSAC method to localisation and pose estimation of a multi-camera system placed on top of a car. Furthermore, other works have opted to expand the margin of a multi-view scenario using two or more cameras in a structure where the information can be fully exploited [9,10]. However, these works employ a more elaborate procedure by working with the geometry of the multiple views and combining computer vision techniques to achieve the result.

Fortunately, deep learning has changed the perspective such that researchers use convolutional neural networks (CNN) and learning techniques [18]. Therefore, works such as [22,25] creating multi-camera relocalisation systems using CNN for pedestrian pose estimation. Likewise, in [6,7] use the networks AlexNet and ResNet for multi-view camera pose estimation achieving promising results for robotics. However, these systems need to generate a dataset a configure the architecture to get the highest accuracy in the output results. For example, in [5,11,29] they develop a suitable dataset to perform this kind of task with a multi-camera system, allowing object detection and relocation and autonomous navigation.

As a result, architectures have been developed to solve the relocalisation problem. PoseNet [15] has become one of the most widely used networks for position estimation using both terrestrial and aerial RGB imagery [8]. In [4] used PoseNet architecture with UAVs to estimate the GPS coordinates from aerial images in outdoor scenarios. Other works use the PoseNet network as a basis for developing new approaches in different zones [2,20] by improving the network as in [24,30] or using it for street and road location [3]. However, using a learning model requires a large dataset, leading to higher computational costs and longer training time.

Due to the limitations in deep learning, a new technique has emerged to perform online training as further information arrives. The work [21] developed a method called latent replay implemented in a MobileNet network, in which information is trained continuously without losing previous knowledge. To address the problem of multi-camera relocalisation using deep learning and avoid creating a large dataset. We propose to use the latent replay method with a dataset created with a multi-camera so that the training is performed in mini-batches, constantly updating the model when new information arrives. Thus, it will allow the system to relocate using multiple images of the scenario.

3   Methodology

The methodology used for this work consisted of 2 stages. The first stage describes the generation of the dataset from images captured with a multi-camera and their association with GPS coordinates. For the second stage, we train the CNN continually using the mini-batches generated from the navigation.

3.1   Dataset Generation

The images and the GPS data have been obtained with the Robot Operating System (ROS) using the sensor multi-camera Guidance from the DJI company

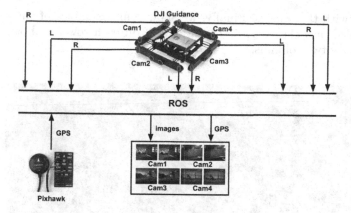

Fig. 2. Configuration for dataset collection using ROS, a multi-camera, and GPS information from the Pixhawk flight controller.

and the Pixhawk flight controller. DJI Guidance consists of five sensors with IMU, ultrasonic, and velocity data attached to two cameras (one left and another right) that support grey-scale images with 320 × 240 resolution. The Pixhawk is an autopilot flight controller used for navigation with UAVs, allowing read telemetry data, altitude, pose, GPS, and control of the rotors. In this way, we placed four sensors in a square structure to view the front, back, left, and right of a scenario and mounted the Pixhawk with a GPS device on the structure. Thus, we carried out the dataset generation through ROS, connecting the system to a ground control station. We show a diagram of the set-up for dataset generation in Fig. 2, where we receive the separate images and the GPS information, associating it to each of them.

For the collection of the dataset, we decided to put the multi-camera on top of a car while a pilot drives on the city's streets, capturing thus the training and testing data. For continual learning, we have divided the dataset into mini-batches that represents a section along the path. Each section is a mini-batch with five classes for the first section and two classes for every one of the following. On the other hand, we obtained each class by advancing approximately 50 m from the last class, obtaining at the end of the trajectory 23 classes within ten mini-batches. In this way, we can train CNN while collecting new information and creating new classes. Finally, the images captured were rescaled to 128 × 128 and we converted the GPS information from decimals (*latitude, longitude*) to metres (*x, y*). The final dataset consists of 9200 grey-scale images for training and 8648 for testing.

Hence, we generate a new class once the car moves 50 m forward, storing 50 frames per camera view. Since the multi-camera has eight views, we collected 400 frames from the video stream in total per class. In Fig. 3 we show the training dataset collection path in Google Earth with the GPS coordinates like waypoints. It is worth mentioning that each section count with two or more GPS coordinates associated with the images. Finally, we show some examples of the dataset in Fig. 4 divided into classes 1, 5, 12, 18 and 23 in each row and both cameras' views in every two columns.

Fig. 3. Training trajectory into Google Maps where each waypoint represents a GPS coordinate.

3.2 Continual Learning

We based the continual learning on the work [21] training the CNN for a multi-camera relocalisation approach using mini-batches and classes like reference points. The CNN is a MobileNet based on ImageNet applying the latent replay method in a network layer storing the essential patterns of each class. In this way, when new data arrives, the previous patterns are combined with the new ones, repeating the information of the last classes to CNN. Thus, we train the CNN with the first mini-batch while generating the new ones and apply a data augmentation in each mini-batch to improve the training.

The classes consist of a centroid that expressed a mean GPS with two o more positions encompassing the coordinates within that section. Therefore by creating a mini-batch, we saved the mean coordinate in a text file to annotate its association with the group of images. For a better understanding of this labelling per group, in Table 1 we show the class index and its corresponding mean GPS coordinate. In addition, we draw in Fig. 5 the centroids that represent the classes expressed like mean GPS coordinates, indicating the position of each group along the trajectory. Likewise, we draw the training trajectory in red and the test trajectory in green. The red and green squares indicate the captured coordinates.

Fig. 4. Example images of the 23 classes in the training trajectory. Each row denotes the following classes: 1, 6, 12, 18, and 23, and consist of 8 images, 2 for each camera.

For continual learning, we have used the following parameters: 1) SGD Optimiser; 2) Batch size: 128; 3) Epochs: 10; 4) Cross-Entropy as loss function; 5) Learning Rate: 0.001. Besides, to maintain more information about the classes, we have extended the number of patterns injected in the latent replay layer in the *pool6* layer, set up with 1500 and 3500 patterns. We argue that increase the number of patterns could improve the accuracy of the output result when an image arrives, recovering the mean GPS coordinates belonging to the image. Finally, we have two evaluation models to show the efficacy of continual learning in a multi-camera relocalisation approach.

4   Experiments and Results

We present the experiments and results obtained with the two models created from the mini-batches and GPS coordinates in a testing trajectory. Thus, the evaluation consisted of two experiments: 1) test all the images in raw; 2) test with a voting method using the eight images sequences to determine the correct class. Finally, we show the result in a graph, drawing the GPS coordinates recovered using the best model to represent the relocalisation in a trajectory section, and we show the performance time. All the tests were carried out using the grey-scale images captured with the Guidance multi-camera and ROS communication.

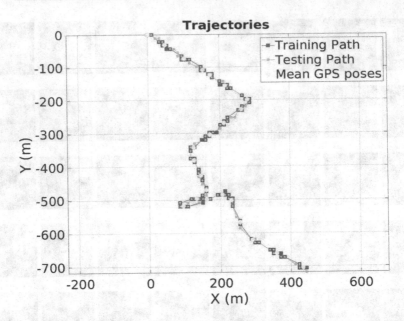

Fig. 5. Training and Testing trajectories plotted. The red squares represent the training path, green for the testing path, and yellow circles represent the Mean GPS coordinate. The training path has a traversed length of 1.4 km, testing path of 1.3 km. Thus, the total traversed length was 2.7 km. (Color figure online)

We perform the continual learning and experimental process on Ubuntu 18.04, using Pytorch 1.1.0, Opencv 3, and Nvidia Geforce 960M.

For the first experiment, we assessed the CNN with 8648 images of the testing trajectory. All the images were processed and passed to the CNN randomly, obtaining an index indicating the class belonging in the output. This index has a mean GPS coordinate that represents a section along the path in metres. Thereby, at the end of the evaluation, we save the output results in a text file and compare them w.r.t the ground truth labels. In this way, whether the output result is the same as the ground truth, we consider it correct classification and draw the GPS coordinate in a graph. Otherwise, we draw a black square in the plot to indicate the missing pose and, therefore, the lost localisation at that point. Finally, we present the accuracy results for the testing trajectory using both models in Table 2.

For the second experiment, we carried out a voting method to compare the output results for each camera sequence and determine the correct label. We argued that a voting method could improve the final result by having more consults about a chunk along the trajectory using the eight views of the multi-camera. Hence, we want to take advantage of the multi-camera system to relocalise ourselves using a query method on which class it belongs to. For the latter, we combine all the images belonging to each of the cameras by combining their classes in a single sequence, having eight sequences corresponding to the eight

Table 1. Distribution of classes' index associated with 23 labels as discrete GPS coordinate expressed in metres.

Index	Mean GPS coordinate
0	x: 0.42, y: 0.0
1	x: 47.5, y: −31.67
2	x: 101.13, y: −74.16
3	x: 152.84, y: −108.47
4	x: 173.32, y: −124.10
5	x: 215.34, y: −156.67
6	x: 278.88, y: −199.34
7	x: 261.95, y: −229.07
8	x: 211.18, y: −269.56
9	x: 171.55, y: −304.70
10	x: 125.61, y: −343.40
11	x: 126.00, y: −381.24
12	x: 144.91, y: −427.63
13	x: 159.83, y: −470.73
14	x: 132.57, y: −496.30
15	x: 95.000, y: −518.36
16	x: 160.57, y: −498.40
17	x: 205.14, y: −484.70
18	x: 230.68, y: −501.86
19	x: 254.95, y: −567.50
20	x: 290.11, y: −620.68
21	x: 359.54, y: −661.55
22	x: 435.28, y: −704.17

cameras. In this way, we have 1081 consults in every camera, that is, 1081 set of 8 images: 1081 consults × 8 views = 8648 images.

Afterwards, we ran the model 8 times for each sequence and compared the results at the end of the evaluation, choosing the class corresponding through the voting method. Therefore, the most repeat index will be the final class corresponding to the set of eight views. In contrast, if there are two or more repeat indices, we take that output as a draw and add it to the incorrect results by not clearly classifying the network output. In Fig. 6, we show the evaluation procedure for each sequence and the voting method at the end of it.

Table 2. Result of the evaluation using all the images to obtain the GPS position for each of them. The number of labels represents the correct results w.r.t the ground truth positions.

Model	Images	Correct label	Accuracy
1	8648	6108	0.7063
2	8548	6239	**0.7214**

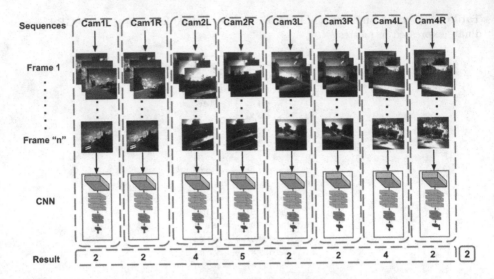

Fig. 6. Voting method to determine the label corresponding to a series of 8 images. The most repeated result will be the final label indicating the final class.

We show the accuracy result using both models in Table 3. For this evaluation, we take into account the correct consults with the voting method and show the incorrect consults and those that are tied. In addition, for a better understanding of the advantage of relocalisation with a multi-camera system, we plotted the GPS coordinates using the best model and the results of experiments 1 and 2 in Fig. 7. In this way, we present with green squares the recovered pose and with black squares the missing poses.

Table 3. Results using eight sequences with a set of 8 images per consult. We took into account the eight images of the multi-camera system with a voting method to determine the correct classification and get the GPS coordinate. The correct labels indicate the number of times the voting method obtained the corresponding coordinate for the section. Model 1 consists of the training with 1500 patterns and 2 with 3500 patterns.

Model	Consults	Correct label	Incorrect	Draw	Accuracy
1	1081	807	199	75	0.7405
2	1081	842	147	92	**0.7789**

Besides, in Fig. 7, we can appreciate that the section corresponding to classes 17, 18, 19 and 20 present some GPS coordinates missing for both experiments. However, we recovered some of those missing coordinates with the voting method by taking advantage of the multiple views in that part of the trajectory, increasing the percentage to 4%. We argue these sections present most of the missing

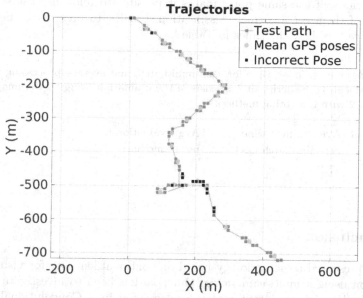

(a) Results of experiment 1 without a voting method

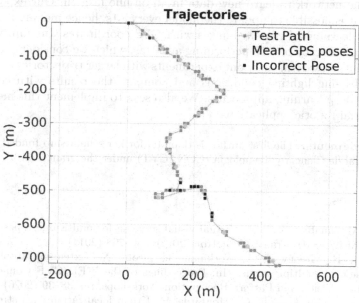

(b) Results of experiment 2 with a voting method

Fig. 7. GPS coordinates retrieved using the full imagery and queries with a voting method. A video illustrating this process can be found at video link.

labels due to the similarity of the scenarios with the previous classes. Still, having more views of the same point makes it possible to relocalise the system if at least 5 of them get a correct classification. Finally, to present the network's performance, we show the times in Table 4.

Table 4. Average training time for each mini-batch, and average processing time for image classification. Experiment 1 consists of the evaluation using all the images, and experiment 2 with the voting method.

Experiment	Avg. training time per mini-batch [sec]	Avg. evaluation time per frame [ms]	Fps
1	87.13	18.71	47.75
2	89.42	22.33	50.0

5 Conclusion

We have presented a methodology based on continual learning for visual place recognition using a multi-camera rig. Our goal has been to investigate the use of a method known as latent replay implemented in a Convolutional Neural Network, which enables the network to update its weights in certain layers, thus allowing the network to learn new data in an on-line fashion. Our experiments show that it is possible to update the model every 80 s, hence making it possible for on-line learning of camera images with GPS coordinates. In sum, we have achieved an accuracy of 0.78 performing at 50 fps, which we consider promising.

In the future, we will carry out experiments with larger trajectories and under more challenging lighting conditions and compare the results with computer vision and deep learning approaches. We also seek to implement this methodology for aerial robotic applications.

Acknowledgments. The first author is thankful for her scholarship funded by Consejo Nacional de Ciencia y Tecnología (CONACYT) under the grant 727018.

References

1. Assa, A., Janabi-Sharifi, F.: Virtual visual servoing for multicamera pose estimation. IEEE/ASME Trans. Mechatron. **20**(2), 789–798 (2014)
2. Blanton, H., Greenwell, C., Workman, S., Jacobs, N.: Extending absolute pose regression to multiple scenes. In: Proceedings of the IEEE/CVF Conference on Computer Vision and Pattern Recognition Workshops, pp. 38–39 (2020)
3. Bresson, G., Yu, L., Joly, C., Moutarde, F.: Urban localization with street views using a convolutional neural network for end-to-end camera pose regression. In: 2019 IEEE Intelligent Vehicles Symposium (IV), pp. 1199–1204. IEEE (2019)
4. Cabrera-Ponce, A.A., Martinez-Carranza, J.: Aerial geo-localisation for MAVs using PoseNet. In: 2019 Workshop on Research, Education and Development of Unmanned Aerial Systems (RED UAS), pp. 192–198. IEEE (2019)

5. Cai, Z., Yu, C., Zhang, J., Ren, J., Zhao, H.: Leveraging localization for multi-camera association. arXiv preprint arXiv:2008.02992 (2020)
6. Charco, J.L., Sappa, A.D., Vintimilla, B.X., Velesaca, H.O.: Camera pose estimation in multi-view environments: from virtual scenarios to the real world. Image Vis. Comput. **110**, 104182 (2021)
7. Charco, J.L., Vintimilla, B.X., Sappa, A.D.: Deep learning based camera pose estimation in multi-view environment. In: 2018 14th International Conference on Signal-Image Technology & Internet-Based Systems (SITIS), pp. 224–228. IEEE (2018)
8. Clark, R., Wang, S., Markham, A., Trigoni, N., Wen, H.: VidLoc: a deep spatio-temporal model for 6-DoF video-clip relocalization. In: Proceedings of the IEEE Conference on Computer Vision and Pattern Recognition, pp. 6856–6864 (2017)
9. Frahm, J.-M., Köser, K., Koch, R.: Pose estimation for multi-camera systems. In: Rasmussen, C.E., Bülthoff, H.H., Schölkopf, B., Giese, M.A. (eds.) DAGM 2004. LNCS, vol. 3175, pp. 286–293. Springer, Heidelberg (2004). https://doi.org/10.1007/978-3-540-28649-3_35
10. Geppert, M., Liu, P., Cui, Z., Pollefeys, M., Sattler, T.: Efficient 2d-3d matching for multi-camera visual localization. In: 2019 International Conference on Robotics and Automation (ICRA), pp. 5972–5978. IEEE (2019)
11. Grenzdörffer, T., Günther, M., Hertzberg, J.: YCB-M: a multi-camera RGB-D dataset for object recognition and 6dof pose estimation. In: 2020 IEEE International Conference on Robotics and Automation (ICRA), pp. 3650–3656. IEEE (2020)
12. Harmat, A., Trentini, M., Sharf, I.: Multi-camera tracking and mapping for unmanned aerial vehicles in unstructured environments. J. Intell. Robot. Syst. **78**(2), 291–317 (2015)
13. Hee Lee, G., Pollefeys, M., Fraundorfer, F.: Relative pose estimation for a multi-camera system with known vertical direction. In: Proceedings of the IEEE Conference on Computer Vision and Pattern Recognition, pp. 540–547 (2014)
14. Kasyanov, A., Engelmann, F., Stückler, J., Leibe, B.: Keyframe-based visual-inertial online slam with relocalization. In: 2017 IEEE/RSJ International Conference on Intelligent Robots and Systems (IROS), pp. 6662–6669. IEEE (2017)
15. Kendall, A., Grimes, M., Cipolla, R.: PoseNet: a convolutional network for real-time 6-DOF camera relocalization. In: Proceedings of the IEEE International Conference on Computer Vision, pp. 2938–2946 (2015)
16. Le, T., Gjevestad, J.G.O., From, P.J.: Online 3D mapping and localization system for agricultural robots. IFAC-PapersOnLine **52**(30), 167–172 (2019)
17. Lee, G.H., Li, B., Pollefeys, M., Fraundorfer, F.: Minimal solutions for the multi-camera pose estimation problem. Int. J. Robot. Res. **34**(7), 837–848 (2015)
18. Muñoz-Salinas, R., Yeguas-Bolivar, E., Saffiotti, A., Medina-Carnicer, R.: Multi-camera head pose estimation. Mach. Vis. Appl. **23**(3), 479–490 (2012)
19. Mur-Artal, R., Montiel, J.M.M., Tardos, J.D.: ORB-SLAM: a versatile and accurate monocular slam system. IEEE Trans. Robot. **31**(5), 1147–1163 (2015)
20. Ott, F., Feigl, T., Loffler, C., Mutschler, C.: ViPR: visual-odometry-aided pose regression for 6DoF camera localization. In: Proceedings of the IEEE/CVF Conference on Computer Vision and Pattern Recognition Workshops, pp. 42–43 (2020)
21. Pellegrini, L., Graffieti, G., Lomonaco, V., Maltoni, D.: Latent replay for real-time continual learning. arXiv preprint arXiv:1912.01100 (2019)

22. Remelli, E., Han, S., Honari, S., Fua, P., Wang, R.: Lightweight multi-view 3D pose estimation through camera-disentangled representation. In: Proceedings of the IEEE/CVF Conference on Computer Vision and Pattern Recognition, pp. 6040–6049 (2020)
23. Sewtz, M., Luo, X., Landgraf, J., Bodenmüller, T., Triebel, R.: Robust approaches for localization on multi-camera systems in dynamic environments. In: 2021 7th International Conference on Automation, Robotics and Applications (ICARA), pp. 211–215. IEEE (2021)
24. Valada, A., Radwan, N., Burgard, W.: Deep auxiliary learning for visual localization and odometry. In: 2018 IEEE International Conference on Robotics and Automation (ICRA), pp. 6939–6946. IEEE (2018)
25. Xu, Y., Roy, V., Kitani, K.: Estimating 3D camera pose from 2D pedestrian trajectories. In: 2020 IEEE Winter Conference on Applications of Computer Vision (WACV), pp. 2568–2577. IEEE (2020)
26. Yang, A.J., Cui, C., Bârsan, I.A., Urtasun, R., Wang, S.: Asynchronous multi-view slam. arXiv preprint arXiv:2101.06562 (2021)
27. Yang, S., Yi, X., Wang, Z., Wang, Y., Yang, X.: Visual slam using multiple RGB-D cameras. In: 2015 IEEE International Conference on Robotics and Biomimetics (ROBIO), pp. 1389–1395. IEEE (2015)
28. Yang, Y., Tang, D., Wang, D., Song, W., Wang, J., Fu, M.: Multi-camera visual slam for off-road navigation. Robot. Auton. Syst. **128**, 103505 (2020)
29. Yogamani, S., et al.: WoodScape: a multi-task, multi-camera fisheye dataset for autonomous driving. In: Proceedings of the IEEE/CVF International Conference on Computer Vision, pp. 9308–9318 (2019)
30. Zhang, R., Luo, Z., Dhanjal, S., Schmotzer, C., Hasija, S.: PoseNet++: a CNN framework for online pose regression and robot re-localization (2018)

Facing a Pandemic: A COVID-19 Time Series Analysis of Vaccine Impact

Benjamin Mario Sainz-Tinajero[✉], Dachely Otero-Argote,
Carmen Elisa Orozco-Mora, and Miguel Gonzalez-Mendoza

Tecnologico de Monterrey, School of Engineering and Science, Atizapan de Zaragoza,
52926 Estado de Mexico, Mexico
a01362640@itesm.mx

Abstract. Economics, social encounters, and most importantly, human life was deeply affected by the COVID-19 pandemic. The international state of contingency led vaccine manufacturers worldwide to double their efforts in developing a vaccine that would be influential in contagion rate decrease. This paper offers an overview of the worldwide vaccination process and its impact on confirmed cases. In this work, we present a time series analysis methodology to predict which country group using each of the most popular vaccines will have the less steep curve of confirmed cases in a time window of 21 days. The experiments led to 94% of the data fitting our models on average, leading to a confident suggestion on the vaccine related to the less steep foreseeable contagion slope.

Keywords: COVID-19 · Vaccination · Time series analysis · Data mining

1 Introduction

The capability of governmental entities and organizations to mitigate further effects in natural and human-made disasters could potentially save thousands of lives in the response and recovery phases [11]. To contribute to the analysis of the COVID-19 pandemic containment caused by the Novel Coronavirus SARS-CoV-2, we analyzed three datasets related to the relief and reaction stages of the pandemic. This work focuses on analyzing the number of cases in relation to the global vaccination progress. The mentioned factors are directly influential to the relief of the propagation of the virus, and in this work, we present our approach to the systemic vaccination efforts, arguing its effects on the contagion rate.

The main objective of our proposal is to analyze the behavior of countries using the five vaccines with the most significant worldwide presence. After cleaning and preparing the data, we modeled five time series of confirmed cases, one for each vaccine, and the predictions were used to determine which vaccine is helping the most in *flattening the curve*. This was achieved by training a second set of linear models using our predictions of confirmed cases for the 21 days posterior to the last entry in the collected data and analyzing their coefficients.

© Springer Nature Switzerland AG 2021
I. Batyrshin et al. (Eds.): MICAI 2021, LNAI 13067, pp. 303–314, 2021.
https://doi.org/10.1007/978-3-030-89817-5_23

The rest of the paper is organized as follows. In Sect. 2, we describe multiple approaches to tackle the vaccination problem for predicting and observing its behavior over time. Section 3 goes into detail on the proposed methodology, and the results are further explained and discussed in Sect. 4. In Sect. 5, we point out the final remarks and conclude with the outcomes of this work.

2 Related Work

This development is focused on the relation between COVID-19 confirmed cases and the COVID-19 world vaccination progress data [4,8]. We will analyze and discuss three studies related to the mentioned task that are useful for data-driven decision-making at a micro or even a macro level.

2.1 COVID-19 World Vaccination Progress with Tableau

Sevgi Sy is the author of this work and is available at [12]. In this work, she presents visualizations regarding the vaccine manufacturer used per country. The visualizations suggest that the Pfizer/BioNTech vaccine has been used in different combinations in a high number of countries, and Oxford/AstraZeneca is used in some of the countries with the highest total vaccination figures.

The results of this study show that the countries using Pfizer/BioNTech-Moderna, the most distributed combination, show a downward trend of both cases and deaths after the first half of January 2021. The author also concluded that the countries with the highest daily vaccination applications are the US and the UK. This solution also examined the movement of infection trend after the beginning of the vaccine process and concluded that the daily cases decreased and the deaths remained constant, whereas the number of vaccinations increases rapidly.

2.2 COVID-19: Can We Predict the Future?

Artem Pozdniakov submitted this solution in March 2021 [7]. He started with visualizations of the data by plotting the number of vaccinated citizens per country. He noted that China and the US had more significant quantities of vaccinated people than the rest. In contrast, Indonesia, Argentina, and Ecuador had the lowest vaccinated citizens.

Then, he made a visualization for the number of vaccinated people for each kind of vaccine, and he found that the Pfizer/BioNTech vaccine seems to be the most popular and the most widespread one. Next, he proceeded to predict the weekly quantity of vaccinated people. Finally, he performed a model benchmark, and he compared the results with the original times series and got an acceptable difference.

2.3   COVID-19 EDA: Man Vs. Disease

This solution was submitted by Pawan Bhandarkar in February 2021 and can be found at [1]. The first visualization of this work shows the number of people who received the first vaccine dose; the chart illustrates that the US, China, and the UK are the leading countries. Then, the author points out the main combinations of vaccines that are used around the world. Moderna-Pfizer/BioNTech is the most used combination, followed by Sinopharm/Beijing-Sinopharm/Wuhan-Sinovac.

The vaccination rate is also analyzed in this study, and the US, China, and the UK are the countries with the highest rate. According to the visualizations developed by the author, the number of vaccinations is highly superior to the number of new cases. This solution also presents the popular vaccines by country. Moderna and Pfizer/BioNTech are the most-used vaccines in countries such as the US and Canada. Mexico has been mainly using Pfizer/BioNTech and Oxford/AstraZeneca, while in Europe, Moderna, Pfizer/BioNTech, and Oxford/AstraZeneca are the most popular.

2.4   Remarks

All three solutions make very purposeful use of data and offer sufficient material to develop this work. Each of them provides knowledge on the vaccination progress, even though there are some gaps to be tackled. In the case of the solutions of Sevgi Sy and Bhandarkar, they offer dynamic visualizations that allow us to interact with the information. Nevertheless, these solutions are just focused on data visualization, and it would be useful to see some prediction of vaccine impact on confirmed cases of COVID-19. In the case of the solution of Pozdniakov, we can see an important prediction of the vaccination efforts and their extent over time. Still, it would be an informative complement to see how the number of confirmed COVID-19 cases changes to appreciate the importance and effect of the vaccination per manufacturer.

3   Methodology

This study is based upon the structure of the CRISP-DM methodology (cross-industry standard process for data mining), which breaks up data science tasks into six phases: business understanding, data understanding, data preparation, modeling, evaluation, and deployment [15]. Up to this point, we can establish that the business understanding phase took place in Sect. 2, where we described multiple approaches to tackle similar problems. In the following subsections, we will go into detail about the remaining steps of the study.

3.1   Data Understanding

COVID-19 Daily Updates. Table 1 shows the structural details of the collected data for this work. The first dataset to be used is the COVID-19 daily updates, which contains curated information from the 2019 Novel Coronavirus COVID-19 (2019-nCoV) Data Repository by Johns Hopkins CSSE [8]. This dataset is a convenient compilation of daily updates from the original repository for representing the geographical propagation of the virus from January 2020 up to May 2021 [2]. The attributes of each object are the date, country, number of deaths, recovered, and confirmed cases. This dataset is useful for comparing the response effectiveness of the virus containment policies per country.

Table 1. Datasets used for analysis.

Dataset	Objects	Features	Size (MB)
COVID-19 daily updates	1,578,095	6	72.67
COVID-19 world vaccination progress	6,500	16	1.48

Figure 1 contains ten boxplots with the countries presenting the highest daily standard deviation on confirmed cases. These countries might have unstable relief management during the pandemic. France, India, and the United Kingdom seem to have multiple peaks and days with much larger confirmed cases that range out of the fourth quartile. This dataset intends to capture any underlying structures that will help us determine a pandemic we have not yet understood completely. All of the visualizations in this document were computed using the Seaborn library in the Python programming language along with the Pandas library for data manipulation [9,13,14].

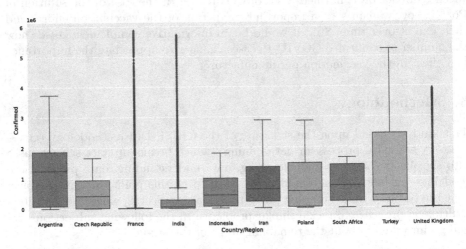

Fig. 1. Box plots of the countries with highest standard deviation of daily confirmed cases.

COVID-19 World Vaccination Progress. The impact of the COVID-19 pandemic implied the inherent challenge of driving research and development to an unprecedented acceleration rate. Vaccine technology platforms entered human clinical testing on March 16th, 2020, only five days after COVID-19 was declared as a pandemic by the World Health Organization [4,16]. The second dataset was also obtained from the COVID-19 World Vaccination Progress repository [4], which collects and merges the raw data from [5,10].

The dataset contains geographical information of 196 countries. The main attributes of this dataset are the number of vaccinations applied per day and the daily vaccinations per hundred, as well as the number of fully vaccinated people considering the immunization scheme. Furthermore, we can find the kind of vaccines applied to the population in each country. Therefore, this dataset is useful for in-depth research on the use of each vaccine per region, relative and absolute vaccination rates, and overall highest-performing countries in this matter.

Figure 2 plots the ten countries with the highest standard deviation regarding their number of vaccinations applied daily, and overall, it has noticeably fewer outliers than Fig. 1. France, India, and Indonesia appear in both Figs. 1 and 2, suggesting that both their spread of the disease and their vaccination efforts have been unstable. We can visualize two countries from Latin America (Brazil and Mexico), five from Europe (France, Germany, Italy, Spain, and the United Kingdom), two from Asia (India and Indonesia) with the remarkable absence of China, where the pandemic started, and the United States. The countries with the highest deviations in this metric are India, and the United States, which are both in the top three countries with the largest populations worldwide [17]. Large countries seem to differ from day to day in their vaccination efforts.

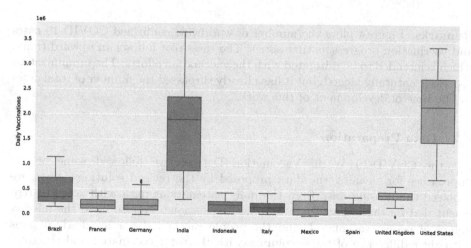

Fig. 2. Box plots of the countries with highest standard deviation in daily vaccinations.

Figure 3 plots the ten countries with the highest ratio between the number of fully vaccinated people (complete schema of either one or two doses) in the territory and its population. Gibraltar shows outstanding results with the Pfizer/BioNTech vaccine, with a ratio over 100%, meaning that they have completed the vaccination schema of a greater number of people than their population size, which is 33,682 [17]. Even though having a higher standard deviation concerning the daily application of vaccines, the United States seems to have already fully vaccinated around 30% of its population, or at least has applied that number of immunizations.

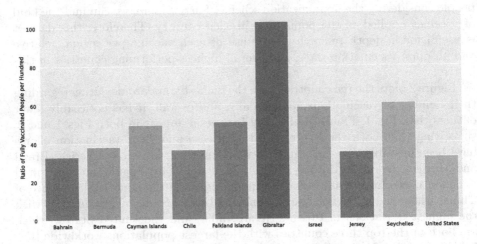

Fig. 3. Countries with the highest complete vaccination vs. population ratio.

Remarks. Figure 4 plots the number of worldwide confirmed COVID-19 cases and vaccination progress as time series. The cases plot follows an upward trend, which is intended to be mitigated with the vaccination efforts. The immunization process is naturally lagged, but it has already surpassed the number of total cases by the time of development of this work.

3.2   Data Preparation

For the COVID-19 World Vaccination Dataset, we followed some of the approaches for cleaning the data proposed by the revised solutions. First, we replaced the missing data from the daily vaccinations data with zeros to represent that on those days there were no vaccinations, we also filled the missing values of people vaccinated, and the people vaccinated per hundred columns with their difference of those columns with the total vaccinated and the total vaccinated per hundred columns mean values.

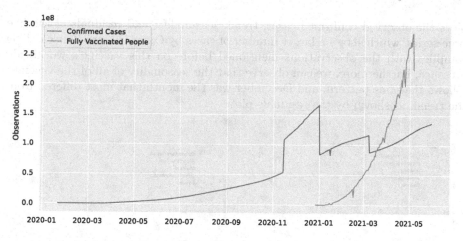

Fig. 4. Vaccination and worldwide COVID-19 confirmed cases time series.

For the COVID-19 Daily Updates Dataset, we proceeded by getting rid of the missing data, and we dropped some of the columns that were not useful for our case study. We also noted that the dataset had information from 2021, but the rows had the country's name missing. Therefore, we got rid of these data and compiled the information with the updated reports from the 2019 Novel Coronavirus COVID-19 (2019-nCoV) Data Repository by Johns Hopkins CSSE [2].

Based upon our resources, the vaccines manufactured by Johnson&Johnson, Moderna, Oxford/Astra-Zeneca, and Pfizer/BioNTech were applied in at least one country in 131 days up to the development of this study, whereas Sinovac is behind them with 114 days applied. The leading vaccines worldwide presented are Oxford/Astra-Zeneca, which is present in 91 countries, Pfizer/BioNTech in 88, Moderna in 41, Sinopharm/Beijing in 26, and Sinovac in 21. Thus, we decided to move forward with the analysis using the following vaccines considering a balance between these two factors:

- Oxford/Astra-Zeneca
- Pfizer/BioNTech
- Moderna
- Johnson&Johnson
- Sinovac

The main purpose of the approach we are presenting is to create five time series of confirmed cases, each for one of the mentioned vaccines to be analyzed, and model them to predict confirmed cases for 21 days. This would let us examine and distinguish the consequences of the presence of a vaccine in the set of countries it is applied. The five time series was formed by separating and summing the total daily cases of countries using the selected vaccines. For instance, the United States uses three different vaccines, and its daily cases are added to three time series: Johnson&Johnson, Moderna, and Pfizer/BioNTech. Figure 5 plots the observed confirmed cases, trend, seasonality, and residuals of the five time series, which shows a larger number of cases by Oxford/Astra-Zeneca. This complies with the observations mentioned before on this vaccine's worldwide presence. Furthermore, we can observe that the seasonality of all of the vaccines follows the same pattern, and December was the month that most differed from the trend, as shown by the residuals plot.

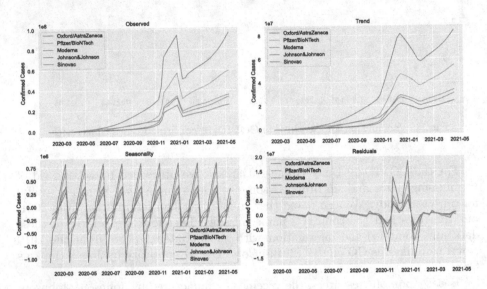

Fig. 5. Weekly decomposition of the time series divided by vaccine and the summed observations per country.

3.3 Modeling

ARIMA Model. Five *Autoregressive Integrated Moving Average* (ARIMA) models were used for training each of the time series. We selected the ARIMA model due to its fitting flexibility and its literature presence. In addition, this model is capable of handling the stochastic nature of time series and is helpful for forecasting with a high degree of accuracy [3].

Experiments. In this phase, we iterated through 729 parameter combinations for training the models. The train test split was held for each of the five time series, splitting them into two parts, where the test set was constituted by the last three weeks in the data, or 21 days. The p (Autoregressive), d (Integrated), q (Moving average), and the first three seasonality order non-negative parameters were tested in a range of 0 to 2 with the seasonality value set to 7. The metrics for comparing the ARIMA models were AIC and BIC, which are to be minimized. The AIC (Akaike's Information Criterion) estimates the relationship between the true likelihood function of the data and the model's fitted likelihood function, with a lower value closer to the truth. BIC is a Bayesian estimate function of a true model's posterior probability, with a lower value considered to be more likely the true model. The resulting best models per vaccine are presented in Table 2. The autoregressive part of the models is set to 0, so a linear regression is not happening. The integrated part will subtract the time series once to make it stationary, and the moving average does not impact the predictions.

Table 2. Best models per vaccine using the training data.

Manufacturer	p	d	q	Seasonality order	AIC	BIC
Oxford/AstraZeneca	0	1	0	(1, 1, 1, 7)	14,941.4585	14,953.8391
Pfizer/BioNTech	0	1	0	(0, 1, 1, 7)	14,119.2497	14,127.4593
Moderna	0	1	0	(0, 1, 1, 7)	13,682.2579	13,690.4674
Johnson & Johnson	0	1	0	(0, 1, 1, 7)	13,413.6959	13,421.9055
Sinovac	0	1	0	(0, 1, 1, 7)	13,926.9017	13,935.1555

4 Results

4.1 Model Evaluation

The best model per vaccine was then tested using the holdout data, which stands for the evaluation phase in the CRISP-DM methodology. The predictions of the last 21 days of the records, or the holdout data, returned the metrics that are presented in Table 3. The forecasts for the last 21 days in the collected data are used to compute the Mean Squared Error, and Root Mean Squared error. The lowest root mean squared value of 198,000 confirmed cases achieved by the Johnson&Johnson implies that the model could predict the average daily confirmed cases within 198 thousand of the actual cases.

Table 3. Evaluation metrics of the ARIMA Models using the holdout data.

ARIMA Model	MSE (10^9)	RMSE ($x10^6$)
Oxford/AstraZeneca	23,635.591	4.862
Pfizer/BioNTech	1,006.123	1.003
Moderna	67.712	0.260
Johnson & Johnson	39.360	0.198
Sinovac	1,809.018	1.345

The observed data are plotted in Fig. 6 from January to the first week of May 2021, followed by the testing data (last three weeks of May) and the ARIMA model's predictions of confirmed cases per vaccine. We included both training and testing data to visualize how the predictions behave in the last 21 days of the records. An additional vertical line is included for differentiating the train test split.

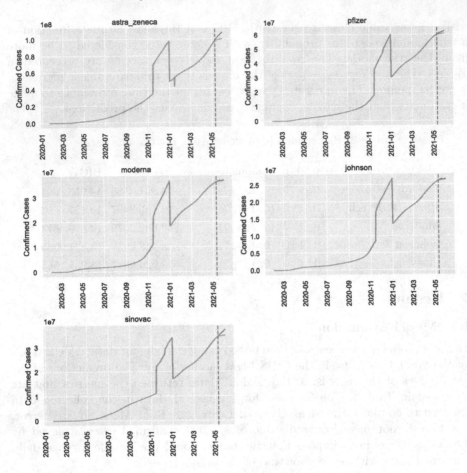

Fig. 6. Predictions of the five ARIMA models.

4.2   Deployment

We used the trained models to predict 21 days after the final entry in the data (i.e., May 30th to June 6th, 2021). The generated time series were used to train a linear regression model for analyzing the slopes in the ARIMA predictions. The linear models were coded using the Scikit-learn library [6]. Table 4 displays the obtained results. All of the linear models returned very high R^2 score values and p-values smaller than 0.05, implying that the coefficients are statistically significant. The smallest coefficient was obtained by the Johnson&Johnson vaccine, suggesting that for the time window of the predictions, it is the vaccine group of the countries with the least steep cases slope. The followed methodology offers a sound approach for finding the vaccine for which countries are prone to see either a higher or lower case rate for three weeks after the data's limit. The source code of this analysis is available in a Python implementation at https:// github.com/benjaminsainz/covid-19-vaccine-tsa.

Table 4. Evaluation Metrics of the Linear Regression models used in the deployment phase.

Linear regression model	Coefficient	R^2 score
Oxford/AstraZeneca	52,397.058	0.941
Pfizer/BioNTech	37,809.104	0.945
Moderna	21,708.588	0.944
Johnson & Johnson	15,075.696	0.943
Sinovac	20,230.047	0.945

5 Conclusion

Our proposal is capable of predicting the slope of the predicted cases of the countries using the vaccines with the greatest presence around the world, aiming to reduce uncertainty by using time series analysis. Thus, the objectives of this work were fulfilled, and we suggest that the vaccine related to the countries with the less steep cases slope according to our predictions is Johnson&Johnson. Our methodology is applicable for any pandemic stage and could be replicated again at a future stage. However, we cannot ensure causality, as the decrease in cases has to do with multiple factors outside the scope of this work. Future work intends to analyze the relation of the most used vaccines and the number of deaths at a macro level.

References

1. Bhandarkar, P.: COVID-19 EDA: man vs disease, February 2021. https://www.kaggle.com/pawanbhandarkar/covid-19-eda-man-vs-disease
2. Dong, E., Du, H., Gardner, L.: An interactive web-based dashboard to track COVID-19 in real time. Lancet Infect. Dis. **20**(5), 533–534 (2020)
3. Ho, S.L., Xie, M.: The use of ARIMA models for reliability forecasting and analysis. Comput. Ind. Eng. **35**(1–2), 213–216 (1998)
4. Le, T.T., et al.: The COVID-19 vaccine development landscape. Nat. Rev. Drug Discov. **19**(5), 305–306 (2020)
5. Mathieu, E., Ritchie, H., Ortiz-Ospina, E.: A global database of COVID-19 vaccinations (2021). https://github.com/owid/covid-19-data/
6. Pedregosa, F., et al.: Scikit-learn: machine learning in Python. J. Mach. Learn. Res. **12**, 2825–2830 (2011)
7. Pozdniakov, A.: COVID-19: can we predict the future? March 2021. https://www.kaggle.com/gpreda/covid-world-vaccination-progress/tasks?taskId=3176
8. Preda, G.: COVID19 daily updates, February 2021. https://www.kaggle.com/gpreda/coronavirus-2019ncov/version/140
9. Reback, J., et al.: chris b1, h vetinari: pandas-dev/pandas: Pandas 1.2.4, April 2021. https://doi.org/10.5281/zenodo.4681666
10. Ritchie, H., et al.: Coronavirus pandemic (COVID-19). Our World in Data (2020). https://ourworldindata.org/coronavirus
11. Rolland, E., Patterson, R.A., Ward, K., Dodin, B.: Decision support for disaster management. Oper. Manage. Res. **3**(1–2), 68–79 (2010)

12. Sevgi, S.: COVID-19 world vaccination progress with Tableau, February 2021. https://www.kaggle.com/sevgisarac/covid-19-world-vaccination-progress-with-tableau
13. Van Rossum, G., Drake, F.L.: Python 3 Reference Manual. CreateSpace, Scotts Valley, CA (2009)
14. Waskom, M.L.: seaborn: statistical data visualization. J. Open Source Softw. **6**(60), 3021 (2021). https://doi.org/10.21105/joss.03021
15. Wirth, R., Hipp, J.: CRISP-DM: towards a standard process model for data mining. In: Proceedings of the 4th International Conference on the Practical Applications of Knowledge Discovery and Data Mining, vol. 1. Springer, London (2000)
16. World Health Organization: Who director-general's opening remarks at the media briefing on COVID-19, 11 March 2020 (2020). www.who.int/director-general/speeches/detail/who-director-general-s-opening-remarks-at-the-media-briefing-on-covid-19--11-march-2020
17. Worldometers.info: Real time world statistics, May 2021. https://www.worldometers.info/

COVID-19 on the Time, Countries Deaths Monitoring and Comparison Dealing with the Pandemic

Juan J. Martínez[✉], Alexander Gelbukh, and Hiram Calvo

Centro de Investigación en Computación, Instituto Politécnico Nacional, Av. Juan de Dios Bátiz S/N, 07738 Mexico City, Mexico
gelbukh@gelbukh.com

Abstract. This paper aims to implement time series normalization methods in order to compare situations for top countries with more deaths due to COVID-19 over the time. In this work, a dashboard set was created using Power BI for analytical dashboards, is tracked the daily data dynamics of the pandemic which is collected and represented graphically. For all data collecting were developed various web scraping scripts mainly based on bash scripting and python which extract data from specific web sites and once the initial inputs are obtained, the transforming process is started. This includes making aggregations, key performance indicators, correlations and mappings giving the facility to use that transformed data for future works. The data has been collected and treated for study from different sources [1–4]. Additionally, all the results and final data after transformations are being published on a daily basis in the following sites [5–8].

Keywords: COVID-19 · Daily COVID deaths · Coronavirus · SARS-CoV-2 · Time series comparison

1 Introduction

Coronavirus disease 2019 (COVID-19) is a coronavirus-borne infection that is caused by critical respirational sickness (SARS-CoV-2). It has spread quickly over the planet, with the year 2020 being the most important. A cluster of pneumonia cases was recorded in Wuhan, Hubei Province, on December 31, 2019, according to the Wuhan Municipal Health Commission in China. WHO (World Health Organization) reported on social media on January 4th, 2020, that a cluster of pneumonia cases – with no deaths – had been detected in Wuhan. Officials confirmed a case of COVID-19 in Thailand on January 13, 2020, the first occurrence outside of China. WHO declared COVID-19 a pandemic on March 11, 2020, after being gravely worried by the frightening levels of spread and intensity, as well as the alarming levels of inaction.

Coughing, sneezing, and chatting are the most common ways for the virus to transmit between people while they are in close proximity. Rather than traveling great distances in the air, the droplets normally fall to the ground or onto surfaces, however they can stay airborne for tens of minutes in some cases. People can get infected by contacting

© Springer Nature Switzerland AG 2021
I. Batyrshin et al. (Eds.): MICAI 2021, LNAI 13067, pp. 315–326, 2021.
https://doi.org/10.1007/978-3-030-89817-5_24

a contaminated surface and then touching their face, albeit this is a rare occurrence. COVID-19 is also known to be most contagious in the first three days after the onset of symptoms, though it can also spread before symptoms occur and from persons who do not have any.

Many data science approaches are used and visualized in this effort to reveal data analytics so that individuals may keep informed and compare virus dynamics across countries.

2   Focus on Deaths Instead of Reported Infections

Since the beginning of the pandemic, we started to track and collect the data from the daily reports for all countries in the world and we identified that many countries did not report real infected people numbers due to particular methodologies for tests or sometimes political implications, this can be clearly identified in the following table.

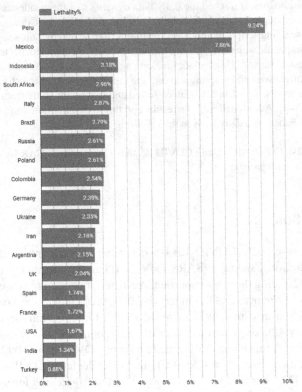

Fig. 1. Lethality per country, top countries with more reported infections.

Taking the example of México and Peru, it is hard to accept a lethality of 9.3% while these countries did not implement a COVID TESTs strategy and basically the reports were based on the people once they arrived to a hospital, comparing data with countries like USA that was more proactive on the cases identification, lethality is very different: 1.8%.

We shown below a quick calculation for the real number of infected people on the time in Mexico, using a theoretical lethality of 2.19% and real deaths (233,622 as of 4th of July).

$$Estimated\ Real\ Infected\ People = \frac{100 \cdot Real\ Deaths}{Estimated\ Real\ Lethality}$$

$$Estimated\ Real\ Infected\ People = \frac{100 * 233,622}{2.19\%}$$

$$Estimated\ Real\ Infected\ People = 10,667,671$$

The result is 420% greater than the official number reported by Mexican Government, (2,540,068) as of July 4th. Using this analysis we came with conclusion of not using the official number of infections per country due to possible inaccuracies. Instead, we use number of deaths as those numbers are more difficult to hide and a better accuracy can be achieved.

3 Methods

In order to normalize the data and get a better comparison, we use as day one per country the day when at least 3 new deaths were reported and, once is homologated, we implemented the following methods.

3.1 Moving Average

A sequence of averages generated utilizing sequential segments of data points over a range of values is known as a moving average. They have a length that specifies how many data points should be included in each average. Moving averages are useful for smoothing time series data, revealing underlying trends, and identifying components for statistical modeling. Smoothing is the technique of reducing random changes in a plot of raw time series data that show as coarseness. It lowers noise in order to highlight the signal, which can include patterns and cycles. Filtering the data is another term used by analysts to describe the smoothing procedure.

The current and past observations for each average are included in one-sided moving averages, for this specific study, we used MA3, MA5 and MA7.

$$MA_3 = \frac{X_{t-2} + X_{t-1} + X_t}{3}$$

$$MA_5 = \frac{X_{t-4} + X_{t-3} + X_{t-2} + X_{t-1} + X_t}{5}$$

$$MA_7 = \frac{X_{t-6} + X_{t-5} + X_{t-4} + X_{t-3} + X_{t-2} + X_{t-1} + X_t}{7}$$

3.2 Plateau

The plateau is defined as the largest interval in which all values at all time positions within the interval are equal to or greater than all values at all time locations outside the interval. The plateau may provide us with more information than the top-k time positions. When we add another constraint to the plateau definition, all of the values in the plateau must be *near enough* to the top-1 value in the entire sequence. The user can specify how close is *close enough* [14].

Having identified the first Plateau for each country, we also perform an evaluation to calculate the duration of the plateau, another comparison method for our analysis.

3.3 Maximum Value

Once data are normalized, we sought to find the Maximum MA_7 and the days each country lasted to have those numbers.

3.4 Economics Information Per Country

3.4.1 GDP

The market value of all the goods and services produced by a country in a given year is referred to as the gross domestic product (GDP). Countries are ranked according to nominal GDP estimates from financial and statistical organisations, which are derived using market or official exchange rates.

3.4.2 Constitutional Form

We used a list of countries by system of government. For our purpose, the top countries analysed are:

Presidential republic: Head of state is the head of government and is independent of legislature.

Constitutional monarchy: Head of state is executive; Monarch personally exercises power in concert with other institutions.

4 Results

Applying our described methods, we obtained the following results:

In Fig. 1, we represent a comparison of the top countries by infected people with the normalized data using MA_7, an homologated top duration of over 500 days.

In order to have a clearer representation for our countries of study, we separated time-series and plot daily new deaths, MA_3, MA_5, MA_7 and the found Plateau. See Figs. 2, 3, 4, 5, 6, 7, 8, 9, 10, 11, 12, 13, 14, 15, 16, 17, 18 and 19.

After having the calculations done, we plot the results for first Plateau, initial day, final day and duration for every country of the study in Fig. 20. Brazil can be recognized as the country with the longest first plateau and the second biggest, that is a Plateau of

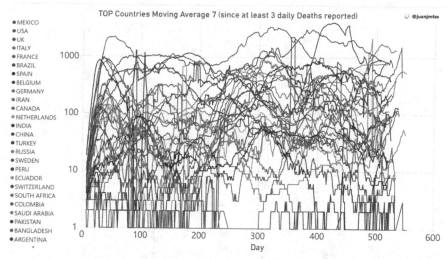

Fig. 2. Top countries Moving Average 7, since first 3 deaths per country.

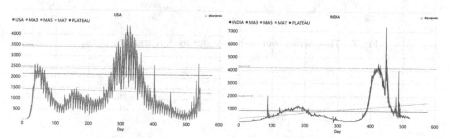

Fig. 3. Time series over daily new deaths. USA and India.

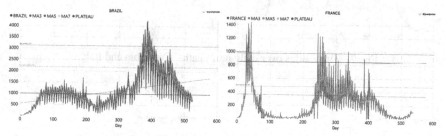

Fig. 4. Time series over daily new deaths. Brazil and France.

992 daily deaths during 106 days. The second longest Plateau belongs to Mexico, with 88 days and 590 daily new deaths.

In Fig. 21, we show the countries that were the best in containing the pandemic effects of deaths evaluating the Maximum number of daily deaths and the time each country lasted to have those Max, also the last day (duration of the pandemic as of 4th

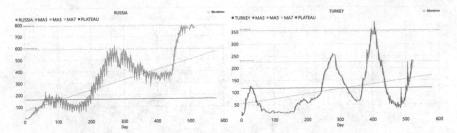

Fig. 5. Time series over daily new deaths. Russia and Turkey.

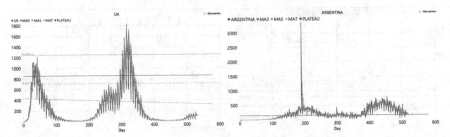

Fig. 6. Time series over daily new deaths. UK and Argentina.

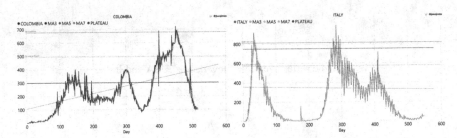

Fig. 7. Time series over daily new deaths. Colombia and Italy.

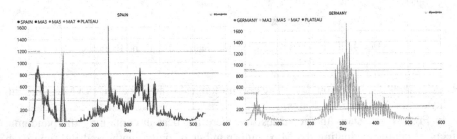

Fig. 8. Time series over daily new deaths. Spain and Germany.

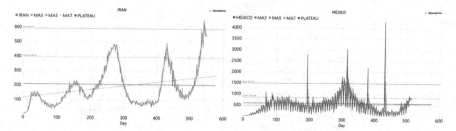

Fig. 9. Time series over daily new deaths. Iran and Mexico.

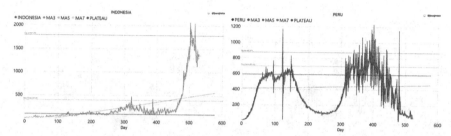

Fig. 10. Time series over daily new deaths. Indonesia and Peru.

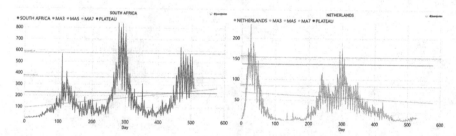

Fig. 11. Time series over daily new deaths. South Africa and Netherlands

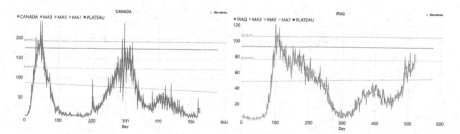

Fig. 12. Time series over daily new deaths. Canada and Iraq.

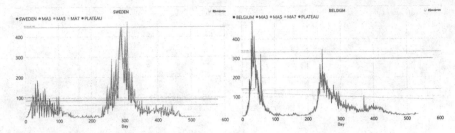

Fig. 13. Time series over daily new deaths. Sweden and Belgium.

Fig. 14. Time series over daily new deaths. Pakistan and Bangladesh.

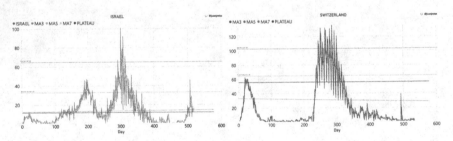

Fig. 15. Time series over daily new deaths. Israel and Switzerland.

Fig. 16. Time series over daily new deaths. Egypt and Kazakhstan.

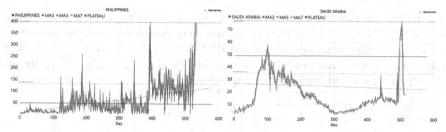

Fig. 17. Time series over daily new deaths. Philippines and Saudi Arabia.

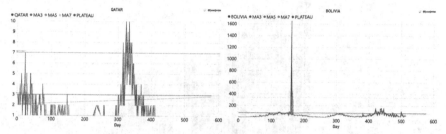

Fig. 18. Time series over daily new deaths. Qatar and Bolivia.

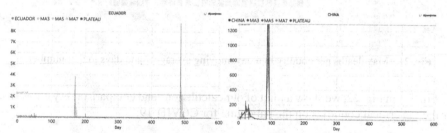

Fig. 19. Time series over daily new deaths. Ecuador and China.

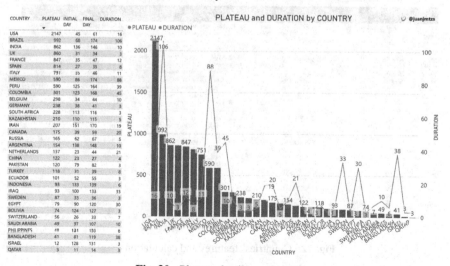

Fig. 20. Plateau details per country.

of July). From the chart we identify Indonesia, Russia and Bangladesh as the countries that lasted the most on having their maximum daily new deaths and low maximums compared with other countries. In contrast USA lasted 313 days to have a maximum number of daily new deaths of 3,432, Brazil 384 days for 3,124 daily new deaths and India 422 days for 4,190 deaths in one single day.

COUNTRY	DAYS TO MAX	MAX MA7	LAST DAY	LAST MA7
IRAN	540	592	546	544
PHILIPPINES	527	398	527	398
INDONESIA	502	1,789	527	1,361
RUSSIA	502	791	511	777
SAUDI ARABIA	499	77	511	18
BANGLADESH	486	252	503	120
ECUADOR	486	2,210	516	115
COLOMBIA	451	678	511	98
INDIA	422	4,190	518	403
TURKEY	401	356	519	232
PAKISTAN	397	144	510	75
PERU	394	874	519	36
BRAZIL	384	3,124	521	698
KAZAKHSTAN	353	2,291	501	934
QATAR	333	7	427	1
MEXICO	314	1,483	513	847
UK	314	1,253	530	104
USA	312	3,425	538	460
GERMANY	301	894	527	3
ISRAEL	299	65	513	16
SWEDEN	288	448	522	4
SOUTH AFRICA	281	578	506	268
SWITZERLAND	261	103	527	1
ARGENTINA	183	789	513	147
BOLIVIA	156	338	510	40
IRAQ	104	106	520	85
SPAIN	99	1,179	532	132
CHINA	85	1,290	548	-3
EGYPT	85	89	518	9
CANADA	44	193	523	15
ITALY	33	814	545	45
BELGIUM	31	333	528	8
FRANCE	29	975	533	96
NETHERLANDS	21	154	527	10

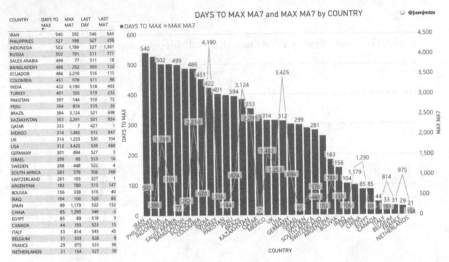

Fig. 21. Max deaths per country in a day, moving average 7 and days to that number.

Finally, in Fig. 22, we show a joint of our calculations and comparison key elements, for the top critical countries on the pandemic for COVID-19.

@juanjmtzs

COUNTRY	PLATEAU	INITIAL DAY	FINAL DAY	DURATION	DAYS TO MAX	LAST DAY	LAST MA7	MAX MA7	GDP (US$million)	CONSTITUTIONAL FORM	CASES	DEATHS	LETHALITY %	POPULATION
USA	2147	45	61	16	312	538	460	3,425	$21,439,453	Republic	38,545,144	645,058	1.67 %	333,214,890
CHINA	122	23	27	4	85	548	-3	1,290	$14,140,163	Republic	94,652	4,636	4.90 %	1,439,323,776
GERMANY	238	38	41	3	301	527	3	894	$3,863,344	Republic	3,876,024	92,478	2.39 %	84,089,414
INDIA	862	136	146	10	422	518	403	4,190	$2,935,570	Republic	32,448,969	434,784	1.34 %	1,395,457,353
UK	860	31	34	3	314	530	104	1,253	$2,743,586	Constitutional Monarchy	6,492,906	131,640	2.03 %	68,292,493
FRANCE	847	35	47	12	29	533	96	975	$2,707,074	Republic	6,619,611	113,311	1.71 %	65,438,227
ITALY	751	35	46	11	33	545	45	814	$1,988,636	Republic	4,484,613	128,751	2.87 %	60,360,383
BRAZIL	992	68	174	106	384	521	698	3,124	$1,847,020	Republic	20,570,891	574,574	2.79 %	214,281,173
CANADA	175	39	59	20	44	523	15	193	$1,730,914	Constitutional Monarchy	1,468,813	26,792	1.82 %	38,119,126
RUSSIA	165	62	67	5	502	511	777	791	$1,637,892	Republic	6,747,087	176,044	2.61 %	146,005,843
SPAIN	814	27	35	8	99	532	132	1,179	$1,397,870	Constitutional Monarchy	4,770,453	83,136	1.74 %	46,775,436
MEXICO	590	86	174	88	314	513	847	1,483	$1,274,175	Republic	3,225,073	253,155	7.85 %	130,475,416
INDONESIA	93	133	139	6	502	527	1,361	1,789	$1,111,713	Republic	3,979,456	126,372	3.18 %	276,817,411
NETHERLANDS	137	23	44	21	21	527	10	154	$902,355	Constitutional Monarchy	1,918,769	17,953	0.94 %	17,178,109
SAUDI ARABIA	49	97	107	10	499	511	18	77	$779,289	Absolute Monarchy	541,994	8,481	1.56 %	35,430,697
TURKEY	118	31	39	8	401	519	232	356	$743,708	Republic	6,215,663	54,533	0.88 %	85,372,527
SWITZERLAND	56	26	33	7	261	527	1	103	$715,360	Republic	752,761	10,941	1.45 %	8,726,741
SWEDEN	87	33	36	3	288	522	4	448	$528,929	Constitutional Monarchy	1,116,584	14,629	1.31 %	10,171,034
BELGIUM	298	34	44	10	31	528	8	333	$517,609	Constitutional Monarchy	1,163,726	25,320	2.18 %	11,647,125
IRAN	207	151	170	19	540	546	544	592	$458,500	Republic	4,677,114	102,038	2.18 %	85,216,720
ARGENTINA	154	138	148	10	183	513	147	789	$445,469	Republic	5,133,831	110,352	2.15 %	45,668,187
ISRAEL	12	128	131	3	299	513	16	65	$387,717	Republic	990,428	6,830	0.69 %	9,326,000
SOUTH AFRICA	228	113	116	3	281	506	268	578	$358,839	Republic	2,690,973	79,421	2.95 %	60,159,918
PHILIPPINES	48	124	130	6	527	527	398	398	$356,814	Republic	1,839,635	31,810	1.73 %	111,241,145
COLOMBIA	301	123	168	45	451	511	98	678	$327,895	Republic	4,889,537	124,216	2.54 %	51,500,501
BANGLADESH	41	81	119	38	486	503	120	252	$317,465	Republic	1,461,998	25,282	1.73 %	166,558,093
EGYPT	79	90	120	30	85	518	9	89	$302,256	Republic	286,352	16,671	5.82 %	104,528,784

Fig. 22. Countries' features and calculations.

5 Conclusions and Future Work

We have presented in this work several methods aimed to normalize time-series and find key comparison elements for tracking daily new deaths, presenting visualisations on the pandemic over the time in the World. Using numbers of deaths allowed us to obtain more accurate statistics on the pandemic behaviour avoiding miss-reports from the countries on contagions. With this time series and data collecting/normalization in conjunction with the data clustering it is easier to identify patterns comparing countries and to use forecasting methods to predict future COVID waves in the world.

As a future work we plan to continue the efforts made up to this point, by increasing the amount of transformed data and improving the dashboards. Moreover, specific work is being done to compare contagion, deaths and recovered curves on a Moving Average of 14 days to avoid peaks on data due to tracking methods in each country; on this path it will be easier to identify real trends and to know, based on real data when a country has overpassed the pandemic or still is going upwards. We worked on having countries classifications on different aspects and look for answers on why some territories had a better handling of the pandemic, or maybe there is hidden data in others, that can explain the reason of high lethality rates.

6 Related Work

Some interesting related works that have been developed by different organizations like CONACYT [13], Scriby [12], Youyang Gu [11] and Institute for Health Metrics [10] and Evaluation, have similar work over data statistics on the pandemic and also different forecasting techniques using machine learning. These are carefully tracking data related to the spread of COVID-19 in the world encouraged by what is being seen in some areas and concerned about what is seen in others. As world moves forward, making decisions based on science, data and facts related to the specific conditions in our communities is needed. Those works are committed to provide accurate, reliable reports to the public and the information presented is updated daily and it is dependent on reporting by numerous agencies across the world.

Acknowledgements. We would like to recognize the great effort from the Center for Systems Science and Engineering at Johns Hopkins University [3] and worldometers.info [4] for collecting all the data from all over the world about the novel Coronavirus COVID-19. Also, we want to acknowledge the work from Mexican Government and all participating people from the Mexican Ministry of Health for the daily tracking and data facilitation for the COVID-19 in Mexico [1, 2]. Without all the work from previous mentioned teams over input data none of the research work presented in this paper would have been achieved.

References

1. Mexican Government: Datos abiertos – dirección general de epidemiología. Obtained from Gobierno de México. https://www.gob.mx/salud/documentos/datos-abiertos-152127 (2020)

2. Mexican Government: Coronavirus (COVID-19)-comunicado técnico diario. Obtained from Gobierno de México. https://www.gob.mx/salud/documentos/coronavirus-covid-19-comuni cado-tecnico-diario-238449 (2020)
3. University, C.F.: COVID-19 dashboard by the center for systems science and engineering (CSSE) at Johns Hopkins University (JHU). Obtained from arcgis.com: https://www.arcgis.com/apps/opsdashboard/index.html#/bda7594740fd40299423467b48e9ecf6 (2020)
4. Worldometer: COVID-19 coronavirus pandemic. Obtained from wordometers.info: https://www.worldometers.info/coronavirus/ (2020)
5. Martínez, J.J.: @JuanJMtzS. Retrieved from Twitter: https://twitter.com/JuanJMtzS (2020)
6. Martínez, J.J.: BI-COVID19. Retrieved from hiketech.com.mx: http://bi-covid19.hiketech.com.mx/ (2020)
7. Martínez, J.J.: BI-COVID19. Retrieved from GitHub: https://github.com/juanjmtzs/BI-COVID19 (2020)
8. Martínez, J.J.: https://www.linkedin.com/in/juanjmtzs/. Retrieved from LinkedIn: https://www.linkedin.com/in/juanjmtzs/detail/recent-activity/shares/ (2020)
9. Mexican Government: Jornada nacional de sana distancia. Retrieved from Gobierno de México: https://www.gob.mx/cms/uploads/attachment/file/541687/Jornada_Nacional_de_Sana_Distancia.pdf (2020)
10. Institute for Health Metrics and Evaluation: COVID-19 Projections. Obtained from covid19.healthdata.org: http://covid19.healthdata.org/united-states-of-america (2020)
11. Gu, Y.: COVID-19 projections using machine learning. Obtained from covid19-projections.com: https://covid19-projections.com/ (2020)
12. Scriby Inc.: COVID-19 Map. Obtained from coronavirus.app: https://coronavirus.app/map (2020)
13. CONACYT: Covid-19 México. Obtained from Gobierno de México: https://coronavirus.gob.mx/datos/ (2020)
14. Wang, M., Wang, X.S.: Finding the Plateau in an Aggregated Time Series. In: Yu, J.X., Kitsuregawa, M., Leong, H.V. (eds.) WAIM 2006. LNCS, vol. 4016, pp. 325–336. Springer, Heidelberg (2006). https://doi.org/10.1007/11775300_28

Linear Structures Identification in Images Using Scale Space Radon Transform and Multiscale Image Hessian

Aicha Baya Goumeidane[1]($\boxtimes$), Nafaa Nacereddine[1], and Djemel Ziou[2,3]

[1] Research Center in Industrial Technologies, CRTI, P.O. BOX 64,
16014 Algiers, Algeria
{a.goumeidane,n.nacereddine}@crti.dz
[2] Département d'informatique, Université de Sherbrooke, Québec, Canada
[3] Shenzhen Institutes of Advanced Technology, Chinese Academy of Sciences,
Shenzhen, China

Abstract. In this paper we propose a stand-alone method to identify lines of different thickness in an image exploiting the scale Space Radon Transform (SSRT) combined to multiscale image Hessian. The proposed approach does not need any prior knowledge about the image content, neither make any assumption about the image to make a decision of the presence or not of a line. This work which consists in seeking information about possible presence of linear structures in an image and exploiting this information while constructing the SSRT space, limits the SSRT computation around precomputed zones. The latter are obtained by multiscale image Hessian. As a consequence, the subsequent maxima detection is done on a restricted transform space freed from unwanted peaks that usually drown the peaks representing lines. Tests done on synthetic and real images have shown that our method highlight the useful maxima of the SSRT permitting to improve SSRT detection of lines of different thickness in an image, while preserving computation time.

Keywords: Scale Space Radon Transform · Multiscale image
Hessian · Line identification

1 Introduction

Detection of linear features by determining their positions and their orientations in images is an important task in many computer vision applications. These applications vary from road network extraction to robot autonomous navigation, including code bar detection or lane marking for vision-based car navigation systems. Frequently, the detection methodology needs to be adapted to the involved application and to the image content. The Radon transform [1] (RT), the Hough transform (HT) [2] and their dozens of extensions have been widely applied to this purpose: They consist in converting the global line detection

© Springer Nature Switzerland AG 2021
I. Batyrshin et al. (Eds.): MICAI 2021, LNAI 13067, pp. 327–340, 2021.
https://doi.org/10.1007/978-3-030-89817-5_25

problem in the image domain into a peak detection problem in the transformations domain [3], even in presence of noise [4]. These peaks positions correspond to the lines parameters. Unfortunately, most of the proposed methods include pre-processing steps, additional tricks, supplementary procedures and predefined thresholds to manage the detection, as often ambiguities regarding the relevant and irrelevant peaks in the transforms spaces rise. This issue have been raised by many authors that have proposed, in turn, how to overcome these ambiguities [5–8]. In fact, applied to images, these transforms correspond to pixels intensity accumulation over the image in all directions. Consequently, peaks may appear in the output signal where there are no linear features. Lastly, Ziou et al. [9] have proposed a new integral transform called Scale Space Radon Transform (SSRT) which is a generalization of the Radon transform. In this transform, the Dirac distribution of the RT is replaced by a Gaussian kernel permitting thus, to handle the thick lines and leading, consequently, to an accurate detection of their centerlines. Nevertheless, the SSRT has inherited from the RT its sensitivity to complex background, especially when the linear structure is quite short compared to the image dimensions. Considering the advantages of the SSRT over the RT and the HT, we propose, in this work, to exploit this transform to detect linear structures in images by demarcating, in the SSRT space, peaks corresponding to linear structures while discarding the others. This operation is achieved through a limitation of the space search in the SSRT domain around precomputed space parameters which permits not only accuracy of the detection/identification, but also a gain in processing time. The precomputed parameters are obtained via multiscale image Hessian computation, as we will see later.

The remainder of this paper is presented as follows. In Sect. 2 we introduce the SSRT transform and recall its principals and advantages. Section 3 is consecrated to the image Hessian. In Sect. 4, we present our line detection framework. Section 5 is dedicated to experiments and results. Finally, conclusion is drawn in Sect. 6.

2 The Scale Space Radon Transform

The Scale Space Radon Transform generalizes the Radon Transform by replacing the matching between an embedded thin structure in an image and the Dirac distribution δ of an implicit parametric shape by a matching between an embedded parametric shape in an image and a kernel depending on a scale parameter. The authors of [9] have argued such replacement by the fact that exploiting the Dirac distribution δ does not handle embedded shapes represented in an image by elongated areas of more than one pixel width. Consequently, the estimated model using RT will be different from the user expectation, and hence, will be inexact. In turn, using the kernel constitutes a trick that allows to the user to control the parametric shape position inside the embedded shape. Moreover, the thickness of the concerned structure can be expressed through the scale parameter and then, the shape detection is reduced to a maxima detection in the Radon space.

When the kernel is a Gaussian along the line parametrized by θ and ρ, then the SSRT computed on an intensity image I is given by [9]

$$SSRT(\rho, \theta) = \int_x \int_y I(x, y) e^{-(x\cos\theta + y\sin\theta - \rho)^2/2\sigma_s{}^2} dx dy \qquad (1)$$

Here, σ_s is the scale space parameter and is related to the line thickness. Furthermore, when the scale parameter $\sigma_s \to 0$, the SSRT reduces to RT, as seen in [9]. On the other hand, the study of the SSRT properties have shown, as for RT, robustness against noise. Moreover, beside performing a good detection, the SSRT permits to recover the linear structure dimensions (length and thickness), when the image is noise-free.

3   The Image Hessian

The image Hessian H_σ, has been used for identifying particular structures centers when the scale σ of the Hessian matches the size of the local structures in images [10]. The Hessian matrix of the intensity image I at a scale σ is equal to

$$H_\sigma = \begin{pmatrix} H_{11}(\sigma) & H_{12}(\sigma) \\ H_{21}(\sigma) & H_{22}(\sigma) \end{pmatrix} = \begin{pmatrix} \sigma^2 I * \frac{\partial^2}{\partial x^2} G(\sigma) & \sigma^2 I * \frac{\partial^2}{\partial x \partial y} G(\sigma) \\ \sigma^2 I * \frac{\partial^2}{\partial y \partial x} G(\sigma) & \sigma^2 I * \frac{\partial^2}{\partial y^2} G(\sigma) \end{pmatrix}. \qquad (2)$$

$G(\sigma)$ is the bivariate Gaussian kernel of standard deviation σ and $*$ is the convolution symbol. The image Hessian components encode objects shape information through, among others, the principal directions provided by H_σ eigenvalues analysis. Let λ_1 and λ_2 be the eigenvalues of H_σ obtained through eigenvalue decomposition as given in [11]

$$\lambda_{1,2} = \frac{1}{2}\left(H_{11} + H_{22} \pm \sqrt{(H_{11} - H_{22})^2 + 4H_{12}^2}\right) \qquad (3)$$

Let $\lambda_1 < \lambda_2$ and let λ_{max} be the maximum eigenvalue derived from λ_1 and λ_2, where $\lambda_{max}(x, y)$ is the maximum absolute value at (x, y). When considering a bright structure on a dark background, as in our study, then λ_1 is negative as depicted in Fig. 1 [12]. Moreover, let $\tilde{\theta}$ be the local directions map derived from $\mathbf{n}(n_x, n_y)$, the unitary vectors field normal to the linear structure, whose components along the x-axis and the y-axis are n_x and n_y respectively and are expressed as

$$[(n_x(x, y), n_y(x, y))] = \left[\frac{H_{11}(x, y), \lambda_{max}(x, y) - H_{11}(x, y)}{\sqrt{H_{12}(x, y)^2 + (\lambda_{max}(x, y) - H_{12}(x, y))^2}}\right] \qquad (4)$$

To get the orientation of the features present in the image, non-maximum suppression procedure is carried out on the direction map $\tilde{\theta}$ by inspecting λ_{max} in order to delete irrelevant angles $\tilde{\theta}(x, y)$ as done in [11]. The remaining directions in $\tilde{\theta}(x, y)$ are the principal directions.

In presence of linear structures and when using an adequate scale and retaining only principal directions $\tilde{\theta}(x, y)$ whose $\lambda_{max}(x, y)$ are above a predefined threshold T, the retained directions will be positioned at the centerline of the linear structure. Before going further with the method explanation, it is worth to note that in all our images the linear structures are bright in dark backgrounds, so when we write λ_{max} it stands for - λ_{max}, since, in this case, the relevant part of λ_{max} (which belongs to λ_1) will be negative as illustrated in Fig. 1 and Fig. 2.

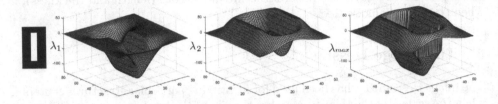

Fig. 1. From left to right, the image to be processed, λ_1, λ_2 and λ_{max}

4   Multiscale Hessian-Based Linear Structure Identification and SSRT Detection

As previously seen, RT, HT, SSRT and all integral transforms have the ability to extract lines from images by transforming the line detection problem in the image domain into peak detection problem in the transform space, which makes them powerful tools for lines detection. However, except the SSRT, the other transforms could not detect centerlines of thick lines without additional processing. Regarding the SSRT, the transform acts on the raw input data with an unsupervised manner in a single step, providing all the detected centerlines parameters, in an elegant way, without any further handling no matter the lines thickness[9].

When exploiting the SSRT in order to detect linear structures, the peaks revealed in the SSRT space are supposed to correspond to lines parameters. However, this is not always true, especially when faced to images with complex backgrounds. In such case, seek some information about the possible presence of linear structures in the image became recommended. This information will be exploited to restrict the SSRT computation to SSRT space zones containing only the useful maxima i.e. those corresponding to the linear structures. Image Hessian is one of the numerous methods that can do the job. However, the image Hessian is associated to a single scale, so what is the most suitable scale we must associate to the Hessian when looking for linear structures informations, and what happens if the structures are of different thickness? Using several Hessian scales can respond to those questions and can give a satisfying solution to this issue. In the following, let us see how the image Hessian of linear structures behaves in scale space.

4.1   Linear Structures Hessian Behaviour in Scale Space

It is known that the Hessian H_σ provides through λ_{max} a high response of the features that match the scale σ [10]. In such case, λ_{max} extrema above the previously introduced threshold T, match the center of the studied features. This behaviour is shown in Fig. 2 for an image representing two linear structures of five (5) and of thirteen (13) pixels width. The Hessian H_σ for σ_1 equals 2.5 and σ_2 equal to 6.5 are computed on this image and λ_{max} is illustrated for each scale in Fig. 2b and Fig. 2c. This figure demonstrates clearly that the linear structure Hessian has a high response only when the scale is an appropriate one, i.e. the scale match the structure thickness, and this response drops when the scale is large compared to this thickness or became improper (for our study) when the scale is small (the maxima are near the boundaries). Indeed, σ_1 which is equal to 2.5 does not lead to a good response for the large line, conversely to σ_2 that seems to give a correct response for both lines. This suggests that the high scale can be used for the thin and thick lines. However, to handle both responses simultaneously when considering the large scale, the threshold T used to consider only the high part of the responses, has to be low to include both lines responses high parts. This could be a solution for an ideal case: a binary image with an homogeneous background, for example. Nevertheless, in practical cases, images are not binary ones, and backgrounds are often complex, so dropping T value to manage together the two lines Hessian responses, will bring more inconveniences than advantages, compromising thus, the intended results. This indicates that applying Hessians of different scales and gather, scale after scale, only the needed responses parts, would be more likely to reveal the structures present in the images regardless their thickness. The manner of gathering the Hessians responses will be detailed later.

Furthermore, applying Hessian with σ varying from 0.5 to 11 on images containing linear structures of 1 to 23 pixels width, with a step increment equals to 2, has permitted each time to compute the maximum value of λ_{max} in order to observe its evolution when varying scale and structure thickness. This evolution is depicted as a surface on which are highlighted local maxima. Each local maximum expresses the maximum value of the highest Hessian response of a structure thickness, regarding all scales. So, each thickness has an optimal scale that provides the highest Hessian response (optimal in this case), and when moving away from the considered thickness, the maximum responses values decay slowly especially for high scales as shown in Fig. 3a. This suggests that, as shown in Fig. 4 for large scales, high Hessian responses are obtained for a range of thickness values, leading thus to a good linear structures emphasis and then, identification, unlike for small scales. On the other hand, Fig. 3b shows that all optimal responses maximal values are quite similar. This denotes that the threshold T, used to consider only λ_{max} high parts, will be the same for all scales. Consequently, when a particular thickness is handled by considering its Hessian optimal response high part via T, the other thickness will be neglected since their responses are not sufficiently high to exceed this threshold. Lastly, the last view of the aforementioned surface in Fig. 3c shows that we can derive graphically a relationship between a structure thickness and its optimal Hessian scale. Indeed, the best Hessian response for a structure thickness w, has been obtained when $\sigma = w/2$. To summarize, revealing the presence of linear structures of different thickness in an image can be done by considering the optimal Hessian response high part of each structure. Moreover, when no knowledge is available about the structures and their thickness and when expecting reaching automatically all structures optimal responses, the involvement of several Hessian scales must be considered, and then the corresponding Hessian responses must be examined separately.

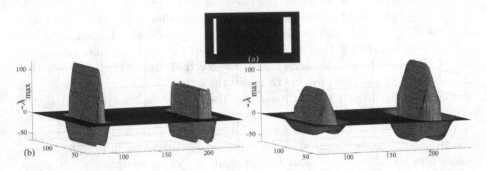

Fig. 2. Image Hessian response for the image in (a), with $\sigma = 2.5$ in (b) and $\sigma = 6.5$ in (c)

4.2   Multiscale Hessian Application

In the following, we will see how the Hessian responses of several scales are used to derive the needed information; namely the structures orientations θ and locations ρ. Suppose that the involved scales in the Hessians computations are $\sigma_{min}, \ldots, \sigma_{max}$. To begin, let us consider that our structures are bright on dark backgrounds. So, after computation of $\lambda_{max}^{\sigma_{min}}, \ldots, \lambda_{max}^{\sigma_{max}}$ and the principal directions $\tilde{\theta}^{\sigma_{min}}, \ldots, \tilde{\theta}^{\sigma_{max}}$ for all scales, only principal directions $\tilde{\theta}(x, y)$ whose $\lambda_{max}(x, y)$ is above a predefined threshold T i.e. $\lambda_{max}^{\sigma_i}(x, y) > T$, where $\sigma_i \in \{\sigma_{min}, \ldots, \sigma_{max}\}$ are kept. We recall that the role of T is to favour the structure responses which thickness matches σ_i and neglect the others. Afterwards, a histogram $h(\tilde{\theta})$ of the retained directions of all scales, is constructed and smoothed. The structures directions are computed as the arguments $\tilde{\theta}_m$ of $h(\tilde{\theta})$ maxima, where $m \in \{1, \ldots, M\}$ with M is the number of the detected maxima and then the number of structures directions. Once the directions are known, the location parameters can be calculated as follows. For each direction $\tilde{\theta}_m$, the corresponding location parameter values ρ can be derived from all principal directions maps of all scales. This is done by searching the coordinates (x_i, y_i)

in $\tilde{\theta}^{\sigma_i}$, with $\sigma_i \in \{\sigma_{min}, \ldots, \sigma_{max}\}$, where; $\tilde{\theta}_m - \Delta\theta \leq \tilde{\theta}^{\sigma_i}(x_i, y_i) \leq \tilde{\theta}_m + \Delta\theta$. Afterwards, the locations parameters set ρ_m for the direction $\tilde{\theta}_m$ is computed as

$$\rho_m = \left\{ \rho_i / \rho_i = x_i \cos \tilde{\theta}_m + y_i \sin \tilde{\theta}_m \right\} \tag{5}$$

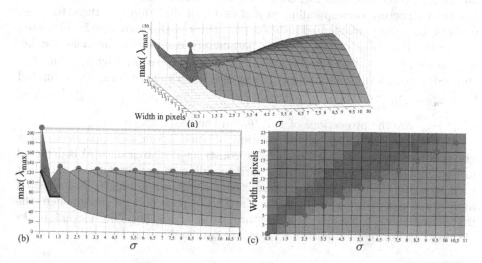

Fig. 3. a, b, and c: Different views of the surface representing the behaviour of λ_{max} with respect to the scale and the structure width

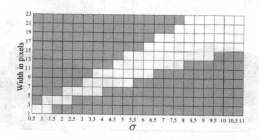

Fig. 4. Desired vs undesired linear structure Hessians responses regarding scales and thickness. In each column, unlike a gray square, a white square stands for a Hessian response exceeding 90% of the optimal response maximum value of the scale represented by the column

Moreover, in presence of parallel structures, one direction can concern more than one structure. Consequently, the set ρ_m contains location parameters of several structures and, consequently, have to be divided into homogeneous classes. As the location parameter values ρ of the same structure are going to be close to each others, as shown in Fig. 5c, where superposed lines corresponding to several location parameters values of one structure are confounded, this operation can be done via k-means procedure, preceded by the determination of k, the classes number. The classes number k is computed as the number of maxima in a histogram constructed with the elements of the set ρ_m. If n classes C_j, $j = 1, \ldots, n$ are found for the direction $\tilde{\theta}_m$, then n linear structures have as orientation $\tilde{\theta}_m$. Let be a structure corresponding to ρ class C_j of direction $\tilde{\theta}_m$, then, to avoid outliers, $\tilde{\rho}_{j,m}$ the precomputed location parameter of this structure is calculated as the median value of the location parameters composing the class C_j, i.e. $\tilde{\rho}_{j,m} = med\{\rho/\rho \in C_j\}$. Finally, for the structure represented by $\tilde{\rho}_{j,m}$ and $\tilde{\theta}_m$, the SSRT transform computation is restricted to the transform space limited by $\left\{\tilde{\theta_m}^T - \Delta\theta \leq \theta \leq \tilde{\theta_m}^T + \Delta\theta\right\}$ and $\left\{\rho_{\tilde{j,m}}^T - \Delta\rho \leq \rho \leq \rho_{\tilde{j,m}}^T + \Delta\rho\right\}$, where $\rho_{\tilde{j,m}}^T / \tilde{\theta_m}^T$ are the projections of $\tilde{\rho}_{j,m} / \tilde{\theta}_m$ in the SSRT space.

We can see in Fig. 5, an illustration of such operations processed on a grayscale image provided in Fig. 5a. The constructed histogram $h(\tilde{\theta})$ is depicted in Fig. 5b, the superposition of lines corresponding to precomputed $\tilde{\theta}$ and their corresponding classes of location parameters ρ are shown in Fig. 5c. The detected lines are illustrated in Fig. 5d, while, the SSRT space and its peaks where the useless ones drown the relevant ones are given in Fig. 5e. The relevant peaks obtained by our method in SSRT zones are depicted in Fig. 5f.

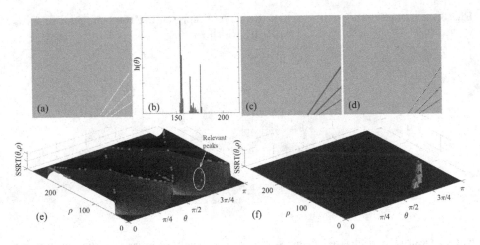

Fig. 5. Illustration of the proposed framework results. a: Example of an image to be processed. b: Histogram $h(\tilde{\theta})$ constructed all the principal directions of the multiscale Hessian. c: Superposition of all lines representing computed directions and all location parameters values belonging to different ρ classes. d: The detected lines in dots. e: SSRT space with all its peaks. f: Relevant SSRT zones and its useful peaks

5 Experiments

To begin, it is worth to note that no numerical results will be given, since the SSRT detection efficiency has already been studied [11]. In fact the proposed method does not modify the SSRT transform in any way but only exploits the useful parts of it. On the other hand, some parameters have to be set. So, $\Delta\theta$ and $\Delta\rho$ have been set equal to 2 and 4 respectively. These values have been chosen empirically, so more investigations should be carried out to find an efficient strategy to choose them. Regarding T, it has been set equal to $0.8 \times \max(\lambda_{max})$, where $\max(\lambda_{max})$ is the maximum value of the computed λ_{max}.

In order to see the results of the proposed method, a set of synthetic and real images containing lines of one or of several pixels widths, have been used. The first experiment is held on a synthetic image consisting in a number of one pixel-width lines of arbitrary orientations and locations. Applying SSRT with a small scale parameter $\sigma_s = 0.1$ provides the SSRT representation given in Fig. 6a where, as reported on the figure, the useful peaks of SSRT, corresponding to the lines present in the image, are almost completely immersed in the irrelevant ones. With our method, relevant SSRT zones have been demarcated as shown in Fig. 6b, and the detection have been carried out successfully on them providing line illustrated in Fig. 6c, while preserving simultaneously processing time.

The proposed method is applied on the second test image illustrated in Fig. 7, which is a synthetic image consisting in thick linear structures. The SSRT is computed only on the predefined zones, obtained as previously proposed and illustrated in Fig. 7b. The detection is achieved as depicted in Fig. 7c, requiring, thus, less computation load, compared to the SSRT in Fig. 7a.

Next experiments are conducted on real images provided in Fig. 8, Fig. 9 and Fig. 10. Here, the deal is to extract from these images the linear structures that compose them even in presence of complex backgrounds. Following the steps of our proposed method has permitted to highlight interesting parts in the SSRT domain, as shown in Fig. 8b, Fig. 9b, and Fig. 10b, as well as detecting only relevant peaks. Examining the entire SSRT spaces in Fig. 8a, Fig. 9a, and Fig. 10a reveals the huge number of peaks composing them, but in spite of that, the useful peaks have been extracted successfully, and detections have been achieved correctly, as shown in Fig. 8c, Fig. 9c and Fig. 10c. Concerning the gain of computation time, it is obvious that computing specific zones of the SSRT space prevents of computing the whole SSRT space which dimension are at least equal to the product of the image diagonal size and 180, the degrees number.

To finish, it is worth noting that when faced to dark lines in a bright background, one has just to invert the images as the SSRT, like the other integral transforms, is a band (line) integration of the pixels intensities over the image. Moreover, it important to mention that the proposed method can be applied for line detection, not only when exploiting the SSRT, but also when considering any integral transforms, to consider only the relevant transform parts and avoid unwanted ones that can only compromise the results.

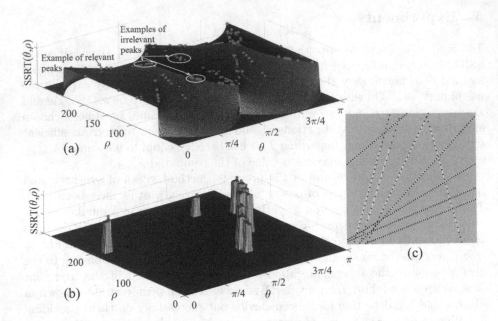

Fig. 6. a: SSRT of the first test image with highlighting of some relevant and irrelevant maxima. b: SSRT computed around precomputed parameters. b: Lines detection on the first test image

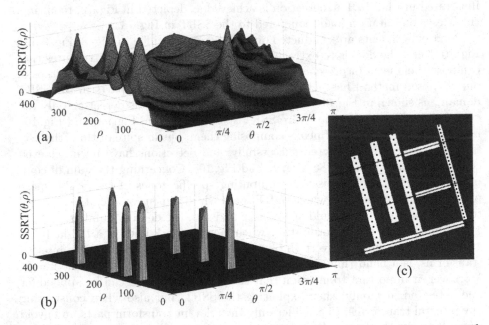

Fig. 7. a: SSRT of the second test image. b: SSRT computed around precomputed parameters. c: Thick lines detection on the second image

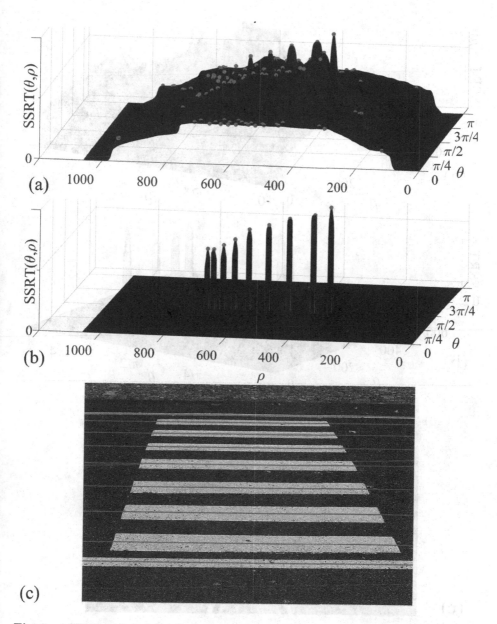

Fig. 8. a: SSRT of the pedestrian crossing image. b: Relevant SSRT zones and useful peaks. c: Detection on the image

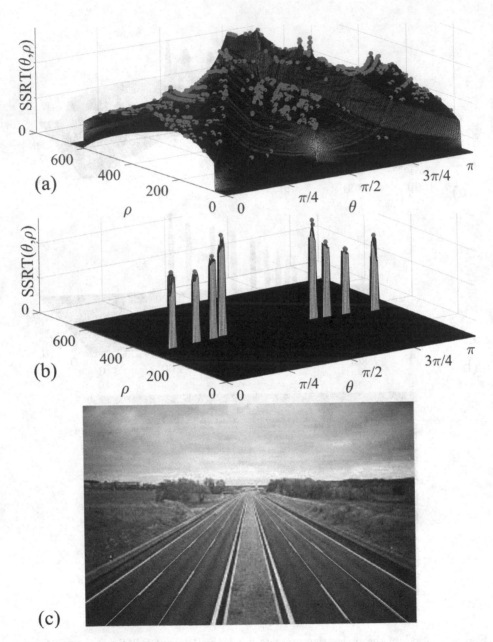

Fig. 9. a: SSRT of the lanes image. b: Relevant SSRT zones and useful peaks. c: Detection on the image

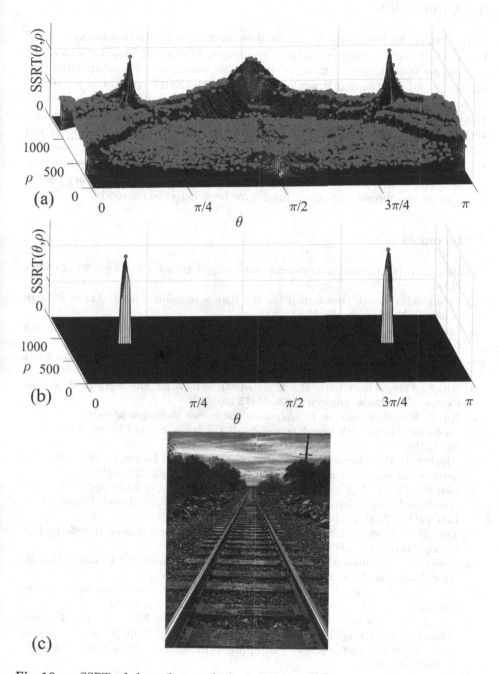

Fig. 10. a: SSRT of the railway rails. b: Relevant SSRT zones and useful peaks. c: Detection on the image

6   Conclusion

In this paper, we have proposed to combine multiscale image Hessian and SSRT transform to detect linear structures in images. Multiscale Hessian, which has as purpose to seek information about possible presence of linear structures in an image and then exploit it while constructing the SSRT space, has permitted to prevent the computation of the whole transform space, which can be very time consuming when faced to large images. In fact, the computation of circumscribed SSRT zones obtained with precomputed transform parameters derived from multiscale Hessian, has allowed, even in complex backgrounds, to overcome the SSRT weakness and extract the relevant transforms peaks corresponding to the linear structures among a huge number of irrelevant ones. Preliminary results seem to be very promising, as all lines have been detected correctly.

References

1. Barrett, H.H.: The radon transform and its applications. Prog. Opt. **21**, 217–286 (1984)
2. Deans, S.R.: Hough transform from the Radon transform. IEEE Trans. Pattern Anal. Mach. Intell. **2**, 185–188 (1981)
3. Liu, W., Zhang, Z., Li, S., Tao, D.: Road detection by using a generalized Hough transform. Remote Sens. **9**(6), 590 (2017)
4. Nacereddine, N., Tabbone, S., Ziou, D.: Robustness of Radon transform to white additive noise: general case study. Electron. Lett. **50**(15), 1063–1065 (2014)
5. Xu, Z., Shin, B.S., Klette, R.: A statistical method for line segment detection. Comput. Vis. Image Underst. **138**, 61–73 (2015)
6. Xu, Z., Shin, B.S., Klette, R.: Accurate and robust line segment extraction using minimum entropy with Hough transform. IEEE Trans. Image Process. **24**(3), 813–822 (2014)
7. Alpatov, B.A., Babayan, P.V., Shubin, N.Y.: Weighted Radon transform for line detection in noisy images. J. Electron. Imaging **24**(2), 023023 (2015)
8. Liu, W., Zhang, Z., Chen, X., Li, S., Zhou, Y.: Dictionary learning-based Hough transform for road detection in multispectral image. IEEE Geosci. Remote Sens. Lett **14**(12), 2330–2334 (2017)
9. Ziou, D., Nacereddine, N., Goumeidane, A.B.: Scale space Radon transform. IET Image Process. **15**, 2097–2111 (2021)
10. Steger, C.: An unbiased detector of curvilinear structures. IEEE Trans. Pattern Anal. Mach. Intell. **20**(2), 113–125 (1998)
11. Deschenes, F., Ziou, D., Auclair-Fortier, M.-F.: Detection of lines, line junctions and line terminations. Int. J. Remote Sens. **25**(3), 511–553 (2004)
12. Frangi, A.F., Niessen, W.J., Vincken, K.L., Viergever, M.A.: Multiscale vessel enhancement filtering. In: Wells, W.M., Colchester, A., Delp, S. (eds.) MICCAI 1998. LNCS, vol. 1496, pp. 130–137. Springer, Heidelberg (1998). https://doi.org/10.1007/BFb0056195

Deep Neural Networks for Biomedical Image Segmentation: Trends and Best Practices

Cecilia-Irene Loeza-Mejía[1] , Eddy Sánchez-DelaCruz[1]([✉]) ,
and Mirta Fuentes-Ramos[2]

[1] Departamento de Posgrado e Investigación, Tecnológico Nacional de México,
Campus Misantla, Veracruz, Mexico
[2] Universidad Autónoma de Tabasco, Villahermosa, Tabasco, Mexico

Abstract. Biomedical image segmentation is an important process in computer-aided diagnostic systems. Segmentation allows an image to be divided and tagged into anatomical sub-regions such as bones, muscles, blood vessels, and various pathological structures, facilitating image analysis and detecting various diseases. However, it is challenging due to the different features that images present, including noise and contrast. This article provides an overview of deep neural networks for biomedical image segmentation specifically computed tomography, dermoscopy, MRI, ultrasound, and X-ray. Additionally, best practices are discussed to improve the accuracy and sensitivity of results, including in unbalanced datasets.

Keywords: Biomedical image segmentation · Deep neural networks · Deep learning · CAD system

1 Introduction

Biomedical images allow observation of molecular and functional information of the body [1,14], related to specific diseases like infections, obstructions, abnormal pathology, and malignant lesions, which reduce the need for invasive tests, such as biopsy. Every year, around the world, billions of medical images are captured and analyzed [1]. There are different physical processes in the generation of biomedical images, which allows the visualization of various body structures. Also, each acquisition modality is better for certain tissues and pathologies. Due to the relevance of the use of medical images for health care, automatic analysis of biomedical images in Computer-Aided Diagnosis (CAD) systems is an active field and continuous improvement that helps to serve as a second opinion to experts [32,51] which contributes to large-scale medical image analysis and reduces subjectivity.

Segmentation is a highly significant step of CAD systems to provide effective radiological diagnostics [55], increase the sensitivity [33], accuracy [28], and

I. Batyrshin et al. (Eds.): MICAI 2021, LNAI 13067, pp. 341–352, 2021.
https://doi.org/10.1007/978-3-030-89817-5_26

the understanding of structural information [47,55]. Segmentation allows partitioning an image to find the region of interest (ROI) and boundaries (lines, curves, etc.). Each ROI is a connected homogenous subset that possesses similar attributes [14,35], which can be gray levels, contrast, spectral values, or textural properties [3,31,47]. However, segmentation of biomedical images is a challenge [25] due to complex patterns of the image [26,47], irregular shapes of body structures [5–7,11,14,17,21,25], variation in scales [21], blurred boundaries [11], subjectivity of the expert [14,49], and different scanning protocol and image providers. So different types of segmentation are used in biomedical images [6]. Segmentation techniques can be classified depending on the degree of human interaction: manual, semiautomatic and automatic. Semi-automatic and automatic approaches are more common using neural networks.

This paper deals with an overview of different deep neural network architectures of the state-of-the-art for the segmentation of biomedical images specifically computed tomography (CT), dermoscopy, magnetic resonance imaging (MRI), ultrasound, and X-rays images. Also, best practices, are detailed in order to improve the accuracy and recall of the results even in unbalanced datasets.

This paper is organized as follows, in Sect. 2 imaging modalities are exposed, while in Sect. 3 deep neural networks specifically U-Net [36] and hybrid model-based architectures are presented. Section 4 shows the best practices in order to improve the performance of the neural networks. In Sect. 5 is concluded the investigation, also, future needs are explained.

2   Imaging Modalities

There are different physical processes in the generation of biomedical images, which leads to having different modalities of image acquisition. These processes mainly can be based in X-ray transmission [3,12], magnetic fields [11], visible light [11], Gamma-ray emission [11,12], and reflection of ultrasonic waves [12]. Also, there are optical modalities such as endoscopy, microscopy, or photography [11]. In addition, biomedical images can be classified according to the dimensions they have [1,14]:

– Two-dimensional (2D): e.g. 2D ultrasound, dermoscopy, X-rays.
– Three-dimensional (3D): e.g. MRI, CT, 3D ultrasound.

2D techniques can be used for exploratory diagnosis, planning, and control in surgical interventions. On the other hand, 3D techniques can be used for diagnosis, treatment planning [23], simulation [44], manipulation, and analysis. Additionally, biomedical modalities can be classified as structural or functional [18]. Structural modalities include X-rays, CT, MRI, echocardiography, etc., while, functional modalities include positron emission tomography, single-photon emission CT, fluorescence imaging, etc. Table 1 shows biomedical images and their usefulness in various body structures and pathologies.

Also, there are hybrid imaging techniques like the fusion PET-CT and PET-MRI [11]. Because there are different imaging modalities and they have different characteristics, a standard segmentation method cannot be applied [46].

Table 1. Uses of biomedical image

Biomedical	useful in image
CT	Analysis of lung nodules [54]
	Detect the presence of inflammatory diseases
	Detection of calcification, hemorrhage, and bony detail [3,31]
	Early detection of abnormal changes in tissues [3,14,29,33,55]
	Pulmonary parenchyma [3]
	Radiotherapy planning [31]
	3D simulation [44]
Dermoscopy	Melanoma diagnosis [11]
	Skin lesions
	Visualization of deeper details of skin [4]
MRI	Brain tumors [10]
	Cardiac diseases [49]
	Detailed anatomical information [11]
	Early detection of abnormal changes in tissues [3,29,33,55]
	Fine blood vessels [11]
	Image-guided interventions
	Organs of the chest and abdomen
	Pelvic organs
	Planning and treatment monitoring [23]
	Soft tissues as cartilage, brain [3,14,41], etc.
	Surgical planning
	Uterine fibroids
Ultrasound	Creating images of live tissue
	Detection of breast abnormalities [3]
	Examination of the abdomen [12]
	Heart disease [11]
	Monitor the fetus [14]
	Needle biopsies
	Soft tissues as heart and blood vessels
	Urinary tract
X-rays	Dental problems
	Detection of foreign objects
	Fracture detection [14]

2.1   Characteristics of Images

To characterize biomedical images, different properties such as image size, brightness, contrast, histogram, radiometric resolution (image depth), spatial resolution, and spectral resolution can be used. Table 2 shows the characteristics of different biomedical images.

Table 2. Characteristics of images

Biomedical image characteristics	
CT	Cross-sectional images [14]
	Gaussian noise
	High resolution [18]
	Low signal-to-noise ratio
	Low-contrast soft tissue (particularly the brain) [31,33]
	Poor sensitivity and specificity [47]
Dermoscopy	It contains hair-like artifacts [53]
MRI	Cross-sectional images
	Higher contrast in comparison with CT
	High signal-to-noise ratio [39]
	Provides morphological and functional information
	Rician distribution of noise
Ultrasound	Inherent speckle noise [7]
	Low contrast [7]
	Poor transmission through bone or air [55]
X-rays	Good contrast between bone, soft tissue, lung, and air
	Poor differentiation between soft tissues
	Poisson noise

3   Deep Learning Methods

When developing deep neural networks for biomedical image segmentation, two main approaches are followed: i) convolutional network-based architectures for biomedical image segmentation (U-Net) [2,7,10,21,22,29,43,45,47], and ii) emerging neural networks including hybrid approaches [13,22,27,34,53,56,56].

3.1   U-Net Based

U-Net [36] is a fully convolutional network for imaging segmentation. It is a symmetric architecture consisting of multi-scale Encoder and Decoder parts. U-Net is one of the most popular network in the state-of-the-art literature [2,7,10,21,22,29,43,45,47]. Some variants have been developed in order to improve accuracy [22] and feature extraction such as U-Net [32], Improved Attention U-Net [2], Attention-based nested segmentation network (ANU-Net) [25], MultiResUNet [21], and USE-Net [37].

3.2   Emerging Neural Networks and Hybrid Approaches

Currently, different investigations report variations of the ResNet [16], DenseNet [20], and U-Net [36] architectures to develop novel hybrid models, where the results obtained with the original architectures are improved [10,22]. By the integration of ResNet and DenseNet, a Dense Residual U-net (DRU-net) [22] was developed using U-net as a base structure. Also, U-Net and Res-Net have been integrated into an architecture called Res-Unet [53]. Likewise, approaches using U-Net and DenseNet have been integrated into Adaptive Fully Dense U-Net (AFD-UNet) [47]. In addition, new models have surged like ResNet-based models (i.e. Cascaded Dual-Pathway Residual Network (CDP-ResNet) [27]), ensemble deep learning methods (i.e. Ensemble-A, Ensemble-L and Ensemble-S in [13]), edge attention-based architectures (i.e. ET-Net [56]), and deep clustering architectures (i.e. Y-Net [34]).

3.3   Analysis of Approaches

Deep neural networks offer an important automatic technique for the segmentation of biomedical images. Table 3 shows a comparison of neural networks and their applications in the segmentation of CT, dermoscopy, MRI, ultrasound, and X-rays images including the organs that have been segmented. On the other hand, Fig. 1 shows the pipeline of deep neural networks incorporating best practices. Data augmentation and preprocessing must be performed before it enters into the neural network. Moreover, network blocks, ensemble deep learning, loss function, and weighted sampling strategy must be configured when developing the deep neural network model.

The use of neural networks offers great advantages in the segmentation of biomedical images. However, there are many challenges. In the U-Net architecture, several limitations have been observed, compared to hybrid and emerging approaches, which may be since the U-Net literature has been analyzed more. For example, U-Net cannot identify small areas [47] or extract semantic information in the pancreas [26]. Also, U-Net makes false predictions on noisy images with no clear ROI limits [21]. AFD-UNet has only been analyzed in the segmentation of liver tumors. ANU-Net and Attention U-Net have been only analyzed in CT and MRI images. CDP-ResNet has only been applied in CT images. DRU-net and MultiResUNet have been only analyzed in dermoscopy and MRI images. ET-Net has been used in CT and X-rays. Improved Attention U-Net has been applied in dermoscopy and ultrasound images. Res-Unet and Ensemble-A have been only analyzed in dermoscopy images. USE-Net has been only analyzed in MRI images. Therefore, hybrid architectures should continue to be analyzed in other organ segmentation tasks and different biomedical imaging modalities. Furthermore, it is necessary to reduce the number of parameters even when integrating network blocks.

Table 3. Advantages and best practices applied in Deep Neural Networks

Deep neural Network	Advantages	Best practices applied to improve performance	Applications	
			Biomedical imaging	*Segmented structures*
AFD-UNet [47]	It uses effectively shallow and deep features adaptively even with complex boundary, and it is better compared to U-Net [47]	Loss function Network blocks	CT	Liver tumor [47]
ANU-Net [25]	It can increase the weight of the ROI and suppress the irrelevant tissue [25]	Loss function Network blocks	CT MRI	Liver tumor [25] Abdominal Organs [25]
Attention U-Net [32]	It can be applied in dense predictions and obtain sufficient semantic context [32]	Network blocks	CT MRI	Abdominal Organs [32] Knee menisci [8]
CDP-ResNet [27]	It can extract local texture information of lung nodules [27]	Weighted sampling strategy	CT	Lung nodules [27]
DRU-net [22]	It is efficient in terms of memory and training time. In addition, it overcomes U-Net and Improved Attention U-Net accuracy [22]	Network blocks	Dermoscopy MRI	Skin lesions [22] Brain
Ensemble-A [13]	It outperforms the sensitivity of U-Net [13]	Ensemble deep learning	Dermoscopy	Skin lesions [13]
ET-Net [56]	It integrates edge detection and object segmentation. Also, it is better compared to U-Net [56]	Loss function Network blocks	CT X-rays	Lung [56] Lung [56]
Improved Attention U-Net [2]	It is more suitable for the segmentation of small lesions compared to Attention U-Net [2]. Also, it gets better recall compared to U-Net and DRU-net [22]	Loss function Network blocks	Dermoscopy Ultrasound	Skin lesions [2,22] Breast lesion [2]
MultiRes UNet [21]	It outperforms U-Net results [29] in less number of training epochs and a smaller number of parameters. Also, it is more reliable against outliers and noise [21]	Network blocks	Dermoscopy MRI	Skin lesions [21] Brain tumor [21]
Res-Unet [53]	It outperforms the Jaccard Index and Dice coefficient of U-Net [53]	Network blocks	Dermoscopy	Skin lesions [53]
U-Net [36]	It allows capturing the context [36] and presents efficient use of GPU [32], [22,49]	Data augmentation	CT	Liver tumor [47] Lung [56] Lung-tissue [24] Pancreas [26]
			Dermoscopy	Skin lesions [4,22], [13,34]
			MRI	Brain [22] Brain tumor [10,45] [48] Left ventricle [49] Prostate zones [37]
			Ultrasound	Arterial walls [52] Breast mass [7]
			X-rays	Lung [56]
USE-Net [37]	It outperforms the Dice coefficient of U-Net [37]	Network blocks	MRI	Prostate zones [37]

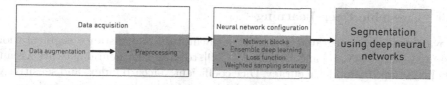

Fig. 1. Pipeline for the segmentation of biomedical images using deep neural networks incorporating best practices

4 Best Practices

The best practices that we identify to improve segmentation performance can be divided into two types:

1. Best practices focused on the image including data augmentation and preprocessing.
2. Best practices related to the neural network including the use of ensemble deep learning, implementation of network blocks, choice of the loss function, and application of weighted sampling strategy.

4.1 Data Augmentation

The performance of convolutional neural networks depends on having large datasets [4,5,14]. However, it is not always possible to acquire more samples [21], therefore data augmentation techniques are used, which allow generating new images without collecting new samples. Data augmentation can help overcome overfitting problems [4,38,40,42] and improve model performance [21]. Some data augmentation techniques like the traditional techniques, mixing images, and those based on neural networks can be found in [42].

4.2 Preprocessing

It is recommended to apply preprocessing techniques before performing segmentation because images are often affected by noise and tissue movement [25]. The artifacts of the image impact segmentation accuracy [41]. Also, preprocessing allows us to discard non-relevant areas and accentuate image features for subsequent segmentation (i.e. in dermoscopy images preprocessing is applied to remove the hairs) [53].

Feature enhancement has been shown to help achieve better results in segmentation [30,41,48]. Besides, preprocessing has been shown to facilitate learning for the model [4]. Moreover, regions of interest are often in a small area of the image [2,10,47]. Furthermore, most neural network architectures do not use the same number of neurons or have limited input size [14] mainly in pretrained architectures, it is necessary to resize the images, in order to use certain architectures. In addition, it may be necessary to resize the images to the GPU memory (i.e. [21,29]).

4.3   Ensemble Deep Learning

Goyal *et al.* [13] integrated semantic segmentation and instance segmentation using Mask R-CNN [15] and DeeplabV3C [9], to show that neural network assembly helps improve model accuracy, Dice coefficient, Jaccard index, and sensitivity, thus, surpassing the results obtained by U-Net in skin lesion segmentation.

4.4   Network Blocks

Incorporating network blocks can help improve network performance [19,22]. In the literature Attention Gates (AGs), DenseNet blocks, Residual blocks, and Squeeze-and-Excitation (SE) block in U-Net (i.e. [2,25,32]) has been used, showing effective generalization between different datasets [19,32,50]. Also, AGs allow detection of relevant spatial information [2], improve sensitivity, and highlight salient features [32], even if different training sizes are used [32]. Furthermore, Weighted Aggregation Module has been used in emerging neural networks such as ET-Net to aggregate multi-scale information and edge-attention representations [56].

4.5   Loss Function

Choosing the loss function is important for improving performance [47,56], especially when dealing with unbalanced datasets [2]. By means focal Tversky loss function in [2] accuracy and recall balance were improved. The use of hybrid loss function as presented in [25] (combining advantages of soft dice coefficient loss, focal loss, and pixel-wise binary cross-entropy loss) allowed to furnish smooth gradient and solve the class imbalance. Also, Lovász-Softmax loss has been used for class imbalance problems [56]. In addition, BCE loss and Dice loss have also been combined in [47].

4.6   Weighted Sampling Strategy

Recent approaches focusing on the segmentation of lung nodules used an improved weighted sampling strategy to select training samples based on the edge in order to improve the robustness of the model [27].

5   Conclusion and Future Directions

Medical image processing is a multi-disciplinary field of research. The segmentation of biomedical images requires constant changes, in order to improve the feature extraction and classification in CAD systems. In this paper, neural networks based on U-Net and hybrid approaches were analyzed. In addition, best practices to improve model performance, such as data augmentation, ensemble deep learning, network blocks, loss function, preprocessing, and weighted sampling strategy were presented.

The use of hybrid approaches and ensemble deep learning has significantly improved the results obtained by the original architectures. However, there is still a long way to go to meet the challenges. Future work should focus on the following areas:

- Analyze the feasibility of reducing the number of parameters in network blocks.
- Apply hybrid methods in different organ segmentation tasks and biomedical imaging modalities to compare their performance.
- Compare different architectures on the same dataset.
- Consider variations (i.e. shape, texture) of ROI even if there is poor contrast between regions.
- Decrease the computational cost in ensemble deep learning.
- Develop pre-trained architectures that can be used in different types of biomedical images.
- Implement algorithms to obtain the best set of hyper-parameters (i.e. learning rate, number of layers).
- Implement a weighted sampling strategy in the segmentation of different organs.
- Integrate neural network blocks to improve image features.
- Propose loss functions appropriate to different imaging modalities.

Acknowledgement. Consejo Nacional de Ciencia y Tecnología (CONACyT), Mexico.

References

1. Abdallah, Y.M.Y., Alqahtani, T.: Research in medical imaging using image processing techniques. In: Medical Imaging-Principles and Applications. IntechOpen (2019)
2. Abraham, N., Khan, N.M.: A novel focal Tversky loss function with improved attention U-Net for lesion segmentation. In: 2019 IEEE 16th International Symposium on Biomedical Imaging (ISBI 2019), pp. 683–687. IEEE (2019)
3. Acharya, T., Ray, A.K.: Image Processing: Principles and Applications. Wiley, Hoboken (2005)
4. Al-Masni, M.A., Al-Antari, M.A., Choi, M.T., Han, S.M., Kim, T.S.: Skin lesion segmentation in dermoscopy images via deep full resolution convolutional networks. Comput. Methods Programs Biomed. **162**, 221–231 (2018)
5. Asaeikheybari, G., Green, J., Qian, X., Jiang, H., Huang, M.C.: Medical image learning from a few/few training samples: melanoma segmentation study. Smart Health **14**, 100088 (2019)
6. Ben Rabeh, A., Benzarti, F., Amiri, H.: Segmentation of brain MRI using active contour model. Int. J. Imaging Syst. Technol. **27**(1), 3–11 (2017)
7. Byra, M., et al.: Breast mass segmentation in ultrasound with selective kernel U-Net convolutional neural network. Biomed. Sig. Process. Control **61**, 102027 (2020)
8. Byra, M., et al.: Knee menisci segmentation and relaxometry of 3D ultrashort echo time cones MR imaging using attention U-Net with transfer learning. Magn. Reson. Med. **83**(3), 1109–1122 (2020)

9. Chen, L.C., Papandreou, G., Kokkinos, I., Murphy, K., Yuille, A.L.: DeepLab: semantic image segmentation with deep convolutional nets, atrous convolution, and fully connected CRFs. IEEE Trans. Pattern Anal. Mach. Intell. **40**(4), 834–848 (2017)
10. Daimary, D., Bora, M.B., Amitab, K., Kandar, D.: Brain tumor segmentation from MRI images using hybrid convolutional neural networks. Procedia Comput. Sci. **167**, 2419–2428 (2020)
11. Deserno, T.M.: Biomedical Image Processing. Springer, Heidelberg (2011). https://doi.org/10.1007/978-3-642-15816-2
12. Dougherty, G.: Digital Image Processing for Medical Applications. Cambridge University Press, Cambridge (2009)
13. Goyal, M., Oakley, A., Bansal, P., Dancey, D., Yap, M.H.: Skin lesion segmentation in dermoscopic images with ensemble deep learning methods. IEEE Access **8**, 4171–4181 (2019)
14. Haque, I.R.I., Neubert, J.: Deep learning approaches to biomedical image segmentation. Inform. Med. Unlocked **18**, 100297 (2020)
15. He, K., Gkioxari, G., Dollár, P., Girshick, R.: Mask R-CNN. In: Proceedings of the IEEE International Conference on Computer Vision, pp. 2961–2969 (2017)
16. He, K., Zhang, X., Ren, S., Sun, J.: Deep residual learning for image recognition. In: Proceedings of the IEEE Conference on Computer Vision and Pattern Recognition, pp. 770–778 (2016)
17. Hesamian, M.H., Jia, W., He, X., Kennedy, P.: Deep learning techniques for medical image segmentation: achievements and challenges. J. Digit. Imaging **32**(4), 582–596 (2019). https://doi.org/10.1007/s10278-019-00227-x
18. Histed, S.N., Lindenberg, M.L., Mena, E., Turkbey, B., Choyke, P.L., Kurdziel, K.A.: Review of functional/anatomic imaging in oncology. Nucl. Med. Commun. **33**(4), 349 (2012)
19. Hu, J., Shen, L., Sun, G.: Squeeze-and-excitation networks. In: Proceedings of the IEEE Conference on Computer Vision and Pattern Recognition, pp. 7132–7141 (2018)
20. Huang, G., Liu, Z., Van Der Maaten, L., Weinberger, K.Q.: Densely connected convolutional networks. In: Proceedings of the IEEE Conference on Computer Vision and Pattern Recognition, pp. 4700–4708 (2017)
21. Ibtehaz, N., Rahman, M.S.: MultiResUNet: rethinking the U-Net architecture for multimodal biomedical image segmentation. Neural Netw. **121**, 74–87 (2020)
22. Jafari, M., Auer, D., Francis, S., Garibaldi, J., Chen, X.: DRU-Net: an efficient deep convolutional neural network for medical image segmentation. In: 2020 IEEE 17th International Symposium on Biomedical Imaging (ISBI), pp. 1144–1148. IEEE (2020)
23. Jarrett, D., Stride, E., Vallis, K., Gooding, M.J.: Applications and limitations of machine learning in radiation oncology. Br. J. Radiol. **92**(1100), 20190001 (2019)
24. Kayal, S., Dubost, F., Tiddens, H.A., de Bruijne, M.: Spectral data augmentation techniques to quantify lung pathology from CT-images. In: 2020 IEEE 17th International Symposium on Biomedical Imaging (ISBI), pp. 586–590. IEEE (2020)
25. Li, C., et al.: ANU-Net: attention-based nested U-Net to exploit full resolution features for medical image segmentation. Comput. Graph. **90**, 11–20 (2020)
26. Li, F., Li, W., Shu, Y., Qin, S., Xiao, B., Zhan, Z.: Multiscale receptive field based on residual network for pancreas segmentation in CT images. Biomed. Sig. Process. Control **57**, 101828 (2020)
27. Liu, H., et al.: A cascaded dual-pathway residual network for lung nodule segmentation in CT images. Physica Med. **63**, 112–121 (2019)

28. Loeza Mejía, C.I., Biswal, R.R., Rodriguez-Tello, E., Ochoa-Ruiz, G.: Accurate identification of tomograms of lung nodules using CNN: influence of the optimizer, preprocessing and segmentation. In: Figueroa Mora, K.M., Anzurez Marín, J., Cerda, J., Carrasco-Ochoa, J.A., Martínez-Trinidad, J.F., Olvera-López, J.A. (eds.) MCPR 2020. LNCS, vol. 12088, pp. 242–250. Springer, Cham (2020). https://doi.org/10.1007/978-3-030-49076-8_23

29. Lou, A., Guan, S., Loew, M.H.: DC-UNet: rethinking the U-Net architecture with dual channel efficient CNN for medical image segmentation. In: Medical Imaging 2021: Image Processing, vol. 11596, p. 115962T. International Society for Optics and Photonics (2021)

30. Mo, J., Zhang, L., Wang, Y., Huang, H.: Iterative 3D feature enhancement network for pancreas segmentation from CT images. Neural Comput. Appl. 32, 12535–12546 (2020). https://doi.org/10.1007/s00521-020-04710-3

31. Nanthagopal, A.P., Rajamony, R.S.: A region-based segmentation of tumour from brain CT images using nonlinear support vector machine classifier. J. Med. Eng. Technol. 36(5), 271–277 (2012)

32. Oktay, O., et al.: Attention U-Net: learning where to look for the pancreas. arXiv preprint arXiv:1804.03999 (2018)

33. Padma, A., Sukanesh, R.: Combined texture feature analysis of segmentation and classification of benign and malignant tumour CT slices. J. Med. Eng. Technol. 37(1), 1–9 (2013)

34. Pathan, S., Tripathi, A.: Y-Net: biomedical image segmentation and clustering. arXiv preprint arXiv:2004.05698 (2020)

35. Pratt, W.K.: Digital Image Processing. Wiley, Hoboken (2001)

36. Ronneberger, O., Fischer, P., Brox, T.: U-Net: convolutional networks for biomedical image segmentation. In: Navab, N., Hornegger, J., Wells, W.M., Frangi, A.F. (eds.) MICCAI 2015. LNCS, vol. 9351, pp. 234–241. Springer, Cham (2015). https://doi.org/10.1007/978-3-319-24574-4_28

37. Rundo, L., et al.: USE-Net: incorporating squeeze-and-excitation blocks into U-Net for prostate zonal segmentation of multi-institutional MRI datasets. Neurocomputing 365, 31–43 (2019)

38. Setio, A.A.A., et al.: Pulmonary nodule detection in CT images: false positive reduction using multi-view convolutional networks. IEEE Trans. Med. Imaging 35(5), 1160–1169 (2016)

39. Sharma, N., Aggarwal, L.M.: Automated medical image segmentation techniques. J. Med. Phys./Assoc. Med. Phys. India 35(1), 3 (2010)

40. Suresh, S., Mohan, S.: NROI based feature learning for automated tumor stage classification of pulmonary lung nodules using deep convolutional neural networks. J. King Saud Univ. Comput. Inf. Sci. (2019)

41. Tahir, B., et al.: Feature enhancement framework for brain tumor segmentation and classification. Microsc. Res. Tech. 82(6), 803–811 (2019)

42. Tajbakhsh, N., Jeyaseelan, L., Li, Q., Chiang, J.N., Wu, Z., Ding, X.: Embracing imperfect datasets: a review of deep learning solutions for medical image segmentation. Med. Image Anal. 63, 101693 (2020)

43. Tomita, N., Jiang, S., Maeder, M.E., Hassanpour, S.: Automatic post-stroke lesion segmentation on MR images using 3D residual convolutional neural network. NeuroImage Clin. 27, 102276 (2020)

44. Uchida, M.: Recent advances in 3D computed tomography techniques for simulation and navigation in hepatobiliary pancreatic surgery. J. Hepatobiliary Pancreat. Sci. 21(4), 239–245 (2014)

45. Venkateswarlu Isunuri, B., Kakarla, J.: Fast brain tumour segmentation using optimized u-net and adaptive thresholding. Automatika: časopis za automatiku, mjerenje, elektroniku, računarstvo i komunikacije **61**(3), 352–360 (2020)

46. Wadhwa, A., Bhardwaj, A., Verma, V.S.: A review on brain tumor segmentation of MRI images. Magn. Reson. Imaging **61**, 247–259 (2019)

47. Wang, E.K., Chen, C.M., Hassan, M.M., Almogren, A.: A deep learning based medical image segmentation technique in Internet-of-Medical-Things domain. Futur. Gener. Comput. Syst. **108**, 135–144 (2020)

48. Wang, H., Wang, G., Xu, Z., Lei, W., Zhang, S.: High- and low-level feature enhancement for medical image segmentation. In: Suk, H.-I., Liu, M., Yan, P., Lian, C. (eds.) MLMI 2019. LNCS, vol. 11861, pp. 611–619. Springer, Cham (2019). https://doi.org/10.1007/978-3-030-32692-0_70

49. Wu, B., Fang, Y., Lai, X.: Left ventricle automatic segmentation in cardiac MRI using a combined CNN and U-Net approach. Comput. Med. Imaging Graph. **82**, 101719 (2020)

50. Wu, Y.: Deep convolutional neural network based on densely connected squeeze-and-excitation blocks. AIP Adv. **9**(6), 065016 (2019)

51. Xie, H., Yang, D., Sun, N., Chen, Z., Zhang, Y.: Automated pulmonary nodule detection in CT images using deep convolutional neural networks. Pattern Recogn. **85**, 109–119 (2019)

52. Yang, J., Faraji, M., Basu, A.: Robust segmentation of arterial walls in intravascular ultrasound images using Dual Path U-Net. Ultrasonics **96**, 24–33 (2019)

53. Zafar, K., et al.: Skin lesion segmentation from dermoscopic images using convolutional neural network. Sensors **20**(6), 1601 (2020)

54. Zhang, G., Yang, Z., Gong, L., Jiang, S., Wang, L., Zhang, H.: Classification of lung nodules based on CT images using squeeze-and-excitation network and aggregated residual transformations. La Radiol. Med. **125**, 374–383 (2020). https://doi.org/10.1007/s11547-019-01130-9

55. Zhang, Z., Sejdić, E.: Radiological images and machine learning: trends, perspectives, and prospects. Comput. Biol. Med. **108**, 354–370 (2019)

56. Zhang, Z., Fu, H., Dai, H., Shen, J., Pang, Y., Shao, L.: ET-Net: a generic edge-aTtention guidance network for medical image segmentation. In: Shen, D., et al. (eds.) MICCAI 2019. LNCS, vol. 11764, pp. 442–450. Springer, Cham (2019). https://doi.org/10.1007/978-3-030-32239-7_49

Evolutionary and Metaheuristic Algorithms

Mexican Stock Return Prediction with Differential Evolution for Hyperparameter Tuning

Ramón Hinojosa Alejandro[1]([✉]) [ID], Luis A. Trejo[2] [ID], Laura Hervert-Escobar[1],
Neil Hernández-Gress[1], and Enrique González N.[2] [ID]

[1] Tecnologico de Monterrey, Campus Mty, Monterrey, N.L., Mexico
a01382300@itesm.mx, {laura.hervert,ngress}@tec.mx
[2] Tecnologico de Monterrey, Campus Edo Mex, Lopez Mateo, Edo Mex, Mexico
ltrejo@tec.mx, a00457801@itesm.mx

Abstract. Technical analysis aims to predict market movement by examining historical data through statistical procedures. Nevertheless, it is sensitive to the parameter it is working with. An optimization problem is defined to tune technical analysis parameters by minimizing an error metric for stock return prediction. Differential Evolution is a metaheuristic that provides good solutions to an optimization problem, searching for the optimal combination of parameters for technical analyzers to predict Mexican stock returns. For the application of the metaheuristic, an objective function based on a Random Forest prediction is used.

The literature has proven the use of different macroeconomic variables (MEV) to determine expected returns, such as the Capital Asset Pricing Model (CAPM) or different Arbitrage Pricing Theories (APT). This paper considers the influence of macroeconomic factors on stock prices; it is approached with a Granger-causality test on the different sector indexes of the Mexican stock exchange, to see the relationship they hold.

Instead of supervising the error from the machine learning models, it is proposed to analyze their performance in a more realistic scenario, by simulating a portfolio. Constructing a diversified portfolio is a smart way to allocate your money parting from the expected returns computed, still, other relevant factors may alter its performance.

This work shows the performance of different portfolios constructed from the same expected return computations, reaching excess returns over the benchmarks of the 12% in the 3 years analyzed.

Keywords: Metaheuristics · Hyperparameter tuning · Causality · Stock market returns · Forecast

1 Introduction

Understanding the behavior of the stock market is a subject of high relevance, and different approaches has been taken, from econometrics methodologies [1],

Supported by Tecnologico de Monterrey and Conacyt.

I. Batyrshin et al. (Eds.): MICAI 2021, LNAI 13067, pp. 355–368, 2021.
https://doi.org/10.1007/978-3-030-89817-5_27

machine learning [2], sentiment analysis [3], the application of deep learning [4] and neural networks [5]. As much as good advances have been made, it remains a growing topic.

One of the main questions to ask when forming a machine learning model is: which attributes to consider when making a prediction (market data, macroeconomic factors, social media sentiment, etc.). The present paper will approach the Macroeconomic Variables (MEV) and Technical Analysis (TA) attributes.

In the financial area, MEV have proven themselves to influence stock market movements [6]. It can be seen in econometrics, methodologies for stock return prediction that includes MEV as factors of influence, as the Capital Asset Pricing Model (CAPM) [7] or different Arbitrage Pricing Theories (APT) [8] like the Fama-French three-factor and the five-factor model.

The MEV to consider for the present project are the Gross Domestic Product (GDP), the inflation rate (represented by "Unidad de Inversion" (UDI) and the Consumer Price Index (CPI)), Yields on Treasury Certificates (CETES), Unemployment Rate (UR), the International Reserves (IR), the exchange rate (USDMXN), and the Oil price. The economy of market partners can also affect the stock index of a country, henceforth, the stock index of the main import and export partners of Mexico are considered. According to the World Bank by 2018 [9], they were the USA (S&P 500), China (SSE), Canada (S&P/TSX), Japan (Nikkei 225), and Germany (DAX).

A Granger-causality is a statistical method that determines if a time series is capable of forecasting another. It is used in the present work to detect if a change in a MEV produces a change in the sector indexes of the Mexican exchange. A research question here to answer is to see how differently the MEV affect companies of different areas.

TA is another financial tool that can be used to predict market movements. Their profitability is extremely sensitive to the input parameters. Search and optimization algorithms, like metaheuristics, tackle this problem. Comparative papers as [10,11] highlight the performance of the Differential Evolution (DE) algorithm when compared against other metaheuristics.

To truly test the performance of the expected returns computed, a portfolio is constructed following the Modern Portfolio theory [12]. Results are documented in terms of final accumulated return and Sharpe ratio.

This paper first addresses the Granger-causality of the MEV to the Mexican sector indexes, followed by the application of the metaheuristic for hyperparameter tuning, and at the end, we construct a portfolio parting from the methodology proposed and compared against two benchmarks: the S&P/IPC BMV (MEXBOL) and the free-risk rate of return.

2   Related Work

This paper approaches three main subjects: the effect of MEV on the stock market, machine learning for stock return prediction, and metaheuristic for hyperparameter tuning.

2.1   Macroeconomic Variables Effect on the Mexican Stock Market

Between the relation of the MEV and their corresponding stock market, there are two main lines of investigation based on the causality direction, MEV cause the stock market or the other way around. Here, it is being tested for the causality direction in which the MEV cause the stock market.

Many papers analyze the first commented line of investigation, like [13, 14] on the China and USA's stock market, [15, 16] in Turkey, [17] in Nigeria, [18, 19] in India, [20, 21] in China, to mention some. The results gathered from each of them are specific to their country, time under analysis, and methodology taken, yet important attention is taken to the MEV considered. The most common MEV analyzed are the inflation, exchange, and interest rates, industrial production, GDP, and money supply.

The oil price is another important MEV that has been heavily studied, [22–26] are some examples.

The unemployment rate is another variable with a strong relationship with the stock market [19, 27]. The results of [28] for the USA's market tell that when the unemployment rate starts to increase, higher levels of risk and returns are seen in the market. Unemployment is used as an economic indicator to tell the probability of recession in a country [29] considering the use of monetary policies to regulate and incentivize the market activities, it becomes of high relevance for the construction of robust models that can anticipate these market movements.

International reserves and foreign indexes are considered because of the international influence on a stock market. Benson [30] have shown how developed and emerging markets react to the USA market index (S&P 500). Anand [31] show how the international influence, in terms of exchange rates and foreign investment, is related to the Pakistan Stock Market.

2.2   Machine Learning for Stock Return Prediction

Strader [32] review papers of the last 20 years and categorized them according to the machine learning methodology; common observations and limitations were extracted for the corresponding categories. Neural networks have proven better for the numerical prediction of stock market indexes. Support Vector Machines, for categorical prediction of stocks rising or falling. Genetic Algorithms, for better input (parameter setting) or portfolio selection for higher returns. Hybrid models can help with these limitations, nevertheless, they become so complex that their process is not efficient in practice.

Cervelló-Royo [33] compare different machine learning techniques for the prediction of the technology NASDAQ index. Random Forest (RF), Deep Learning, Gradient Boosting Machine, and Generalized Linear Models altogether with Technical Indicators were used, where RF outperformed them all in terms of average accuracy. Nabipour [34] work with technical analysis as attributes for machine learning techniques, from which RF was one of them, nevertheless, the Long short-term Memory (LSTM) showed more accurate results.

Kusuma [35] and Selvin [36]'s conclusions matched those stated by Strader [32], where neural networks outperformed those machine learning methodologies they were compared with, in both cases, Convolutional Neural Networks (CNN) were the ones with better performance.

2.3   Metaheuristics for Hyper-parameter Tuning

TA are statistical methodologies that try to portray the mean market behavior, make a comparison between the present and past, supervise volume traded, or tell price patterns, all these to compute indicators that can help investors tell whether a trend is about to begin or end [37]. Oscillators are TA that compare the actual price, against the mean behavior of the stock or past prices [38].

The main issue is that their performance is linked to the parameters they work with. Zamperin [10] approaches a technical analysis parameter tuning through metaheuristics, where Enhanced Firework (EF), Differential Evolution (DE), and Particle Swarm Optimization (PSO) were tested. EF proved to be the most efficient in the test carried out, followed closely by the DE.

From parameter tuning problems, Pholdee [11] made a comparison between 24 metaheuristics applied to a truss design problem for mass minimization. According to their convergence times and consistency in results, the Covariance Matrix Adaptation Evolution Strategy (CMA-ES) and the DE stood out.

The DE algorithm, as an evolutionary algorithm, is based on the Mutation, Crossover, and Selection steps [39]. From these steps, the following parameters are crucial for its performance, a Crossover factor, Mutation factor, Variant (how to select individuals to be perturbed, number of pairs to consider for crossover, and the crossover type), and Dither (strategy to introduce adaptive weights (F), it allows one per iteration or a different one for each individual). Papers such as [40], applied Neural Networks to search for the best parameters to work with. In optimization problems, other papers as [41] commented how many other works on this subject tend to contradict themselves and show inconsistent results, concluding that the best parameters for an algorithm, are specific to the problem being faced.

We opt for the DE algorithm thanks to the ease of application obtained from using the Python tool Pymoo [42], a multi-objective optimization library.

3   Methodology

The methodology to follow consists of three main stages, the Granger-causality analysis, hyper-parameter tuning for expected return calculation, and the portfolio formation. The three-stage methodology is repeated in a monthly interval. Stock returns are computed each month, from which a portfolio is constructed and balanced. The stages of this methodology are described now.

3.1   Granger-Causality

Granger-causality is used to tell if a change in a predictor variable produces a change in the target one, or in other words if a time series can predict another. It will be used to see how the selected MEV Granger-cause the Mexican sector indexes. Before its computation, the Granger-causality makes certain assumptions about data: it should be stationary, no unit-roots, and no autocorrelation.

To tackle the statistical assumptions, the Augmented Dickey-Fuller (ADF) test is carried out to test unit roots, the Kwiatkowski–Phillips–Schmidt–Shin (KPSS) test for stationarity, and the Durbin-Watson for autocorrelation. Once the data assumptions are satisfied, we compute the corresponding Granger-causality, allowing us to highlight relevant variables. To avoid eliminating important variables from our model, a redundant variables analysis is developed based on the Pearson correlation. A correlation level of 0.8 is considered redundant. We only removed variables that showed a strong correlation with the Granger-causality variables. For those variables that were compared, in which none of them proved causality, we select one randomly.

3.2   Hyper-parameter Tuning

This paper applies DE. The objective function to minimize is the relative test error from a prediction model. It works as follows:

- Our dataset is constructed with the historical data of the Stock to analyze, the MEV based on the sector it belongs to, and the set of nine TAs. It holds monthly observations from 20 years, for a total of 240 instances.
- Our target variable is defined as the Lead Net Return.
- Technical analyzers are initialized randomly, according to the metaheuristic.
- A Random Forest model is trained and tested considering an 80%–20% train-test ratio.
- We run a rollback on the test prediction to get the actual stock price from the predicted returns.
- The relative error is computed to see the performance of the model.
- Based on the test error of the population, it selects the best set of TA parameters and will part from it for the next generations.
- Once all generations have finished, the last Random Forest model with the best TA parameters is used to make the expected return prediction.

The configuration of the DE algorithm will part from the parameters observed in [11] in which DE outperformed 24 metaheuristics and based on the recommendations of [40], the previous parameters are to be modified.

3.3   Portfolio Construction

To test the performance of the calculated expected return, considering only the error obtained after the hyperparameter tuning might be misleading. That is why performance is measured with the profitability obtained by making a portfolio derived from the calculated expected returns. A portfolio is constructed and optimized based on the Modern Portfolio Theory or Mean-Variance portfolio theory.

The portfolio construction follows three steps:

- Portfolio Selection: To reach diversification through industrial sectors, we sort stocks based on the expected return obtained from the DE algorithm. The best ten percent of stocks according to their expected returns form our portfolio.
- Portfolio Optimization: Following the mean-variance methodology, we construct our Efficient Frontier from our expected returns and the exponential covariance. Algorithm 1 reflects how the exponential covariance is computed. Molyboga [43] improve portfolios performances when introducing exponential covariance. The idea is to give more weight to more recent data than past data. [44], following this idea, showed how Exponential Moving Average (EMA) was more capable than Simple Moving Average (SMA) in predicting future prices. By this observation, we argue the application of the exponential covariance in this paper. From the Efficient Frontier, we can optimize our portfolio to different objectives, as the maximization of the Sharpe ratio (MaxSharpe) and the minimization of volatility (MinVol).
- Weights Discretization: The Portfolio Optimization algorithm tells us the participation of each asset in our portfolio in percentage terms. Trying to reach a more realistic scenario, these allocation weights are discretized based on an initial investment. From this, we obtained discrete weights that represent the number of shares to buy for each considered stock.

No other bound constraints are set to the portfolio optimization algorithm to achieve a more efficient allocation. Strategies are compared based on the accumulated return obtained and the Sharpe ratio. The Sharpe ratio tells us the excess return obtained per unit of risk [45], it is helpful to compare different investment strategies. The Sharpe ratio is defined as:

$$Sharpe = \frac{\bar{R}_i - R_f}{StDev(R_i - R_f)} \tag{1}$$

Where $\bar{R}_i$ is the strategy return and R_f, the risk-free rate of return (or CETES).

Computation of expected returns with DE algorithm and the portfolio construction are repeated every month. Every month a new portfolio is formed expecting to adapt better to market changes.

Algorithm 1 Exponential Covariance

Code obtained from PyPortfolio Python Library. For more information see reference [46].

    function EXPCOV(Net, span = 1)   ▷ Where Net - Net returns from asset prices, span - span of exponential weighting function
        Let N = Numer of assets in Net
        Let ECM = Zero Matrix (size: NxN)
        for $i = 0$ to N **do**
            for $j = 0$ to N **do**
                $ECM[i,j] = ECM[j,i] = PairExpCov(N[:,i], N[:,j], span)$
            end for
        end for
        return ECM

    function PAIREXPCOV(X,Y,span)   ▷ Where X and Y - Net returns from assets, span - span of exponential weighting function
        Let $cov = (X - \bar{X}) * (Y - \bar{Y})$
        Let N = Size of X and Y
        Let ewm = Zero array (size: N)
        for $i = 1$ to N **do**
            $ewm[i] = cov[i] * \frac{2}{1+span} + ewm[i-1] * (1 - \frac{2}{1+span})$
        end for
        return Last element in ewm array

---

4 Results

4.1 Data

Data is extracted from two main sources. MEV and the Mexican exchange data as the Mexican stock index (MEXBOL) and the sector indexes (Materials, Industrials, Consumer Discretionary, Consumer Staples, Health Care, Financial, and Telecommunication indexes) are downloaded from the Bloomberg Terminal [47].

There are 145 companies listed on the Mexican Stock Exchange, whose historical data is downloaded using the Yahoo Finance API (Yfinance) [48].

The information gathered is from December 1997 to December 2020, in monthly intervals. There are certain MEV that cannot be retrieved in monthly intervals, but quarterly. For these cases, linear interpolation is performed to fill up the missing information.

4.2 Granger-Causality

All tests performed are done using the statsmodels Python library [49].

First, data assumptions made by the Granger-causality are tackled. Table 1 shows the ADF, KPSS, and Durbin-Watson tests. For ADF and KPSS to be satisfied to a 5% significance level, a second difference is computed. To determine no autocorrelation, the Durbin-Watson test needs to give us lectures between 1.5 to 2.5, in both non-differentiated and second-difference data. We see from the Table 2 that no autocorrelation is present.

Now that our assumptions are satisfied, the Granger-causality is computed. Table 2 resumes the results. In it, we represent with ones those variables that showed Granger-causality with one of the target indexes and with zeros (0) those that do not.

Table 1. Data assumptions. (For the case of ADF and KPSS, values represent T-statistic. *, **, *** for p-values below 10%, 5% and 1%, respectively.)

	Normal data			Second difference		
	ADF	KPSS	Durbin-Watson	ADF	KPSS	Durbin-Watson
MEXBOL	−1.805	0.216**	2.042	−14.036***	0.101	1.928
Consumer discretionary	−7.14***	0.293**	1.975	−9.009***	0.033	1.971
Consumer staples	−16.892***	0.048	1.984	−16.833***	0.047	1.978
Financial	−2.745	0.13*	1.975	−9.764***	0.040	1.971
Health	−1.527	0.234**	2.073	−15.422***	0.105	1.983
Materials	−2.224	0.163**	2.021	−9.213***	0.048	2.074
Telecommunications	−2.416	0.196**	2.046	−14.724***	0.053	1.946
Industrial	−16.567***	0.095	1.982	−16.527***	0.091	1.970
DAX	−2.183	0.277**	2.043	−11.451***	0.040	1.940
Nikkei	−1.464	0.28**	2.011	−13.359***	0.041	1.985
SSE	−4.309***	0.074	2.084	−4.294***	0.074	1.977
TSX	−3.62**	0.057	2.107	−3.613**	0.057	1.929
SP500	0.009	0.375**	2.188	−4.451***	0.087	1.959
CPI	−2.855	0.305**	2.058	−6.935***	0.029	1.974
CETES	−3.254*	0.27**	2.004	−5.655***	0.041	1.981
UR	−2.104	0.27**	2.299	−3.789**	0.095	1.661
RI	−1.558	0.39**	1.900	−3.856**	0.072	2.062
Oil	−1.14	0.259**	2.065	−14.994***	0.076	1.967
Inflation	−1.946	0.243**	2.049	−9.631***	0.024	1.911
USDMXN	−2.246	0.312**	2.021	−9.215***	0.046	2.117
GDP	−2.613	0.32**	1.915	−14.832***	0.042	1.903

Table 2. Granger causality test ("1" represents Granger-causality from the MEV to the Index)

	Financial	Health	Consumer staples	Consumer discretionary	Telecommunications	Materials	Industrial	MEXBOL
SSE	0	0	0	0	0	0	0	0
S&P/TSX	0	1	1	0	0	1	0	0
S&P 500	0	0	0	0	0	0	0	1
Nikkei 225	0	0	0	0	0	0	0	1
DAX	0	0	0	0	0	0	0	1
CETES	0	0	0	0	0	0	0	0
UR	0	0	0	0	0	0	0	0
Oil price	0	1	0	0	0	0	0	1
IR	0	0	0	0	1	0	0	1
Inflation	0	0	0	0	0	0	1	0
GDP	1	1	1	1	0	1	1	0
CPI	0	0	0	0	0	0	0	0
USDMXN	0	0	1	1	1	0	0	1

Table 3 shows the results of the analysis of the redundant variables. Here we see remarked with ones (1) the MEV considered for each of the sector indexes models. With zeros (0) it is represented all the variables that were redundant with the Granger-causal variables or others previously selected in the process.

Table 3. Redundant variables analysis ("1" represents that the corresponding MEV will be part of the ML model according to the sector the stock analyzed belongs to.)

	Financial	Health	Consumer staples	Consumer discretionary	Tele communications	Materials	Industrial	MEXBOL
SSE	1	1	1	1	1	1	1	1
S&P/TSX	0	1	1	0	0	1	0	0
S&P500	1	0	0	0	0	0	0	1
Nikkei 225	0	1	1	1	1	1	1	1
DAX	0	0	0	0	0	0	0	1
CETES	1	1	1	1	1	1	1	1
UR	1	1	1	1	1	1	1	1
Oil price	0	1	0	0	1	0	0	1
IR	0	0	0	0	1	0	0	1
Inflation	0	0	0	0	0	0	1	0
GDP	1	1	1	1	0	1	1	0
CPI	0	0	0	0	0	0	0	0
USDMXN	0	1	1	1	1	1	0	1

4.3 Hyper-parameter Tuning

For the computation of the metaheuristics, we use the Python tool Pymoo [42], a multi-objective optimization library. One of the crucial elements to consider is which parameters to work in the DE algorithm, as it is highly correlated with its performance.

Before proceeding with the corresponding experimentations, we analyze how the error term behaves at each generation. Different sets of parameters are considered, yet the general behavior of them is depicted in Fig. 1, which shows the error term evolution through 200 generations when applied to 14 randomly selected stocks.

An important aspect to consider from this figure is that our models are not overfitting. A good performance with the training dataset, and a bad one with the test dataset, is an indicator of overfitting.

In our model, we see that our model shows a good performance with unseen data.

4.4 Portfolio Construction

For the portfolio optimization, we use the Python tool PyPortfolioOpt [46]. Five years of data are used to compute the exponential covariance matrix.

A distinct set of parameters are tested for the DE algorithm to compute expected returns, resumed in Table 4. We observed in this table that the best in terms of accumulated return and Sharpe ratio is the set of parameters on the second row, with a return on three years of the 39%.

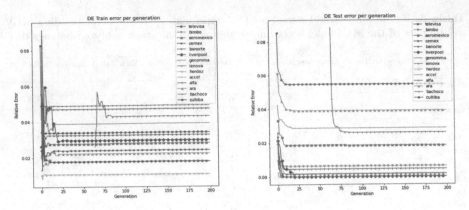

Fig. 1. Relative error per generation on DE (Train and Test, respectively)

Table 4. Performance of DE algorithm at different parameters (Population, Variant, CR, F and Dither are the corresponding DE parameters: Population: Population size. Variant: how to select individuals/number of difference vectors/crossover type. CR: the probability the individual exchanges variable values from the mutant vector. F: Weight of crossover. Dither: Strategy to introduce adaptive weights (F).)

Population					Max sharpe				Min vol			
	Variant	CR	F	Dither	Acc	Net	Std	Sharpe	Acc	Net	Std	Sharpe
100	best/2/bin	0.8	0.7	Scalar	1.272	0.008	0.042	0.357	0.921	−0.001	0.041	0.258
100	**best/2/bin**	**0.8**	**0.1**	**Scalar**	**1.393**	**0.010**	**0.045**	**0.404**	**1.067**	**0.002**	**0.035**	**0.366**
100	best/2/bin	1.0	0.1	Vector	1.249	0.007	0.046	0.340	1.036	0.001	0.029	0.343
100	best/1/bin	1.0	0.1	Vector	1.339	0.009	0.042	0.388	1.063	0.002	0.031	0.350
100	best/1/bin	1.0	0.7	Vector	1.116	0.004	0.039	0.331	1.096	0.003	0.031	0.382

The experiments with low mutation factors showed good results [40] but not all; we only tested high mutation factors [40]; we only tested variants with the format DE/best/n/bin [50,51] where n represents the difference vectors. The number of experimentations performed in terms of DE parameters is not enough to draw statistically significant observations out of them about the most adequate set of parameters to use to tackle the present project.

Figure 2 shows the portfolio accumulated returns compared against the benchmarks when using the best set of DE parameters (seen in bold in Table 4). In this figure, we see the two portfolios, conformed from this strategy (MaxSharpe, MinVol). It is observed that at least the portfolio derived with the MaxSharpe is surpassing the benchmarks, with an excess return of 12.2%.

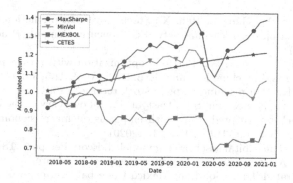

Fig. 2. Portfolio Returns vs Benchmarks. DE parameters: Population: 100. Variant: best/2/bin. CR: 0.8 F: 0.1. Dither: scalar.

5   Conclusion

This paper aims to predict returns of stocks quoted in the Mexican Stock Exchange using macroeconomic variables and technical analysis as input features, applying a machine learning model. Due to the problem of parametrization that technical analysis imposes to make market movement inferences, a tuning algorithm is required. Metaheuristics have proven themselves capable of tuned these parameters, improving the machine learning model's predictions, and minimizing its error.

To test these stock returns predictions, instead of analyzing the error of the machine learning models, a more realistic scenario is proposed to test them: construct an investment strategy by selecting and optimizing a portfolio in monthly intervals, considering the expected returns predictions.

Invest in a strategy on variable income assets, impose on the investor an extra risk to take, by which the expected portfolio exceed return must balance the risk. Results are encouraging as we could beat the MEXBOL and replicate the performance of the free-risk rate of return.

Results are sensitive to the portfolio optimization objectives and configuration, on future work we will research this topic more deeply.

Future work will test different metaheuristics as the Enhanced Firework and a different set of technical analyzers, as volume-based or trend-based analyzers.

References

1. Piamsuwannakit, S., Sriboonchitta, S.: Forecasting risk and returns: CAPM model with belief functions. In: Huynh, V.-N., Kreinovich, V., Sriboonchitta, S., Suriya, K. (eds.) Econometrics of Risk. SCI, vol. 583, pp. 259–271. Springer, Cham (2015). https://doi.org/10.1007/978-3-319-13449-9_18
2. Leung, E., Lohre, H., Mischlich, D., Shea, Y., Stroh, M.: The promises and pitfalls of machine learning for predicting stock returns. J. Fin. Data Sci. 3(2), 21–50 (2021)

3. Shi, Y., Zheng, Y., Guo, K., Ren, X.: Stock movement prediction with sentiment analysis based on deep learning networks. Concurr. Comput. Pract. Exp. **33**(6), e6076 (2021)

4. Ma, Y., Han, R., Wang, W.: Portfolio optimization with return prediction using deep learning and machine learning. Exp. Syst. Appl. **165**, 113973 (2021)

5. Chalupová, K.: Can machines explain stock returns? (2021)

6. Kaviani, M., Fakhrehosseini, S.F., Dastyar, F.: An overview of the importance and why the stock return prediction, with emphasis on macroeconomic variables. J. Acc. Account. Soc. Int. **10**(2), 113–131 (2020)

7. Perold, A.F.: The capital asset pricing model. J. Econ. Perspect. **18**(3), 3–24 (2004)

8. Siddiqui, S.: Arbitrage pricing theory: a review of literature

9. World Integrated Trade Solution: México trade balance, exports and imports by country and region 2018. Accessed 26 May 2020

10. Zamperin, F.: Testing standard technical analysis parameters' efficiency, a meta-heuristic approach. B.S. thesis, Università Ca'Foscari Venezia (2020)

11. Pholdee, N., Bureerat, S.: Comparative performance of meta-heuristic algorithms for mass minimisation of trusses with dynamic constraints. Adv. Eng. Softw. **75**, 1–13 (2014)

12. Markowitz, H.M.: Portfolio Selection. Yale University Press, New Haven (1968)

13. Abbas, G., Wang, S.: Does macroeconomic uncertainty really matter in predicting stock market behavior? A comparative study on China and USA. China Finance Review International (2020)

14. Jin, Z., Guo, K.: The dynamic relationship between stock market and macroeconomy at sectoral level: evidence from Chinese and US stock market. Complexity **2021**, 6645570 (2021)

15. Sivrikaya, A.: Macroeconomic variables and sector-specific returns: evidence from Turkish stock exchange market. Gaziantep University. J. Soc. Sci. **20**(1), 72–89 (2021)

16. Rjoub, H., Türsoy, T., Günsel, N.: Istanbul stock market. Studies in Economics and Finance, The effects of macroeconomic factors on stock returns (2009)

17. Pole, H., Cavusoglu, B.: The effect of macroeconomic variables on stock return volatility in the Nigerian stock exchange market. Asian J. Econ. Fin. Manag. **3**, 32–43 (2021)

18. Syed, A.A.: Symmetric and asymmetric influence of macroeconomic variables on stock prices movement: study of Indian stock market. In: New Challenges for Future Sustainability and Wellbeing. Emerald Publishing Limited (2021)

19. Keswani, S., Wadhwa, B.: Association among the selected macroeconomic factors and Indian stock returns. In: Proceedings Materials Today (2021)

20. Khan, M.K., Teng, J.-Z., Khan, M.I., Khan, M.Z.: Stock market reaction to macroeconomic variables: an assessment with dynamic autoregressive distributed lag simulations. Int. J. Fin. Econ. (2021)

21. Huang, P., Chen, S., Wei, W., Elkassabgi, A.: Influences of macroeconomic variables on stock market in China: an empirical analysis (2019). SSRN 3519674

22. Mayur, M., et al.: Relationship between stock market and economy: empirical evidence from India. Asian J. Emp. Res. **7**(6), 124–133 (2017)

23. Arouri, M.E.H., Nguyen, D.K.: Oil prices, stock markets and portfolio investment: evidence from sector analysis in Europe over the last decade. Ener. Policy **38**(8), 4528–4539 (2010)

24. Hatipoglu, F.B., Uyar, U.: Examining the dynamics of macroeconomic indicators and banking stock returns with Bayesian networks. Bus. Econ. Res. J. **10**(4), 807–822 (2019)

25. Gay, R.D.: Effect of macroeconomic variables on stock market returns for four emerging economies: Brazil, Russia, India, and China. Int. Bus. Econ. Res. J. (Iber) **15**(3), 119–126 (2016)
26. Bermudez Delgado, E., Saucedo, E.: Evidence from Mexico: Nancy Areli Bermudez Delgado, The relationship between oil prices, the stock market and the exchange rate. North Am. J. Econ. Fin. **45**, 266–275 (2018)
27. Celebi, K., Welfens, P.J., et al.: The stock market, labor-income risk and unemployment in the us: empirical findings and policy implications. Technical report, Universitätsbibliothek Wuppertal, University Library (2021)
28. Victoria, A.: Unemployment and aggregate stock returns. J. Bank. Fin. **129**, 106159 (2021)
29. Vrontos, S.D., Galakis, J., Vrontos. I.D.: Modeling and predicting us recessions using machine learning techniques. Int. J. Forecast. **37**(2), 647–671 (2021)
30. Benson, E.D., Kong. S.X.: The influence. Innovations **16**(4), 46–60 (2021)
31. Anand, V., Zhang, J.: International influence on stock markets in Pakistan
32. Strader, T.J., Rozycki, J.J., Root, T.H., John Huang, Y.H.: Machine learning stock market prediction studies: review and research directions. J. Int. Technol. Inf. Manag. **28**(4), 63–83 (2020)
33. Cervelló-Royo, R., Guijarro, F.: Forecasting stock market trend: a comparison of machine learning algorithms. Fin. Market. Val. **6**(1), 37–49 (2020)
34. Nabipour, M., Nayyeri, P., Jabani, H., Mosavi, A., Salwana, E., et al.: Deep learning for stock market prediction. Entropy **22**(8), 840 (2020)
35. Mangir, R., et al.: Using deep learning neural networks and candlestick chart representation to predict stock market. arXiv preprint arXiv:1903.12258 (2019)
36. Selvin, S., Vinayakumar, R., Gopalakrishnan, E.A., Menon, V.K., Soman, K.P.: Stock price prediction using ISTM, RNN and CNN-sliding window model. In: 2017 International Conference on Advances in Computing, Communications and Informatics (ICACCI), pp. 1643–1647. IEEE (2017)
37. McDonagh, S.B.: What drives prices in financial markets? (2020)
38. Schwager, J.D.: Getting Started in Technical Analysis, vol. 19. Wiley, New York (1999)
39. Pant, M., Zaheer, H., Garcia-Hernandez, L., Abraham, A., et al.: Differential evolution: a review of more than two decades of research. Eng. Appl. Artif. Intell. **90**, 103479 (2020)
40. Centeno-Telleria, M., Zulueta, E., Fernandez-Gamiz, U., Teso-Fz-Betoño, D., Teso-Fz-Betoño, A.: Differential evolution optimal parameters tuning with artificial neural network. Mathematics **9**(4), 427 (2021)
41. Wu, G., Shen, X., Li, X., Chen, H., Lin, A., Suganthan, P.N.: Ensemble of differential evolution variants. Inf. Sci. **423**, 172–186 (2018)
42. Blank, J., Deb, K.: pymoo: multi-objective optimization in python. IEEE Access **8**, 89497–89509 (2020)
43. Molyboga, M.: Portfolio management of commodity trading advisors with volatility targeting. J. Invest. Strat. (2018, Forthcoming)
44. Dzikevičius, A., Šaranda, S.: Ema versus SMA usage to forecast stock markets: the case of s&p 500 and OMX Baltic benchmark. Business Theory Pract. **11**(3), 248–255 (2010)
45. Sharpe, W.F.: The sharpe ratio. J. Portfolio Manag. **21**(1), 49–58 (1994)
46. Martin, R.A.: PyPortfolioOpt: portfolio optimization in python. J. Open Sour. Softw. **6**(61), 3066 (2021)
47. Bloomberg Professional: Bloonberg (2012). Accessed 10 Jan 2021

48. Python Software Foundation: Yfinance. Accessed 15 Apr 2021
49. Seabold, S., Perktold, J.: Statsmodels: econometric and statistical modeling with python. In: Proceedings of the 9th Python in Science Conference, Austin, vol. 57, p. 61 (2010)
50. Gämperle, R., Müller, S.D., Koumoutsakos, P.: A parameter study for differential evolution. Adv. Intell. Syst. Fuzzy Syst. Evol. Comput. **10**(10), 293–298 (2002)
51. Mezura-Montes, E., Jesús Velázquez-Reyes, J., Coello, C.A.: A comparative study of differential evolution variants for global optimization. In: Proceedings of the 8th Annual Conference on Genetic and Evolutionary Computation, pp. 485–492 (2006)

Towards a Pareto Front Shape Invariant Multi-Objective Evolutionary Algorithm Using Pair-Potential Functions

Luis A. Márquez-Vega[1]([✉])([iD]), Jesús Guillermo Falcón-Cardona[2]([iD]),
and Edgar Covantes Osuna[1]([iD])

[1] School of Engineering and Science, Tecnologico de Monterrey, 64849 Monterrey,
Nuevo León, Mexico
a00832536@itesm.mx, edgar.covantes@tec.mx
[2] Department of Applied Mathematics and Systems, UAM Cuajimalpa, 05348
Cuajimalpa de Morelos, Mexico City, Mexico
jfalcon@cua.uam.mx

Abstract. Reference sets generated with uniformly distributed weight vectors on a unit simplex are widely used by several multi-objective evolutionary algorithms (MOEAs). They have been employed to tackle multi-objective optimization problems (MOPs) with four or more objective functions, i.e., the so-called many-objective optimization problems. These MOEAs have shown a good performance on MOPs with regular Pareto front shapes, i.e., simplex-like shapes. However, it has been observed that in many cases, their performance degrades on MOPs with irregular Pareto front shapes. In this paper, we designed a new selection mechanism that aims to promote a Pareto front shape invariant performance of MOEAs that use weight vector-based reference sets. The newly proposed selection mechanism takes advantage of weight vector-based reference sets and seven pair-potential functions. It was embedded into the non-dominated sorting genetic algorithm III (NSGA-III) to increase its performance on MOPs with different Pareto front geometries. We use the DTLZ and DTLZ^{-1} test problems to perform an empirical study about the usage of these pair-potential functions for this selection mechanism. Our experimental results show that the pair-potential functions can enhance the distribution of solutions obtained by weight vector-based MOEAs on MOPs with irregular Pareto front shapes. Also, the proposed selection mechanism permits maintaining the good performance of these MOEAs on MOPs with regular Pareto front shapes.

Keywords: Reference sets · Multi-objective evolutionary algorithms · Pareto front shape invariant performance · Pair-potential functions

The first author acknowledges support given by Tecnológico de Monterrey and Consejo Nacional de Ciencia y Tecnología (CONACYT) to pursue graduate studies under the number CVU 859751.

I. Batyrshin et al. (Eds.): MICAI 2021, LNAI 13067, pp. 369–382, 2021.
https://doi.org/10.1007/978-3-030-89817-5_28

1   Introduction

During the last decades, multi-objective evolutionary algorithms (MOEAs) have been widely used to approximate the solution of complex multi-objective optimization problems (MOPs). Typically, MOEAs have been designed using the Pareto dominance relation[1] as the backbone of their selection mechanisms. In contrast to mathematical programming techniques, population-based meta-heuristics, e.g., MOEAs, can generate multiple approximate solutions in a single execution and do not require information on derivatives which allows them to deal with complex MOPs [2]. An unconstrained MOP can be defined, without loss of generality, as a minimization problem as follows:

$$\min_{\vec{x} \in \Omega} \left\{ f(\vec{x}) := [f_1(\vec{x}), f_2(\vec{x}), \dots, f_m(\vec{x})]^T \right\},$$

where $\vec{x} = [x_1, x_2, \dots, x_n]^T$ is a decision vector, $\Omega \subseteq \mathbb{R}^n$ is the decision space, $f(\vec{x}) \in \Lambda$ is an objective vector with m (≥ 2) mutually conflicting objective functions, and $\Lambda \subseteq \mathbb{R}^m$ is the objective space. The objective function, $f : \Omega \to \Lambda$, maps each decision vector $\vec{x}$ to an objective vector $f(\vec{x})$ and $f_i : \Omega \to \mathbb{R}, i = 1, 2, \dots, m$. Due to the conflicting nature of the objective functions, the solution to a MOP is composed of a set of Pareto optimal solutions, where each one represents a trade-off among the objective functions. In the decision space, this solution set is called the *Pareto set*, and its corresponding image in the objective space is called the *Pareto front*.

Pareto dominance-based MOEAs have demonstrated a good capability to deal with MOPs that include two and three objective functions. However, their performance degrades when tackling MOPs with four or more objective functions, i.e., the so-called many-objective optimization problems (MaOPs). The main reason of this performance degradation is the exponential increase of non-dominated solutions as the objective space increases, resulting in the loss of their selection pressure [11]. Various approaches have been proposed to enhance the convergence and diversity of Pareto dominance-based MOEAs on MaOPs, e.g., decomposition-based MOEAs or indicator-based MOEAs [7,12].

Several decomposition-based MOEAs and indicator-based MOEAs require reference sets to decompose the objective space, define search directions, or calculate quality indicators embedded into the selection mechanisms. A common mechanism to generate such reference sets is the generation of convex weight vectors[2] using the systematic approach proposed by Das and Dennis [3]. The non-dominated sorting genetic algorithm III (NSGA-III) [4] is a well-known MOEA that employs a weight vector-based reference set.

It has been empirically shown that the performance of MOEAs that use a systematically generated weight vector-based reference set depends on the

---

[1] Let $\vec{x}, \vec{y} \in \mathbb{R}^m$. It is said that $\vec{x}$ Pareto-dominates $\vec{y}$ (denoted as $\vec{x} \prec \vec{y}$) if $f_i(\vec{x}) \leq f_i(\vec{y}), \forall i \in \{1, 2, \dots, m\}$ and $\exists j \in \{1, 2, \dots, m\}$ such that $f_j(\vec{x}) < f_j(\vec{y})$.

[2] It is said that a vector $\vec{w} \in \mathbb{R}^m$ is a convex weight vector if $\sum_{i=1}^{m} w_i = 1$ and $w_i \geq 0, \forall i \in \{1, 2, \dots, m\}$.

Pareto front shape [10]. These MOEAs can obtain an outstanding performance on MOPs with regular *Pareto front* shapes, i.e., simplex-like *Pareto fronts*. However, their performance degrades on MOPs with irregular *Pareto front* shapes, e.g., degenerated, discontinuous, or inverted.

In recent years, the so called pair-potential functions have been used in the field of evolutionary multi-objective optimization (EMO) to improve the diversity of MOEAs by maintaining well-distributed solutions. The pair-potential functions can be used as selection mechanisms to promote an invariant performance of MOEAs regardless of the *Pareto front* shape [8,9]. In this paper, we propose a new selection mechanism based on niching and pair-potential functions. This selection mechanism aims to promote an invariant performance of MOEAs that employ weight vector-based reference sets regardless of the *Pareto front* shape. This selection mechanism can be used by weight vector-based MOEAs. However, it requires a previous association procedure between solutions and reference points. In order to investigate the performance of this selection mechanism with different pair-potential functions, it is integrated into NSGA-III.

The remainder of this paper is organized as follows. Section 2 contains a background of basic terms related to the usage of pair-potential functions as selection mechanism. Section 3 presents our proposed approach and Sect. 4 defines the experimental setup. The results are showed and discussed in Sect. 5. Finally, the main conclusions and future work are outlined in Sect. 6.

2   Selection Mechanism Based on Pair-Potential Functions

The pair-potential functions are used to distribute points on a manifold (or discretizing a manifold) [1]. Given a d-dimensional manifold $\mathcal{A}$ in the search space $\mathbb{R}^m$ ($d \leq m$) with a given distribution X and described by some geometric property or by some parametrization, the goal is to generate numerous N points in $\mathcal{A}$ such that they are well-separated and have (nearly) distribution X. In other words, the aim is to generate the smallest population possible that describes the distribution of the full set of elements in $\mathcal{A}$ [8].

The pair-potential functions have been used in the EMO field as diversity indicators and selection mechanisms that aim to generate approximation sets[3] with good diversity regardless of the *Pareto front* shape [8,9]. Mathematically, given an approximation set $\mathcal{A} = \{\vec{a_1}, \vec{a_2}, \ldots, \vec{a_N}\}$, the total energy U using a pair-potential function $\mathcal{K}$ is defined as follows:

$$U(\mathcal{A}) = \sum_{\vec{a_i} \in \mathcal{A}} \left\{ \sum_{\vec{a_j} \in \mathcal{A} \setminus \{\vec{a_i}\}} \mathcal{K}(\vec{a_i}, \vec{a_j}) \right\}.$$

The minimization of the total energy U is related to a better distribution of solutions in the approximation set $\mathcal{A}$. In order to use a pair-potential function as

---

[3] An approximation set $\mathcal{A}$ is a set of objective vectors that aims to approximate the *Pareto front* such that $\forall \vec{a_i}, \vec{a_j} \in \mathcal{A} \mid \vec{a_i} \nprec \vec{a_j}$ and $\vec{a_j} \nprec \vec{a_i}$.

part of a selection mechanism, it has been commonly used the calculation of the individual contributions to the total energy U [8,9]. The individual contribution of a given solution $\vec{a} \in \mathcal{A}$ to U is defined as:

$$C(\vec{a}, \mathcal{A}) = \frac{1}{2}|U(\mathcal{A}) - U(\mathcal{A} \setminus \{\vec{a}\})|,$$

where the term $1/2$ is employed because $\mathcal{K}(\vec{a_i}, \vec{a_j}) = \mathcal{K}(\vec{a_j}, \vec{a_i})$. Using the individual contributions C, it is possible to reduce the cardinality of an approximation set $\mathcal{A}$ by deleting the worst-contributing solution. In this study, the memoization structure proposed by Falcón-Cardona et al. [9] is used to reduce the computational cost of computing the individual contributions.

2.1 Pair-Potential Functions

The selected pair-potential functions are defined in the same way as [8]. In each definition, $\vec{u}$ and $\vec{v}$ are objective vectors, and $||\cdot||$ denotes the Euclidean distance.

Definition 1 (Riesz s-energy). *Given a parameter $s > 0$, it is defined by:*

$$\mathcal{K}_{RSE}(\vec{u}, \vec{v}) = \frac{1}{||\vec{u} - \vec{v}||^s}.$$

Definition 2 (Gaussian α-energy). *Given a parameter $\alpha > 0$, this function is defined by:*

$$\mathcal{K}_{GAE}(\vec{u}, \vec{v}) = e^{-\alpha||\vec{u}-\vec{v}||^2}.$$

Definition 3 (Coulomb's law). *It is defined as follows:*

$$\mathcal{K}_{COU}(\vec{u}, \vec{v}) = \frac{1}{4\pi\epsilon_0} \cdot \frac{q_1 q_2}{||\vec{u} - \vec{v}||^2},$$

where $q_1 = ||\vec{u}||$, $q_2 = ||\vec{v}||$, and $\frac{1}{4\pi\epsilon_0}$ is the Coulomb's constant.

Definition 4 (Pöschl-Teller Potential). *Given $V_1, V_2, \alpha > 0$, it is given by:*

$$\mathcal{K}_{PT}(\vec{u}, \vec{v}) = \frac{V_1}{\sin^2(\alpha||\vec{u} - \vec{v}||)} + \frac{V_1}{\cos^2(\alpha||\vec{u} - \vec{v}||)}.$$

Definition 5 (Modified Pöschl-Teller Potential). *Given the parameters $D, \alpha > 0$, this pair-potential function is defined as follows:*

$$\mathcal{K}_{MPT}(\vec{u}, \vec{v}) = -\frac{D}{\cosh^2(\alpha||\vec{u} - \vec{v}||)}.$$

Definition 6 (General form of the Pöschl-Teller Potential). *Given the parameters $A, B > 0$, this function is given as follows:*

$$\mathcal{K}_{GPT}(\vec{u}, \vec{v}) = \frac{A}{1 + \cos(||\vec{u} - \vec{v}||)} + \frac{B}{1 - \cos(||\vec{u} - \vec{v}||)}.$$

Definition 7 (Kratzer Potential). *Given $V_1, V_2, \alpha > 0$, it is given by:*

$$\mathcal{K}_{KRA}(\vec{u}, \vec{v}) = V_1 \left(\frac{||\vec{u} - \vec{v}|| - 1/\alpha}{||\vec{u} - \vec{v}||} \right)^2 + V_2.$$

3   NSGA-III with Pareto Front Shape Invariant Selection Mechanism

This section is used to explain the implementation of NSGA-III with the new selection mechanism. First, the general framework and the environmental selection are described. Then, the selection mechanism based on niching and pair-potential functions is detailed.

3.1   General Framework

NSGA-III uses a hybrid selection mechanism composed by the Pareto dominance relation and a niching approach. A weight vector-based reference set is used for NSGA-III to perform a niche-preservation strategy and promote diversity. NSGA-III has demonstrated good performance on a wide range of MaOPs. However, it has been found that its performance degrades when tackling MOPs with irregular *Pareto front* shapes [10]. The general framework of NSGA-III with selection mechanism based on niching and pair-potential functions (NSGA-III-$\mathcal{K}$) is presented in Algorithm 1. The implementation of this proposal maintains the general framework of NSGA-III [4].

Algorithm 1. NSGA-III with Pareto front shape invariant selection mechanism

Require: H_1: Number of divisions per objective for the boundary layer; H_2: Number of divisions per objective for the inner layer; $\mathcal{K}$: Pair-potential function.
Ensure: P: Final population.
 1: $W = \text{UniformReferenceSet}(H_1, H_2)$;
 2: $N = |W|$ and $E = \emptyset$;
 3: $P = \text{RandomPopulation}(N)$;
 4: Obtain $\vec{z}_{\min}$ from P;
 5: **while** termination criterion **is not** fulfilled **do**
 6:     $M = \text{RandomMatingSelection}(P, N)$;
 7:     $Q = \text{GenerateOffspring}(M, N)$;
 8:     $R = P \cup Q$;
 9:     Obtain $\vec{z}_{\min\_pop}$ from R;
10:     $z_{\min,i} = \min\{z_{\min,i}, z_{\min\_pop,i}\}, \forall i \in \{1, \ldots, m\}$;
11:     $P, E = \text{EnvironmentalSelection}(R, W, N, \vec{z}_{\min}, E, \mathcal{K})$;
12: **return** P

First, a reference set W is generated using the systematic approach proposed by Das and Dennis [3]. The parameters $H_1, H_2 \in \mathbb{N}$ denote the number of divisions per objective for the boundary and the inner layer, respectively. When only one layer is generated, i.e. $H_1 > 0$ and $H_2 = 0$, the cardinality of the reference set is given by $|W| = C_{m-1}^{H_1+m-1}$, where m is the number of the objective functions. However, when the two layers are required, i.e. $H_1, H_2 > 0$, the cardinality of the reference set is given by $|W| = C_{m-1}^{H_1+m-1} + C_{m-1}^{H_2+m-1}$. The population size is set to $N = |W|$, and an empty matrix E is initialized to contain the extreme points for normalization. Then, a random population P of size N is initialized, and $\vec{z}_{\min}$ is defined using the minimum objective values of the population P.

During the main loop, a random mating selection is used to select the parent population M. The offspring population Q of size N is generated using the simulated binary crossover (SBX) and polynomial mutation [5]. The populations P and Q are combined to generate the population R. Then, the vector $\vec{z}_{\min\_pop}$ is defined using the minimum objective values of the population R, and $\vec{z}_{\min}$ is updated. The environmental selection (described in Algorithm 2) selects the new population P, and the extreme points are updated. If the termination criterion is not met, the algorithm begins its new cycle from the mating selection step.

3.2 Environmental Selection Based on Niching and Pair-Potential

The environmental selection is presented in Algorithm 2. It is similar to the used in NSGA-III. However, a selection mechanism based on niching and pair-potential functions is added. This selection mechanism is employed when the number of non-dominated solutions in the combined population equals or exceeds the population size.

Algorithm 2. EnvironmentalSelection procedure

Require: R: Combined population; W: Reference set; N: Population size; $\vec{z}_{\min}$: Minimum point; E: Extreme points; $\mathcal{K}$: Pair-potential function.
Ensure: P: Selected population; E: Updated extreme points.
1: $P = \emptyset$;
2: $\{F_1, \ldots, F_k\} = \text{NondominatedSort}(R)$;
3: Obtain $\vec{z}_{\max\_front}$ from F_1 and $\vec{z}_{\max\_pop}$ from R;
4: Select l as the minimum number such that: $|\cup_{i=1}^{l} F_i| > N$;
5: $P = \cup_{i=1}^{l-1} F_i$;
6: **if** $|P| < N$ **then**
7:     $K = N - |P|$ and $S = \cup_{i=1}^{l} F_i$;
8:     $S_n, E = \text{Normalize}(S, \vec{z}_{\min}, \vec{z}_{\max\_front}, \vec{z}_{\max\_pop}, E)$;
9:     $\vec{\pi}, \vec{d} = \text{Associate}(S_n, W)$;
10:    **if** $|P| > 0$ **then**
11:        $\rho_i = \sum_{j=1}^{|P|} ((\pi_j = i) \rightarrow 1 : 0), \forall i \in \{1, \ldots, N\}$;
12:        $P = \text{Niching}(K, \vec{\rho}, \vec{\pi}, \vec{d}, W, F_l, P)$;
13:    **else**
14:        $\rho_i = \sum_{j=1}^{|F_l|} ((\pi_j = i) \rightarrow 1 : 0), \forall i \in \{1, \ldots, N\}$;
15:        $P = \text{NichingPairPotential}(N, \vec{\rho}, \vec{\pi}, \vec{d}, F_l, \vec{z}_{\min}, \vec{z}_{\max\_front}, \mathcal{K})$;
16: **return** P, E

First, the population P is set to empty. In the following, the solutions from the combined population R are divided into sub-populations depending on their non-domination levels. The points $\vec{z}_{\max\_front}$ and $\vec{z}_{\max\_pop}$ are defined using the maximum objective values of the first front F_1 and the combined population R, respectively. Then, the front F_l is selected such that $| \cup_{i=1}^{l} F_i | > N$, and the solutions from the first $l-1$ fronts are included in the population P.

If the number of selected solutions in the population P is less than the population size N, the remaining K solutions are selected using the following steps. The first l fronts are combined to create the population S which is normalized using the normalization procedure of NSGA-III. The vectors $\vec{\pi}$ and $\vec{d}$ are defined using the associate procedure of NSGA-III to contain the nearest reference point for each solution and its corresponding perpendicular distance, respectively. If the population P contains at least one solution, then the original steps of NSGA-III are executed. First, the vector $\vec{\rho}$ that contains the number of associated solutions from the first $l-1$ fronts of each reference point is calculated. Then, the niching procedure of NSGA-III is used to select K solutions from the front F_l. However, if the population P is still empty, i.e., $F_1 = F_l$, the vector $\vec{\rho}$ is defined to contain the number of associated solutions from the front F_l of each reference point. Then, the niching procedure with pair-potential functions (described in Algorithm 3) is used to select the population P.

3.3 Niching Procedure with Pair Potential Functions

The niching procedure with the pair-potential functions is presented in Algorithm 3. It uses two different selection mechanisms. The first mechanism selects the nearest solution to each reference point with at least a niche count of one. While, the second mechanism selects the remaining solutions using the pair-potential functions defined in Sect. 2.1 to improve diversity.

First, the population S is set to empty. The vector $\vec{J}$ is generated by including the reference points with at least a niche count of one. Then, the following steps are executed for each reference point j included in $\vec{J}$. The vector $\vec{I}$ is generated by including the solutions that are associated with the reference point j, and the solution from $\vec{I}$ with the minimum perpendicular distance to the reference point j is selected to be included in the population S.

If the number of solutions in the population S is equal to the population size N, the selected population P is set to be equal to the population S. Otherwise, the remaining solutions are selected using the selection mechanism based on the pair-potential functions. The front F_l is normalized using the min-max normalization. Then, the dissimilarity matrix D is calculated using the normalized population $\mathcal{A}$ and a given pair-potential function $\mathcal{K}$. The memoization structure $\vec{r}$ is defined with the method proposed by Falcón-Cardona et al. [9]. The following steps are repeated until $|F_l| = N$. The vector of individual contributions $\vec{C}$ is defined by using the absolute values of the memoization structure $\vec{r}$. Then, the worst solution s in the population $F_l \setminus S$ is detected. In this way, the solutions that are previously included in the population S are protected from being deleted from F_l. The values associated with the worst solution s are deleted from the dissimilarity matrix D and the memoization structure $\vec{r}$ is updated. The worst solution s is also deleted from the front F_l. Finally, when the number of solutions in the front F_l is equal to the population size N, the selected population P is set to be equal to the front F_l.

Algorithm 3. NichingPairPotential procedure

Require: N: Population size; $\vec{p}$: Associated solutions of each reference point; $\vec{\pi}$: Nearest reference point for each solution; $\vec{d}$: Perpendicular distance to nearest reference point for each solution; F_l: Critical front; $\vec{z}_{\min}$: Minimum point; $\vec{z}_{\max\_front}$: Maximum point of first front; $\mathcal{K}$: Pair-potential function.

Ensure: P: Selected population.

1: $S = \emptyset$ and $\vec{J} = \{j \mid \rho_j > 0\}$;
2: **for all** $j \in \vec{J}$ **do**
3:      $\vec{I} = \{i \mid \pi_i = j\}$;
4:      $s = \arg\min_{i \in \vec{I}} d_i$;
5:      $S = S \cup F_{l,s}$;
6: **if** $|S| = N$ **then**
7:      $P = S$;
8: **else**
9:      $\mathcal{A} = (F_l - \vec{z}_{\min})/(\vec{z}_{\max\_front} - \vec{z}_{\min})$;
10:      $D_{ij} = \mathcal{K}(a_i, a_j), \forall a_i, a_j \in \mathcal{A}$ and $i \neq j$;
11:      $r_i = \sum_{j=1}^{|F_l|} D_{ij}, \forall i \in \{1, \ldots, |F_l|\}$;
12:      **while** $|F_l| > N$ **do**
13:          $C_i = |r_i|, \forall i \in \{1, \ldots, |\vec{r}|\}$;
14:          $s = \arg\max_{i \in F_l \setminus S} C_i$;
15:          Delete the row and column associated to s from the dissimilarity matrix D;
16:          Update the memoization structure $\vec{r}$;
17:          $F_l = F_l \setminus F_{l,s}$;
18:      $P = F_l$;
19: **return** P

---

4 Experimental Setup

We present the experimental setup used for the performance assessment of NSGA-III-$\mathcal{K}$ with each pair-potential function as selection mechanism. The test problems, the parameters used for each MOEA, and the performance indicator are presented in this section.

4.1 Test Problems

To assess the performance of the proposed selection mechanism with each pair-potential function, we have selected the DTLZ [6] and DTLZ^{-1} [10] test problems due to their scalability in the objective space and the wide range of *Pareto front* shapes. The DTLZ1–DTLZ4 have regular *Pareto front* shapes from which DTLZ1 has a linear *Pareto front* shape, while the DTLZ2–DTLZ4 have concave *Pareto front* shapes. The rest of test problems includes irregular *Pareto front* shapes. DTLZ5 and DTLZ6 have mostly degenerated *Pareto front* shapes, the DTLZ7 and DTLZ7^{-1} have discontinuous *Pareto front* shapes, the DTLZ1^{-1} has an inverted linear *Pareto front* shape, and the DTLZ2^{-1}–DTLZ6^{-1} have inverted convex *Pareto front* shapes.

Table 1. Parameter settings for each pair-potential function.

Pair-potential function	Parameter values
$\mathcal{K}_{\text{RSE}}$	$s = m - 1$
$\mathcal{K}_{\text{GAE}}$	$\alpha = 512$
$\mathcal{K}_{\text{PT}}$	$V_1 = 5.0,\ V_2 = 3.0,\ \alpha = 0.02$
$\mathcal{K}_{\text{MPT}}$	$D = 1.0,\ \alpha = 25.0$
$\mathcal{K}_{\text{GPT}}$	$A = 1.0,\ B = 1.0$
$\mathcal{K}_{\text{KRA}}$	$V_1 = 5.0,\ V_2 = 3.0,\ \alpha = 0.02$

The number of decision variables for the DTLZ and DTLZ^{-1} test problems is set to $n = m - 1 + k$, where m is the number of objective functions, and k is set to 5 for the DTLZ1, 10 for the DTLZ2–DTLZ6, and 20 for the DTLZ7. The same k values are used for the corresponding DTLZ^{-1} versions. All the test problems are scaled to 3, 5, 8, and 10 objective functions.

4.2 Parameter Settings

The population size N, the number of divisions per objective for the boundary layer H_1 and inner layer H_2 vary depending on the number of objective functions m. Therefore, the tuple (m, H_1, H_2, N) is set as in the original paper of NSGA-III [4]: $(3, 12, 0, 91)$, $(5, 6, 0, 210)$, $(8, 3, 2, 156)$, and $(10, 3, 2, 275)$. The maximum number of generations for the DTLZ1–DTLZ4 is showed in Table 2 [4], while for the DTLZ5–DTLZ7 and DTLZ1^{-1}–DTLZ7^{-1} is included in Table 3. The crossover and mutation probabilities are set to 1 and $1/n$, respectively, where n denotes the number of decision variables, and the crossover and mutation distribution indexes are set to 30 and 20, respectively [4].

Some pair-potential functions require the specification of parameter values. Table 1 summarizes the selected parameter values for each pair-potential function. These parameters values are selected based on the ones proposed by Falcón-Cardona et al. [8]. However, the parameter α of the pair-potential functions $\mathcal{K}_{\text{MPT}}$ and $\mathcal{K}_{\text{KRA}}$ differ from the cited work because they obtain a poor performance in the proposed selection mechanism. In this paper, these two parameters were defined through a preliminary trial-and-error experimentation in order to obtain a better performance.

4.3 Performance Assessment

The hypervolume (HV) indicator [14] is selected to assess the convergence and distribution of the approximation sets obtained by each MOEA. The reference point $\vec{z}_{\text{ref}}$ needed during the HV calculation is defined as $\vec{z}_{\text{ref}} = [1.5, 1.5, \ldots, 1.5]$ in the normalized objective space [13]. Larger values of HV indicate a better performance. All MOEAs are executed for 30 independent runs in each test problem. The one-tailed Wilcoxon test with a significance level of $\alpha = 0.05$ is

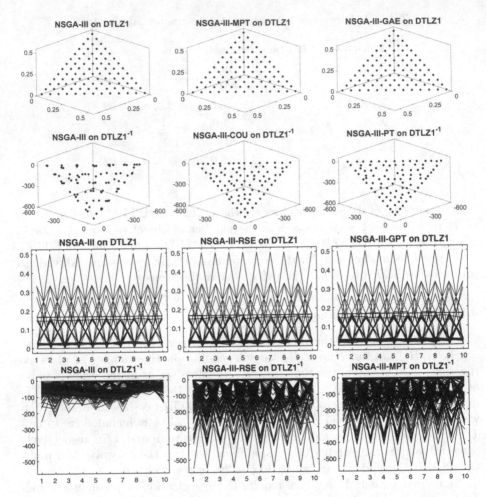

Fig. 1. Approximation sets with the median HV value of NSGA-III and MOEAs with the 1st and 2nd rank on the DTLZ1 and DTLZ1^{-1} with 3 and 10 objectives.

used to compare each version of NSGA-III-$\mathcal{K}$ against NSGA-III. The symbols "+", "−", and "=" indicate if the result is significantly better, significantly worse, or statistically equivalent to the result obtained by NSGA-III. All MOEAs were implemented on Python 3.8. The source code is available at https://github.com/lmarquezvg/NSGA-III-K.

5   Results and Discussion

The performance assessment of MOEAs is given in this section. Each version of NSGA-III-$\mathcal{K}$ is compared against NSGA-III to determine if the use of the proposed selection mechanism based on niching and pair-potential functions is capable to promote an invariant performance regardless the *Pareto front* shape.

Table 2. Mean and standard deviations (in parentheses) of the HV indicator on MOPs with regular *Pareto front* shapes. The two best mean values per MOP are highlighted in grayscale, where the darker tone corresponds to the best mean value. The symbols "+", "−", and "=" indicate that the result is significantly better, significantly worse, or statistically equivalent to the result obtained by NSGA-III based on the one-tailed Wilcoxon test using a significance level of $\alpha = 0.05$. The superscripts represent the obtained rank in the comparison.

MOP	Obj.	Gen.	NSGA-III	NSGA-III-RSE	NSGA-III-GAE	NSGA-III-COU	NSGA-III-PT	NSGA-III-MPT	NSGA-III-GPT	NSGA-III-KRA
DTLZ1	3	400	3.175577e+00^5 (2.980116e-03)	3.176038e+00^7 = (1.826034e-03)	3.176357e+00^2 = (1.459748e-03)	3.176127e+00^6 = (1.529506e-03)	3.176100e+00^3 = (2.043837e-03)	3.176436e+00^8 (1.366416e-03)	3.176254e+00^3 = (2.144405e-03)	3.176100e+00^5 = (1.872642e-03)
	5	600	7.562174e+00^4 (1.810539e-04)	7.562211e+00^6 = (1.482115e-04)	7.562227e+00^3 = (1.178189e-04)	7.562152e+00^8 = (2.527862e-04)	7.562177e+00^5 = (2.041152e-04)	7.362242e-00^1 (1.964118e-04)	7.562179e+00^0 = (1.634752e-04)	7.562233e+00^4 = (1.568861e-04)
	8	750	2.562621e+01^5 (8.142481e-05)	2.562621e+01^2 = (6.139096e-05)	2.562621e+01^3 = (6.911925e-05)	2.562622e+01^4 = (5.333046e-05)	2.562621e+01^6 = (6.491374e-05)	2.562621e+01^6 = (7.319004e-05)	2.562619e+01^7 = (6.694997e-05)	2.562619e+01^1 = (5.719597e-05)
	10	1000	5.766448e+01^5 (7.943768e-06)	5.766448e+01^4 = (6.512587e-06)	5.766448e+01^4 = (8.709883e-06)	5.766448e+01^3 = (6.432675e-06)	5.766448e+01^6 = (7.427814e-06)	5.766448e+01^7 = (8.304548e-06)	5.766448e+01^2 + (8.276820e-06)	5.766448e+01^3 + (6.102572e-06)
DTLZ2	3	250	2.791140e+00^5 (1.344799e-04)	2.791149e+00^2 = (7.876595e-05)	2.791144e+00^8 = (9.430047e-05)	2.791146e+00^9 = (9.083346e-05)	2.791132e+00^0 = (9.707114e-05)	2.791122e+00^7 = (1.482088e-04)	2.791085e+00^6 = (1.538171e-04)	2.791159e+00^0 = (8.266148e-05)
	5	350	7.304333e+00^5 (1.840647e-04)	7.304281e+00^4 = (2.476045e-04)	7.304343e+00^7 = (1.640247e-04)	7.304383e+00^3 = (1.714986e-04)	7.304298e+00^0 = (1.731407e-04)	7.304369e+00^2 = (1.832655e-04)	7.304334e+00^6 = (1.684122e-04)	7.304331e+00^0 = (1.823997e-04)
	8	500	2.550185e+01^5 (3.886465e-04)	2.550184e+01^4 = (3.076401e-04)	2.550171e+01^6 = (3.323231e-04)	2.550196e+01^1 = (4.331393e-04)	2.550177e+01^7 = (3.163166e-04)	2.550192e+01^2 = (2.822520e-04)	2.550181e+01^6 = (2.745163e-04)	2.550181e+01^6 = (3.695267e-04)
	10	750	5.760274e+01^5 (1.393948e-04)	5.760275e+01^2 = (1.013387e-04)	5.760273e+01^3 = (1.522928e-04)	5.760279e+01^1 = (1.323197e-04)	5.760268e+01^6 = (1.484282e-04)	5.760270e+01^7 = (1.431923e-04)	5.760273e+01^6 = (1.163551e-04)	5.760275e+01^2 = (1.214666e-04)
DTLZ3	3	1000	2.810352e+00^5 (4.066996e-03)	2.809489e+00^6 = (5.243501e-03)	2.811294e+00^2 = (3.651869e-03)	2.811603e+00^7 = (3.053299e-03)	2.810540e+00^3 = (3.583411e-03)	2.810705e+00^4 = (5.160424e-03)	2.811117e+00^8 = (3.846387e-03)	2.810212e+00^7 = (4.360882e-03)
	5	1000	7.302804e+00^5 (2.106571e-03)	7.303134e+00^6 = (2.124160e-03)	7.302494e+00^0 = (2.891471e-03)	7.303416e+00^3 = (1.277905e-03)	7.303625e+00^2 (1.513639e-03)	7.303189e+00^6 = (2.196343e-03)	7.303385e+00^4 = (2.420360e-03)	7.303459e+00^2 + (9.221803e-03)
	8	1000	2.562686e+01^2 (1.016394e-04)	2.562289e+01^4 = (2.113306e-02)	2.562073e+01^6 − (3.254554e-02)	2.562687e+01^1 − (7.050230e-05)	2.560363e+01^5 − (9.947627e-02)	2.560885e+01^7 − (9.726683e-02)	2.562245e+01^6 − (1.681090e-02)	2.562656e+01^7 − (4.827674e-04)
	10	1500	5.766462e+01^5 (6.789106e-06)	5.766462e+01^6 = (8.603661e-06)	5.766463e+01^1 = (8.466723e-06)	5.766463e+01^2 = (4.901325e-06)	5.766462e+01^7 = (1.063501e-05)	5.766462e+01^7 = (7.143842e-06)	5.766462e+01^3 = (9.613209e-06)	5.766462e+01^7 − (8.448628e-06)
DTLZ4	3	600	2.696569e+00^5 (4.107718e-01)	2.801681e+00^2 = (1.555989e-02)	2.750559e+00^4 = (2.956073e-01)	2.785048e+00^3 = (1.066882e-01)	2.750489e+00^4 = (2.955931e-01)	2.750541e+00^7 = (2.956030e-01)	2.80441e+00^1 (4.376422e-04)	2.750557e+00^5 = (2.956060e-01)
	5	1000	7.340938e+00^5 (1.388998e-04)	7.340936e+00^6 = (1.472632e-04)	7.340946e+00^0 = (1.055228e-04)	7.340947e+00^3 = (1.578714e-04)	7.340957e+00^2 = (1.019578e-04)	7.340906e+00^7 = (1.549399e-04)	7.340971e+00^1 (1.148422e-04)	7.340931e+00^0 = (1.936923e-04)
	8	1250	2.549580e+01^2 (1.549123e-04)	2.549581e+01^1 = (1.633841e-21)	2.549579e+01^6 = (1.334356e-04)	2.549577e+01^8 = (2.007572e-04)	2.549578e+01^7 = (1.423594e-04)	2.549579e+01^4 = (1.867019e-04)	2.549579e+01^4 = (1.833206e-04)	2.549579e+01^0 = (2.441980e-04)
	10	2000	5.759736e+01^5 (9.640730e-05)	5.759738e+01^1 = (7.645598e-05)	5.759735e+01^6 = (1.070718e-04)	5.759737e+01^3 = (1.052665e-04)	5.759734e+01^6 = (1.065170e-04)	5.759736e+01^7 = (1.219648e-04)	5.759736e+01^4 = (1.046005e-04)	5.759737e+01^0 = (1.000552e-04)
+/-/=				1/1/14	0/1/15	0/0/16	0/2/14	0/1/15	1/2/13	3/3/10

The results of the selected MOEAs in terms of HV on MOPs with regular *Pareto front* shapes are shown in Table 2. It can be seen that all versions of NSGA-III-$\mathcal{K}$ are capable of obtaining an equivalent performance to NSGA-III on most MOPs. This behavior is possible due to their primary selection mechanism to select the nearest solution to each reference point with at least a niche count of one. This selection mechanism permits that all reference points can contribute to the selection of its nearest solution on MOPs with regular *Pareto front* shapes. However, on the DTLZ3 with 8 and 10 objectives, various MOEAs perform significantly worse than NSGA-III. These results can be due to the multi-modality of the MOP, which promotes the emergence of dominance resistant solutions. Such solutions can be selected by the pair-potential functions due to their high level of diversity. This observation is supported by the fact that on the DTLZ2 and DTLZ4, i.e., MOPs without multi-modality, all the MOEAs perform statistically equivalent to NSGA-III. NSGA-III-COU obtains the best results on MOPs with regular *Pareto front* shapes, and it is the only MOEA capable of being statistically equivalent to NSGA-III on all MOPs.

The performance assessment of MOEAs in terms of HV on MOPs with irregular *Pareto front* shapes is shown in Table 3. It shows that all the versions of NSGA-III-$\mathcal{K}$ perform significantly better than NSGA-III on most MOPs. Additionally, NSGA-III obtains the last rank on 27 out of 40 MOPs. These results are due to the fact that not all reference points have solutions associated with

Table 3. Mean and standard deviations (in parentheses) of the HV indicator on MOPs with irregular *Pareto front* shapes. The two best mean values per MOP are highlighted in grayscale, where the darker tone corresponds to the best mean value. The symbols "+", "−", and "=" indicate that the result is significantly better, significantly worse, or statistically equivalent to the result obtained by NSGA-III based on the one-tailed Wilcoxon test using a significance level of $\alpha = 0.05$. The superscripts represent the obtained rank in the comparison.

MOP	Obj.	Gen.	NSGA-III	NSGA-III-RSE	NSGA-III-GAE	NSGA-III-COU	NSGA-III-PT	NSGA-III-MPT	NSGA-III-GPT	NSGA-III-KRA
DTLZ5	3	400	1.683065e+00^{8} (3.003386e-03)	1.699980e+00^{4} + (5.762528e-04)	1.696195e+00^{7} + (5.847770e-04)	1.700047e+00^{3} + (4.606680e-04)	1.699919e+00^{5} + (4.970190e-04)	1.697058e+00^{6} + (6.243073e-04)	1.700063e+00^{2} + (4.394574e-04)	1.700105e+00^{1} + (4.510313e-04)
	5	600	7.271004e+00^{2} (5.693000e-03)	7.259936e+00^{6} − (5.264967e-03)	7.255926e+00^{8} − (7.616037e-03)	7.284951e+00^{1} + (2.723032e-03)	7.265152e+00^{5} − (4.559885e-03)	7.255937e+00^{7} − (4.570385e-03)	7.265815e+00^{3} − (4.349501e-03)	7.265523e+00^{4} − (4.355205e-03)
	8	750	2.441108e+01^{2} (3.445466e-02)	2.409257e+01^{8} − (9.435674e-02)	2.411584e+01^{7} − (1.047563e-01)	2.456996e+01^{1} + (1.767660e-02)	2.422908e+01^{3} − (6.711108e-02)	2.413111e+01^{6} − (8.453526e-02)	2.420593e+01^{5} − (5.726898e-02)	2.421006e+01^{4} − (6.842454e-02)
	10	1000	5.483313e+01^{2} (8.335605e-02)	5.422963e+01^{7} − (2.186670e-01)	5.411770e+01^{8} − (1.986003e-01)	5.531009e+01^{1} + (2.935791e-02)	5.465657e+01^{3} − (1.197149e-01)	5.426811e+01^{6} − (1.328445e-01)	5.464194e+01^{5} − (1.331660e-01)	5.464843e+01^{4} − (1.058191e-01)
DTLZ6	3	400	2.362658e+00^{8} (1.846698e-03)	2.373543e+00^{3} + (4.008124e-05)	2.371253e+00^{7} + (2.882926e-04)	2.373476e+00^{5} + (5.331024e-05)	2.373543e+00^{4} + (3.954001e-05)	2.371959e+00^{6} + (2.156646e-04)	2.373547e+00^{2} + (4.998409e-05)	2.373548e+00^{1} + (5.249706e-05)
	5	600	7.568151e+00^{2} (7.604395e-04)	7.567464e+00^{6} = (2.256934e-03)	7.567092e+00^{8} − (2.027593e-03)	7.570797e+00^{1} + (4.557752e-04)	7.567361e+00^{7} = (3.470327e-03)	7.567697e+00^{5} = (2.079136e-03)	7.567958e+00^{3} = (1.395505e-03)	7.567912e+00^{4} = (2.023821e-03)
	8	750	2.549315e+01^{2} (3.148201e-02)	2.544180e+01^{5} − (6.611674e-02)	2.545647e+01^{3} − (5.700224e-02)	2.552946e+01^{1} + (2.816517e-03)	2.539963e+01^{7} − (6.822166e-02)	2.545154e+01^{4} − (5.215940e-02)	2.540228e+01^{6} − (7.535575e-02)	2.538590e+01^{8} − (7.751317e-02)
	10	1000	5.729264e+01^{2} (5.099834e-02)	5.713326e+01^{5} − (2.326677e-01)	5.721923e+01^{3} − (1.804132e-01)	5.739462e+01^{1} + (5.442550e-03)	5.705625e+01^{7} − (2.042287e-01)	5.719957e+01^{4} − (1.510127e-01)	5.703727e+01^{8} − (2.556414e-01)	5.705735e+01^{6} − (2.378984e-01)
DTLZ7	3	400	2.380405e+00^{8} (4.842439e-03)	2.398290e+00^{1} + (3.213980e-03)	2.395450e+00^{4} + (2.869740e-03)	2.386454e+00^{6} + (6.754157e-02)	2.385976e+00^{7} + (6.620726e-02)	2.394787e+00^{5} + (4.176402e-03)	2.397648e+00^{3} + (3.566738e-03)	2.398148e+00^{2} + (3.279279e-03)
	5	600	5.198370e+00^{7} (3.295134e-02)	5.198769e+00^{6} = (1.343136e-02)	5.207086e+00^{2} = (1.465077e-02)	5.277876e+00^{1} + (1.280768e-02)	5.206297e+00^{3} = (1.708775e-02)	5.200456e+00^{5} = (1.776545e-02)	5.202415e+00^{4} = (1.703395e-02)	5.194385e+00^{8} = (2.061225e-02)
	8	750	1.753352e+01^{2} (4.198670e-02)	1.722826e+01^{6} − (6.770429e-02)	1.765887e+01^{1} + (5.000517e-02)	1.729897e+01^{5} − (3.786583e-02)	1.713589e+01^{7} − (6.732135e-02)	1.732681e+01^{4} − (7.109251e-02)	1.742654e+01^{3} − (7.865710e-02)	1.713324e+01^{8} − (6.627246e-02)
	10	1000	4.161526e+01^{2} (1.360459e-01)	4.074876e+01^{6} − (1.476264e-01)	4.166514e+01^{1} = (1.460388e-01)	4.092649e+01^{4} − (8.229699e-02)	4.063510e+01^{8} − (1.593349e-01)	4.144889e+01^{3} − (1.782307e-01)	4.078496e+01^{5} − (9.215879e-02)	4.068056e+01^{7} − (1.560742e-01)
$DTLZ1^{-1}$	3	400	1.639953e+00^{8} (4.016034e-03)	1.684223e+00^{4} + (2.605900e-03)	1.682769e+00^{7} + (2.914343e-03)	1.684790e+00^{1} + (1.974968e-03)	1.684470e+00^{2} + (2.048041e-03)	1.683013e+00^{6} + (1.622414e-03)	1.683944e+00^{5} + (2.300191e-03)	1.684447e+00^{3} + (2.413731e-03)
	5	600	8.575874e-01^{8} (3.416623e-02)	1.112416e+00^{5} + (4.091570e-03)	1.069450e+00^{7} + (1.058851e-02)	1.140662e+00^{1} + (1.546987e-03)	1.136320e+00^{2} + (2.049798e-03)	1.082581e+00^{6} + (7.653301e-03)	1.135439e+00^{4} + (2.755802e-03)	1.136177e+00^{3} + (2.918011e-03)
	8	750	1.071626e-01^{8} (7.139890e-03)	2.850547e-01^{1} + (2.191552e-03)	2.653132e-01^{7} + (3.081039e-03)	2.754732e-01^{3} + (3.014487e-03)	2.735758e-01^{6} + (2.088246e-03)	2.781008e-01^{2} + (1.947575e-03)	2.744208e-01^{4} + (2.914339e-03)	2.737868e-01^{5} + (2.425082e-03)
	10	1000	3.247164e-02^{8} (2.313260e-03)	1.145618e-01^{1} + (1.019499e-03)	1.070459e-01^{3} + (1.002758e-03)	1.050904e-01^{4} + (1.255150e-03)	1.041448e-01^{6} + (1.350572e-03)	1.129353e-01^{2} + (1.048430e-03)	1.042948e-01^{5} + (1.858401e-03)	1.040354e-01^{7} + (1.602635e-03)
$DTLZ2^{-1}$	3	250	2.426107e+00^{8} (3.854343e-03)	2.455639e+00^{4} + (1.946594e-03)	2.454733e+00^{6} + (3.739631e-03)	2.457116e+00^{1} + (3.286721e-03)	2.455195e+00^{5} + (2.028939e-03)	2.454387e+00^{7} + (3.462336e-03)	2.456002e+00^{2} + (1.457478e-03)	2.455784e+00^{3} + (2.492471e-03)
	5	350	2.350007e+00^{8} (3.189458e-02)	2.586880e+00^{1} + (7.237966e-03)	2.572632e+00^{4} + (1.733438e-02)	2.584350e+00^{2} + (4.704734e-03)	2.500410e+00^{7} + (8.163761e-03)	2.582274e+00^{3} + (1.133009e-02)	2.507910e+00^{5} + (8.714111e-03)	2.501008e+00^{6} + (7.778844e-03)
	8	500	8.195578e-01^{8} (7.927094e-02)	1.050752e+00^{3} + (1.444336e-02)	1.142808e+00^{1} + (2.099845e-02)	9.315582e-01^{4} + (3.567355e-02)	8.311273e-01^{7} = (6.438544e-02)	1.129489e+00^{2} + (2.659085e-02)	8.343061e-01^{6} = (6.575584e-02)	8.564420e-01^{5} = (7.922644e-02)
	10	750	4.406968e-01^{4} (8.861395e-02)	6.122128e-01^{3} + (1.595939e-02)	6.856783e-01^{2} + (1.883502e-02)	4.312420e-01^{5} = (2.968891e-02)	4.045091e-01^{8} − (3.912023e-02)	6.867218e-01^{1} + (1.126337e-02)	4.186998e-01^{6} − (4.788076e-02)	4.105392e-01^{7} − (5.276687e-02)
$DTLZ3^{-1}$	3	1000	2.427100e+00^{8} (3.978260e-03)	2.456995e+00^{7} + (1.454871e-03)	2.458462e+00^{2} + (1.934791e-03)	2.459080e+00^{1} + (1.553236e-03)	2.457349e+00^{5} + (1.498918e-03)	2.458230e+00^{3} + (1.819823e-03)	2.457031e+00^{6} + (1.442687e-03)	2.457489e+00^{4} + (1.158182e-03)
	5	1000	2.300383e+00^{8} (2.136280e-02)	2.581276e+00^{2} + (8.662465e-03)	2.535248e+00^{4} + (1.303254e-02)	2.589319e+00^{1} + (4.575343e-03)	2.504495e+00^{7} + (7.735019e-03)	2.554507e+00^{3} + (1.627775e-02)	2.514435e+00^{5} + (6.662934e-03)	2.506279e+00^{6} + (6.968694e-03)
	8	1000	7.163381e-01^{8} (6.495046e-02)	1.046078e+00^{3} + (2.341057e-02)	1.046145e+00^{2} + (2.966509e-02)	9.331763e-01^{4} + (3.268142e-02)	8.055261e-01^{7} + (3.290860e-02)	1.086491e+00^{1} + (2.263674e-02)	8.218731e-01^{6} + (3.639250e-02)	8.282479e-01^{5} + (4.361378e-02)
	10	1500	4.022272e-01^{5} (8.523433e-02)	6.065511e-01^{3} + (1.049319e-02)	6.391393e-01^{2} + (1.887857e-02)	4.367740e-01^{4} = (3.044940e-02)	3.978547e-01^{6} = (3.373736e-02)	6.649189e-01^{1} + (1.500163e-02)	3.978459e-01^{7} = (2.601526e-02)	3.903164e-01^{8} = (4.044816e-02)
$DTLZ4^{-1}$	3	600	2.427846e+00^{8} (3.801265e-03)	2.457787e+00^{4} + (1.078293e-03)	2.458750e+00^{2} + (1.478288e-03)	2.458517e+00^{3} + (2.027693e-03)	2.457741e+00^{5} + (1.316283e-03)	2.458752e+00^{1} + (2.072326e-03)	2.457324e+00^{7} + (1.117478e-03)	2.457413e+00^{6} + (1.352871e-03)
	5	1000	2.361561e+00^{8} (2.449103e-02)	2.593589e+00^{2} + (6.171817e-03)	2.587644e+00^{3} + (1.347211e-02)	2.586898e+00^{4} + (4.832960e-03)	2.499648e+00^{7} + (9.379624e-03)	2.597186e+00^{1} + (9.653957e-03)	2.508008e+00^{5} + (8.994388e-03)	2.499813e+00^{6} + (8.040674e-03)
	8	1250	7.101595e-01^{8} (1.742561e-01)	1.050636e+00^{3} + (1.276098e-02)	1.133452e+00^{1} + (2.939362e-02)	9.493025e-01^{4} + (4.663841e-02)	8.331453e-01^{7} + (6.029250e-02)	1.119698e+00^{2} + (3.439785e-02)	8.556684e-01^{5} + (6.878305e-02)	8.370736e-01^{6} + (6.520487e-02)
	10	2000	3.415418e-01^{8} (8.473208e-02)	6.072786e-01^{3} + (1.426179e-02)	6.724087e-01^{1} + (1.654860e-02)	4.336076e-01^{4} + (2.922868e-02)	3.903247e-01^{5} + (3.361700e-02)	6.710877e-01^{2} + (1.108425e-02)	3.757644e-01^{7} + (2.282619e-02)	3.881267e-01^{6} + (3.473286e-02)
$DTLZ5^{-1}$	3	400	2.476107e+00^{8} (3.315611e-03)	2.510958e+00^{3} + (2.431047e-03)	2.504371e+00^{7} + (2.853644e-03)	2.509060e+00^{5} + (2.135497e-03)	2.511617e+00^{2} + (1.505994e-03)	2.506988e+00^{6} + (1.993247e-03)	2.510729e+00^{4} + (1.542846e-03)	2.511703e+00^{1} + (2.109414e-03)
	5	600	2.714893e+00^{8} (2.456552e-02)	2.979104e+00^{5} + (9.291639e-03)	2.912514e+00^{7} + (1.254716e-02)	3.009631e+00^{1} + (5.321107e-03)	2.979314e+00^{4} + (4.893081e-03)	2.934682e+00^{6} + (1.438410e-02)	2.979494e+00^{3} + (5.674395e-03)	2.982068e+00^{2} + (5.881932e-03)
	8	750	1.122540e+00^{8} (1.337096e-01)	1.872579e+00^{2} + (1.639879e-02)	1.821866e+00^{4} + (2.872548e-02)	1.825822e+00^{3} + (1.950487e-02)	1.731168e+00^{6} + (1.177111e-02)	1.888242e+00^{1} + (1.790679e-02)	1.743051e+00^{5} + (1.572238e-02)	1.731132e+00^{7} + (1.757934e-02)
	10	1000	7.599818e-01^{8} (8.859381e-02)	1.385056e+00^{2} + (1.886237e-02)	1.357671e+00^{3} + (2.177749e-02)	1.289846e+00^{4} + (1.857520e-02)	1.194430e+00^{7} + (2.013369e-02)	1.413946e+00^{1} + (1.792777e-02)	1.211922e+00^{5} + (1.063207e-02)	1.205323e+00^{6} + (1.101863e-02)
$DTLZ6^{-1}$	3	400	2.438891e+00^{8} (2.556978e-03)	2.469193e+00^{4} + (7.604818e-04)	2.469215e+00^{3} + (9.169312e-04)	2.469330e+00^{2} + (9.349019e-04)	2.469075e+00^{7} + (6.006657e-04)	2.469594e+00^{1} + (9.545435e-04)	2.469103e+00^{6} + (8.713162e-04)	2.469137e+00^{5} + (9.121320e-04)
	5	600	2.457585e+00^{8} (1.804186e-02)	2.689606e+00^{3} + (4.061632e-03)	2.699722e+00^{2} + (6.046173e-03)	2.674587e+00^{4} + (3.332989e-03)	2.621122e+00^{6} + (5.695655e-03)	2.700654e+00^{1} + (5.334639e-03)	2.624983e+00^{5} + (2.872699e-03)	2.620958e+00^{7} + (4.393912e-03)
	8	750	9.196447e-01^{8} (6.252599e-02)	1.222941e+00^{3} + (1.925354e-02)	1.340432e+00^{1} + (2.023835e-02)	1.111650e+00^{4} + (2.885494e-02)	1.037305e+00^{6} + (4.106375e-02)	1.304040e+00^{2} + (2.550192e-02)	1.048056e+00^{5} + (3.848979e-02)	1.026896e+00^{7} + (3.535104e-02)
	10	1000	4.657110e-01^{8} (7.896088e-02)	7.613810e-01^{3} + (1.127991e-02)	8.776534e-01^{1} + (1.742538e-02)	5.841050e-01^{4} + (1.951415e-02)	5.574816e-01^{6} + (2.606097e-02)	8.463911e-01^{2} + (1.610230e-02)	5.620053e-01^{5} + (2.336188e-02)	5.564723e-01^{7} + (2.842876e-02)
$DTLZ7^{-1}$	3	400	2.834353e+00^{8} (3.937099e-03)	2.843035e+00^{2} + (1.826106e-03)	2.839792e+00^{7} + (3.298823e-03)	2.841139e+00^{6} + (3.301679e-03)	2.842890e+00^{3} + (2.689247e-03)	2.841451e+00^{5} + (2.050558e-03)	2.843361e+00^{1} + (1.606716e-03)	2.842784e+00^{4} + (2.208112e-03)
	5	600	6.123502e+00^{8} (1.624968e-01)	6.220762e+00^{5} + (1.241462e-02)	6.185174e+00^{7} + (2.464440e-02)	6.319688e+00^{1} + (8.205647e-03)	6.238808e+00^{2} + (1.998119e-02)	6.202463e+00^{6} + (2.425379e-02)	6.233244e+00^{4} + (1.642926e-02)	6.236887e+00^{3} + (1.283589e-02)
	8	750	1.735203e+01^{2} (5.865975e-01)	1.639168e+01^{5} − (1.117788e-01)	1.625531e+01^{7} − (1.380380e-01)	1.785960e+01^{1} + (1.532403e-01)	1.646296e+01^{4} − (1.365410e-01)	1.626340e+01^{6} − (1.183856e-01)	1.621840e+01^{8} − (3.666152e-01)	1.648245e+01^{3} − (1.334969e-01)
	10	1000	3.717341e+01^{3} (2.195385e-01)	3.709311e+01^{7} = (1.795264e-01)	3.712378e+01^{5} = (1.899615e-01)	3.903706e+01^{2} + (2.492819e-01)	3.712388e+01^{4} = (2.295007e-01)	3.701863e+01^{8} − (1.948909e-01)	3.905638e+01^{1} + (9.024978e-01)	3.710414e+01^{6} = (2.049699e-01)
+/−/=				29/8/3	30/7/3	36/2/2	26/9/5	29/9/2	27/9/4	26/9/5

them. Using NSGA-III-$\mathcal{K}$, only some reference points select its nearest solutions. The remaining solutions are chosen using a pair-potential function aiming to improve diversity. However, it is observed that in the MaOPs versions of the DTLZ5–DTLZ7, some MOEAs perform significantly worse than NSGA-III. These results can be due to the presence of dominance resistant solutions. These MOPs promote that the selection mechanism with some pair-potential functions presents difficulties related to convergence to the *Pareto front*. The best MOEA on MOPs with irregular *Pareto front* shapes is NSGA-III-COU. Also, it is capable of significantly outperform NSGA-III on 36 out of 40 MOPs. Additionally, NSGA-III-COU was the only MOEA with significantly better performance than NSGA-III on the MaOPs versions of the DTLZ5 and DTLZ6. This result suggests that the definition of $\mathcal{K}_{COU}$ helps to improve the convergence to the *Pareto front* of NSGA-III-$\mathcal{K}$ on MaOPs.

Figure 1 shows the approximation sets of some MOEAs on the DTLZ1 and DTLZ1^{-1} with 3 and 10 objectives. It can be seen that the versions of NSGA-III-$\mathcal{K}$ are capable of maintaining the same performance as NSGA-III on the DTLZ1. Also, they obtain a better distribution of solutions on the DTLZ1^{-1}. Especially, NSGA-III fails to fully cover the *Pareto front* of the DTLZ1^{-1} with 10 objectives. However, with the use of pair-potential functions it is possible to deal with this drawback. Additional approximation sets can be visualized as supplementary material at https://github.com/lmarquezvg/NSGA-III-K.

6   Conclusions and Future Work

In this paper, we have included a new selection mechanism based on niching and pair-potential functions into NSGA-III to promote an invariant performance regardless of the geometry of the *Pareto front*. We have performed an experimental comparison of seven pair-potential functions to assess their capability to be used on this selection mechanism. Our experimental results showed that all versions of NSGA-III-$\mathcal{K}$ are capable of preserving the good performance of NSGA-III on MOPs with regular *Pareto front* shapes. Also, they significantly improve the performance of NSGA-III on MOPs with irregular *Pareto front* shapes. However, some pair-potential functions perform significantly worse than NSGA-III on some MOPs with irregular *Pareto front* shapes that include 5, 8, or 10 objectives. This is mainly due to the presence of dominance resistant solutions. NSGA-III-COU obtained the best results on MOPs with regular and irregular *Pareto front* shapes. This result suggests that $\mathcal{K}_{COU}$ is capable of dealing with dominance resistant solutions, and it is the most convenient pair-potential function to be used on this selection mechanism.

As future work, we are interested in improving the convergence capability of the selection mechanism based on niching and pair-potential functions on MaOPs. Also, we want to validate our proposed selection mechanism on a wider variety of test problems and assess it against more state-of-the-art MOEAs. We aim to propose an external archive with this selection mechanism, which can be used to adapt the reference set of any MOEA that employs reference sets.

References

1. Borodachov, S.V., Hardin, D.P., Saff, E.B.: Discrete Energy on Rectifiable Sets. SMM, Springer, New York (2019). https://doi.org/10.1007/978-0-387-84808-2
2. Coello, C., Veldhuizen, D., Lamont, G.: Evolutionary Algorithms for Solving Multi-Objective Problems. Genetic and Evolutionary Computation, 2nd edn. Springer, New York (2007). https://doi.org/10.1007/978-0-387-36797-2
3. Das, I., Dennis, J.E.: Normal-Boundary intersection: a new method for generating the pareto surface in nonlinear multicriteria optimization problems. SIAM J. Optim. **8**(3), 631–657 (1998). https://doi.org/10.1137/S1052623496307510
4. Deb, K., Jain, H.: An evolutionary many-objective optimization algorithm using reference-point-based nondominated sorting approach, Part I: solving problems with box constraints. IEEE Trans. Evol. Comput. **18**(4), 577–601 (2014). https://doi.org/10.1109/TEVC.2013.2281535
5. Deb, K., Pratap, A., Agarwal, S., Meyarivan, T.: A fast and elitist multiobjective genetic algorithm: NSGA-II. IEEE Trans. Evol. Comput. **6**(2), 182–197 (2002). https://doi.org/10.1109/4235.996017
6. Deb, K., Thiele, L., Laumanns, M., Zitzler, E.: Scalable test problems for evolutionary multiobjective optimization. In: Abraham, A., Jain, L., Goldberg, R. (eds.) Evolutionary Multiobjective Optimization: Theoretical Advances and Applications, pp. 105–145. Springer, London (2005). https://doi.org/10.1007/1-84628-137-7_6
7. Falcón-Cardona, J.G., Coello Coello, C.A.: Indicator-Based multi-objective evolutionary algorithms: a comprehensive survey. ACM Comput. Surv. **53**(2), 1–35 (2020). https://doi.org/10.1145/3376916
8. Falcón-Cardona, J.G., Covantes Osuna, E., Coello Coello, C.A.: An overview of pair-potential functions for multi-objective optimization. In: Ishibuchi, H., et al. (eds.) EMO 2021. LNCS, vol. 12654, pp. 401–412. Springer, Cham (2021). https://doi.org/10.1007/978-3-030-72062-9_32
9. Falcón-Cardona, J.G., Ishibuchi, H., Coello Coello, C.A.: Riesz s-energy-based reference sets for multi-objective optimization. In: 2020 IEEE Congress on Evolutionary Computation (CEC), pp. 1–8. IEEE (2020). https://doi.org/10.1109/CEC48606.2020.9185833
10. Ishibuchi, H., Setoguchi, Y., Masuda, H., Nojima, Y.: Performance of decomposition-based many-objective algorithms strongly depends on pareto front shapes. IEEE Trans. Evol. Comput. **21**(2), 169–190 (2017). https://doi.org/10.1109/TEVC.2016.2587749
11. Ishibuchi, H., Tsukamoto, N., Nojima, Y.: Evolutionary many-objective optimization: a short review. In: 2008 IEEE Congress on Evolutionary Computation (CEC), pp. 2419–2426 (2008). https://doi.org/10.1109/CEC.2008.4631121
12. Li, B., Li, J., Tang, K., Yao, X.: Many-Objective evolutionary algorithms: a survey. ACM Comput. Surv. **48**(1), 1–35 (2015). https://doi.org/10.1145/2792984
13. Pang, L.M., Ishibuchi, H., Shang, K.: NSGA-II with simple modification works well on a wide variety of many-objective problems. IEEE Access **8**, 190240–190250 (2020). https://doi.org/10.1109/ACCESS.2020.3032240
14. Zitzler, E.: Evolutionary Algorithms for Multiobjective Optimization: Methods and Applications. Ph.D. thesis, ETH Zurich, Switzerland (1999)

Endowing the MIA Cloud Autoscaler with Adaptive Evolutionary and Particle Swarm Multi-Objective Optimization Algorithms

Virginia Yannibelli[1]([⊠]), Elina Pacini[2,3,4], David Monge[2], Cristian Mateos[1], and Guillermo Rodriguez[1]

[1] ISISTAN (UNICEN-CONICET), Tandil, Buenos Aires, Argentina
{virginia.yannibelli,cristian.mateos,
guillermo.rodriguez}@isistan.unicen.edu.ar
[2] ITIC, UNCUYO, Mendoza, Argentina
{epacini,dmonge}@uncu.edu.ar
[3] Facultad de Ingeniería, UNCuyo, Mendoza, Argentina
[4] CONICET, Buenos Aires, Argentina

Abstract. PSE (Parameter Sweep Experiments) applications represent a relevant class of computational applications in science, engineering and industry. These applications involve many computational tasks that are both resource-intensive and independent. For this reason, these applications are suited for Cloud environments. In this sense, Cloud autoscaling approaches are aimed to manage the execution of different kinds of applications on Cloud environments. One of the most recent approaches proposed for autoscaling PSE applications is MIA, which is based on the multi-objective evolutionary algorithm NSGA-III. We propose to endow MIA with other multi-objective optimization algorithms, to improve its performance. In this respect, we consider two well-known multi-objective optimization algorithms named SMS-EMOA and SMPSO, which have significant mechanic differences with NSGA-III. We evaluate MIA endowed with each of these algorithms, on three real-world PSE applications, considering resources available in Amazon EC2. The experimental results show that MIA endowed with each of these algorithms significantly outperforms MIA based on NSGA-III.

Keywords: Parameter Sweep Experiments · Cloud autoscaling · Multi-objective evolutionary algorithm · Multi-objective particle swarm optimization · CloudSim · Amazon EC2

1 Introduction

PSE (Parameter Sweep Experiments) applications represent an important class of computational applications in science, engineering, industry and technology [1]. These applications are aimed to explore all the possible results of a given simulated computational model by varying the parameter settings of this model. For example, the behavior of an airfoil can be explored by executing its model many times, each time with a different

© Springer Nature Switzerland AG 2021
I. Batyrshin et al. (Eds.): MICAI 2021, LNAI 13067, pp. 383–400, 2021.
https://doi.org/10.1007/978-3-030-89817-5_29

parameter setting, where the parameters include speed, angle attack, and shape. Then, a PSE application is inherent to a given computational model which has several parameters. Each parameter has a range of possible values. Then, the application contains as many tasks as different parameter settings for the model. Each task has a different parameter setting associated with it, and implies executing the model with its parameter setting. Then, each task produces an output from the model, according to its associated parameter setting. The set containing the outputs of all the tasks represent the result of the PSE application.

These tasks are usually resource-intensive, which means that they require a large amount of computing resources and time to be executed. However, these tasks are independent and can be executed in parallel. For these reasons, PSE applications are naturally appropriate for distributed environments such as Cloud environments [2]. In this respect, the potential speedup that can be reached by executing these applications across Cloud environment resources is really significant.

Cloud environments provide access to computational resources, such as instances of virtual machines (VMs), under a pay-per-use scheme. These environments provide a wide variety of types of VMs. In this respect, instances of different types of VMs have a different hardware and software configuration, and a different monetary cost. Besides, the instances can be acquired under different acquisition models: on-demand model or spot model. The on-demand instances are more expensive than the spot instances. However, unlike the spot instances, the on-demand instances are stable. This means that on-demand instances are not subject to failures that can end abruptly the tasks assigned, and then negatively impact on the execution of the application. Therefore, the execution of an application on a Cloud environment (i.e., the computing time and monetary cost inherent to the execution) depends on the number and type of instances of VMs acquired.

Based on the above-mentioned facts, to execute an application on a Cloud environment, it is necessary to determine the number and type of instances of VMs to be requested to the Cloud provider. This should be done in such a way that the computing time and monetary cost inherent to the execution are optimized. This problem is recognized in the literature as a multi-objective NP-Hard optimization problem. Cloud autoscaling approaches are aimed to address this problem [3, 4]. These approaches scale up and down the virtual infrastructure acquired (i.e., instances of VMs acquired) according to the workload of the application (i.e., number of tasks to be executed), and schedule such workload on the acquired infrastructure.

In the literature, some Cloud autoscaling approaches have been proposed to execute different kinds of applications such as workflow applications [5–7]. However, very few autoscaling approaches have been reported for executing PSE applications on Cloud environments. To the best of our knowledge, one of the most recent approaches is the autoscaler MIA [8], which is based on the known multi-objective evolutionary algorithm named NSGA-III [11]. This autoscaler utilizes only on-demand instances of VMs to execute the tasks of a given PSE, in order to avoid tasks failures. Besides, this autoscaler considers two relevant optimization objectives: the minimization of the makespan (i.e., computing time) and the minimization of the monetary cost. Although this autoscaler has reached an excellent performance regarding the two optimization objectives considered, its performance depends greatly on the multi-objective optimization algorithm utilized.

Thus, the replacement of the used algorithm NSGA-III by some other multi-objective optimization algorithm can impact significantly on the performance of this autoscaler.

In this paper, we propose to endow the autoscaler MIA with others multi-objective optimization algorithms, with the aim of improving its performance regarding the optimization objectives considered. To do that, we consider two well-known multi-objective optimization algorithms, which have significant differences with algorithm NSGA-III currently used by the autoscaler. One of these algorithms is an adaptive multi-objective evolutionary algorithm named SMS-EMOA [9]. The other algorithm is a multi-objective particle swarm optimization algorithm named SMPSO [10]. We consider these new algorithms mainly because have been shown to be competitive with traditional multi-objective evolutionary algorithms, across a wide variety of NP-Hard problems, including scheduling and resource assignment problems [9, 10].

The remainder of the paper is organized as follows. In Sect. 2, we describe the multi-objective autoscaling problem addressed, and its mathematical formulation. In Sect. 3, we describe in detail the autoscaler MIA, including a description of its current optimization algorithm NSGA-III. In Sect. 4, we describe the two multi-objective optimization algorithms considered to endow the autoscaler MIA. In Sect. 5, we present in detail the computational experiments developed to evaluate the autoscaler MIA endowed with each of the algorithms, and also a comparative analysis of the results obtained. In Sect. 6, we present relevant related works. Finally, in Sect. 7, we present the conclusions of this work, and also future works.

2 Multi-objective Autoscaling Problem

PSE applications are composed by many computational tasks that are both resource-intensive and independent, which means that these tasks require a large amount of computing resources and time, but can be executed in parallel. Therefore, these applications are suited for Cloud environments.

Public cloud environments (e.g., Amazon EC2, Google Cloud, Microsoft Azure) provide access to instances of a wide variety of types of VMs. Instances of different types of VMs have a different hardware and software configuration (e.g., different number of processors, processing power, memory size, memory speed, disk size, etc.), and also a different monetary cost. Besides, the instances can be acquired under different acquisition models, which determine their monetary cost and behavior.

Here, the utilization of the acquisition model named on-demand is considered. This is because the instances acquired under this model are not subject to failures which can suddenly end the execution of the tasks assigned. Under this model, the instances are acquired for one computing hour. For simplicity, instances acquired under this model will be referred as on-demand instances.

The multi-objective autoscaling problem addressed here implies two problems that are interrelated. The first problem is determining a scaling plan detailing the number and type of on-demand instances to request to the Cloud provider for the next hour, in order to execute the tasks of the PSE application. The second problem is scheduling the tasks of the application on the acquired virtual infrastructure. These problems must be addressed in such a way that the predefined optimization objectives are reached. In this respect, two relevant optimization objectives are considered: the minimization of the makespan (i.e., computing time) and the minimization of the monetary cost. Note that these two problems must be addressed every one hour during the execution of the application, since the instances are acquired for one computing hour. Each one-hour period is called *autoscaling stage*. Therefore, the virtual infrastructure is scaled up or scaled down at the beginning of each autoscaling stage according the workload of the application (i.e., the number of tasks to be executed), and this workload is scheduled on the virtual infrastructure, so that the objectives are achieved.

2.1   Mathematical Formulation of the Multi-objective Autoscaling Problem

Considering I, the set of available types of VMs in the Cloud, and $n = |I|$, the number of these types, a scaling plan X is represented as a vector $x^{od} = (x_1^{od}, x_2^{od}, ..., x_n^{od})$. The position i ($i \in [1, n]$) of this vector contains a feasible number of instances of type i to request from the Cloud provider for the current autoscaling stage (i.e., the next hour). The limit is imposed by the Cloud provider.

Then, given T, the set of application's tasks considered for the current autoscaling stage, the multi-objective problem inherent to this stage is defined by Eqs. (1–3), and is subject to constraints (Eqs. (4–5)) which define the current state of the application execution as well as the current state of the cloud infrastructure load.

In Eq. (1), *makespan*(X) represents the estimated computing time of executing the tasks in T on the instances detailed in X. Considering T and X, a scheduling algorithm named ECT (earliest completion time) is used to estimate this time. Via this algorithm, each task is scheduled to the instance which promises the earliest completion time. This term *makespan*(X) is defined by Eq. (2), where $ST(t)$ represents the start time of the task t, and d_t represents the duration of the task t. The values of these terms are estimated by the mentioned scheduling algorithm.

Then, the term *cost*(X) represents the monetary cost of acquiring all the instances detailed in X. This term is defined by Eq. (3), where x_i^{od} represents the number of instances of type i detailed in X, and *price*$_i$ represents the monetary cost of one instance of type i for one computing hour.

$$min\ (makespan(X),\ cost(X)). \tag{1}$$

$$makespan(X) = \max_{t \in T}\{ST(t) + d_t\} - \min_{t \in T}\{ST(t)\}. \tag{2}$$

$$cost(X) = \sum_{i=1}^{n} x_i^{od} \times price_i. \tag{3}$$

Several constraints are considered as part of the problem. The constraints defined by Eq. (4) determine the feasible numbers of instances that could be requested for each type i of VM. In this equation, X_i^{max} and X_i^{min} are the upper and lower bounds of the number of instances of type i for the current autoscaling stage. Specifically, X_i^{min} refers to the number of running instances of type i which are executing at least one task assigned during a previous autoscaling stage. Note that the Cloud provider cannot terminate instances that are executing tasks assigned during previous autoscaling stages. The term X_i^{max} refers to the number of available instances of type i in the Cloud for the current autoscaling stage.

The constraint in Eq. (5) determines that at least one instance must be acquired in the current autoscaling stage.

$$X_i^{min} \leq x_i^{od} \leq X_i^{max}. \tag{4}$$

$$\sum_{i=1}^{n} x_i^{od} \geq 1. \tag{5}$$

3 Multi-objective Autoscaler MIA

The autoscaler MIA [8] is aimed to address the problem described in Section [2]. In this problem, during the execution of a PSE application, a different autoscaling stage begins every one hour. The autoscaler MIA addresses the multi-objective autoscaling problems inherent to these autoscaling stages.

In order to solve the multi-objective autoscaling problem of each autoscaling stage, the autoscaler MIA follows three sequential phases. In the first phase, the problem is addressed by using the well-known multi-objective evolutionary algorithm NSGA-III [11], to acquire an approximation to the optimal Pareto set. In the second phase, one solution is selected from the obtained Pareto set, according to a predefined selection criterion. In the third phase, the scaling plan represented by such solution is applied to define the virtual infrastructure, and the tasks are scheduled on this infrastructure. The phases of the autoscaler MIA are exposed in Fig. 1 (left), and are described below. This figure also shows the two algorithms considered to endow MIA (right).

3.1 First Phase

The algorithm NSGA-III [11] is applied in this phase. NSGA-III is based on the well-known multi-objective evolutionary algorithm NSGA-II, and incorporates a selection process based on reference points. This process is aimed to preserve both the diversity and well-distribution of the resulting Pareto set.

This algorithm starts creating a random initial population with s solutions. In this case, each solution represents a feasible scaling plan, and is encoded like the vector x^{od} mentioned in Sect. 2.1.

At each generation t, the algorithm creates an offspring population with s solutions by applying the variation operators SBX (Simulated Binary Crossover) and PM (Polynomial Mutation) to randomly selected solutions from the current population P_t.

After the offspring population is created, the algorithm combines the current and offspring populations, and then selects s solutions from this combined population, to create a new population P_{t+1} for the next generation. To select such solutions, first the solutions in the combined population are grouped according to their non-domination levels as $\{F_1, F_2, \ldots\}$. Then, each level is selected one at a time to create the new population P_{t+1}, beginning from F_1, until the size of P_{t+1} is equal to s or higher than s. When the size of P_{t+1} is equal to s, the next generation starts from P_{t+1}. On the other hand, when the size of P_{t+1} is higher than s, the last level selected F_l is not fully included in P_{t+1}. In this respect, the solutions from level F_1 to level F_{l-1} are included in P_{t+1}, and the k remaining solutions ($k = s - |F_1 \bigcup \ldots \bigcup F_{l-1}|$) are selected from level F_l.

In order to select the k remaining solutions from the level F_l, NSGA-III utilizes a selection process based on reference points. Such process starts creating a set of reference points widely and uniformly distributed on the normalized hyperplane that corresponds to the optimization objectives of the problem addressed by the algorithm. Then, the process emphasizes the selection of solutions from F_l which are associated with each one of these reference points. Thus, this process promotes the selection of diverse and well-distributed non-dominated solutions, with the aim of preserving the diversity and distribution of the new population P_{t+1}.

The algorithm ends its execution once a predetermined number of evaluations is reached. Then, the algorithm provides the Pareto set (i.e., the set of non-dominated solutions) of the last generation as the obtained result.

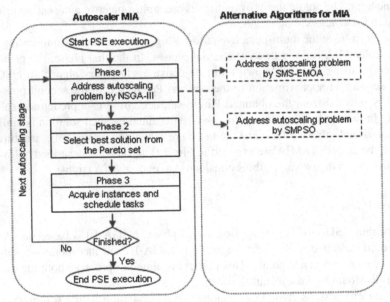

Fig. 1. Phases of the autoscaler MIA (left), and algorithms considered to endow MIA (right).

3.2 Second Phase

In this phase, one solution is selected from the Pareto set provided by the first phase. Specifically, the solution of this set which minimizes the distance to an ideal solution is selected. Here, the ideal solution is such whose makespan and cost is equal to 0. Then, the distance of each solution of the Pareto set to this ideal solution is calculated by using the well-known L_2-norm metric. This allows analyzing simultaneously the makespan and the cost of each solution of the Pareto set, and thus considering the trade-off of each solution of this set between both optimization objectives. Note that the makespan and cost of each solution are calculated by Eqs. (2)-(3), respectively.

3.3 Third Phase

In this phase, the solution selected by the second phase is considered, to define the virtual infrastructure. Specifically, the number of instances detailed in such solution for each type in I is acquired. After that, the tasks in T are scheduled on the acquired instances, by using the scheduling algorithm named ECT. Recall that I refers to the set of available types of VMs in the Cloud, and T refers to the set of tasks considered for the current autoscaling stage.

4 Alternative Multi-objective Optimization Algorithms for MIA

The performance of the autoscaler MIA depends on the multi-objective optimization algorithm used during the first phase of this autoscaler. Thus, exploring alternatives to the currently used NSGA-III algorithm can impact significantly on the performance of this autoscaler.

We propose to endow the autoscaler MIA with others multi-objective optimization algorithms, in order to improve its performance regarding the optimization objectives considered. In this respect, we consider two well-known multi-objective optimization algorithms, which have significant differences with NSGA-III. One of these algorithms is an adaptive multi-objective evolutionary algorithm named SMS-EMOA [9]. The other algorithm is a multi-objective particle swarm optimization algorithm named SMPSO [10]. A description of these algorithms is presented below.

4.1 Algorithm SMS-EMOA

The algorithm SMS-EMOA ($\mathscr{S}$metricselection EMOA) is an adaptive steady-state multi-objective evolutionary algorithm, which features a selection operator based on the hypervolume metric (or $\mathscr{S}$metric) combined with the concept of non-dominated sorting. This operator is aimed to maximize the dominated hypervolume within the optimization process. The hypervolume rewards the convergence towards the Pareto front as well as the representative distribution of solutions along the front. Thus, by this selection operator, the algorithm population evolves to a well-distributed set of solutions, thereby focusing on interesting regions of the Pareto front.

This algorithm starts creating a random initial population with s solutions. In this case, each solution represents a feasible scaling plan, and is encoded like the vector x^{od} mentioned in Sect. 2.1.

At each generation t, a new solution is created by applying the variation operators SBX and PM to a pair of randomly selected solutions from the current population P_t. Then, the selection operator is applied to determine if this new solution will become a member of the next population P_{t+1} for the next generation. By using this operator, the new solution will be considered for the next population, if replacing another solution leads to a higher quality of the population with respect to the hypervolume metric.

The selection operator starts considering the s solutions of the current population P_t plus the new solution created. Then, these $s + 1$ solutions are sorted according to non-domination levels as $\{F_1, F_2, \ldots F_v\}$. After that, one solution is discarded from the worst level F_v, in order to obtain the s solutions for the new population P_{t+1}. Specifically, the solution $x \in F_v$ that minimizes Eq. (6) is eliminated.

In Eq. (6), the value of the term $\Delta_{\mathscr{S}}(x, F_v)$ represents the contribution of x to the hypervolume metric value of its corresponding level F_v. Thus, the selection operator considers the contribution of each candidate solution to the hypervolume metric value, and keeps those solutions which maximize the hypervolume metric value of the population.

The algorithm ends its execution when a predefined number of evaluations is reached. After that, the algorithm provides the Pareto set (i.e., set of non-dominated solutions) of the last generation as the obtained result.

$$\Delta_{\mathscr{S}}(x, F_v) = \mathscr{S}(F_v) - \mathscr{S}(F_v - \{x\}). \tag{6}$$

4.2 Algorithm SMPSO

The algorithm SMPSO (Speed-constrained Multi-objective PSO) is based on the well-known algorithm OMOPSO, and is characterized by the incorporation of a velocity constriction procedure, to control the velocity of the solutions (i.e., particles) of the population (i.e., swarm), and thus to get that the solutions move effectively through the search space. Other important features of SMPSO include the PM operator as turbulence factor, and an external archive, named leaders archive, to store the non-dominated solutions found during the search.

The algorithm starts creating a random initial population with s solutions, and the leaders archive with the non-dominated solutions in the population. In this case, each solution represents a feasible scaling plan, and is encoded like the vector x^{od} mentioned in Sect. 2.1.

In each generation t, the velocities of the s solutions in the current population P_t are calculated first, by Eq. (7). In this equation, $\vec{v}_j(t)$ and $\vec{v}_j(t-1)$ represent the velocity of solution j ($j \in [1, s]$) at t and $t - 1$, respectively. The term $\vec{x}_j$ represents to solution j, $\vec{x}_{pj}$ is the best solution that $\vec{x}_j$ has viewed, $\vec{x}_{gj}$ is the best solution that the entire population has viewed, w is the inertia weight of the solution and controls the trade-off between global and local experience, r_1 and r_2 are two uniformly distributed random numbers in the range [0, 1], and C_1 and C_2 are random numbers in the range [1.5, 2.5] and control the effect of the personal and global best solutions. Then, the resulting velocities are multiplied by a constriction coefficient χ, to control the velocity of the solutions. This coefficient χ is calculated by Eq. (8), where φ is calculated by Eq. (9). After that, the accumulated velocity $v_{j,i}$ of each variable i of each solution j is further bounded by the velocity constriction equation shown in Eq. (10), where $delta_i$ is calculated by Eq. (11). In this equation, X_i^{max} and X_i^{min} are the upper and lower bounds of the variable i.

Once calculated the velocities of the solutions, the positions of the solutions are calculated (i.e., the solutions are updated) considering their velocities, by Eq. (12). Then, the PM operator is applied on the solutions as turbulence factor, with a given probability. The resulting solutions are evaluated, and the solutions' memory is updated for the next generation. Besides, the leaders archive is updated for the next generation, in order to preserve the non-dominated solutions found during the search.

The algorithm ends its execution when a predefined number of evaluations is reached. Then, the algorithm provides the leaders archive of the last generation as the Pareto set found.

$$\vec{v}_j(t) = w \cdot \vec{v}_j(t-1) + C_1 \cdot r_1 \cdot \left(\vec{x}_{pj} - \vec{x}_j\right) + C_2 \cdot r_2 \cdot \left(\vec{x}_{gi} - \vec{x}_j\right) \tag{7}$$

$$\chi = \frac{2}{2 - \varphi - \sqrt{\varphi^2 - 4\varphi}} \tag{8}$$

$$\varphi = \begin{cases} C_1 + C_2 & \text{if } C_1 + C_2 > 4 \\ 0 & \text{if } C_1 + C_2 \leq 4 \end{cases} \tag{9}$$

$$v_{j,i} = \begin{cases} delta_i & \text{if } v_{j,i} > delta_i \\ -delta_i & \text{if } v_{j,i} \leq -delta_i \\ v_{j,i} & \text{otherwise} \end{cases} \tag{10}$$

$$delta_i = \frac{X_i^{max} - X_i^{min}}{2} \tag{11}$$

$$\vec{x}_j(t) = \vec{x}_j(t-1) + \vec{v}_j(t) \tag{12}$$

5 Computational Experiments

We present below the computational experiments developed to evaluate the autoscaler MIA endowed with each of the multi-objective optimization algorithms described in Sect. 4. First, we present the PSE applications, and also the on-demand instances of VMs, used in these experiments. Then, we detail the experimental settings defined. Finally, we present and analyze the results obtained.

5.1 PSE Applications

We used two real-world PSE applications named Cruciform and Plate3D. Moreover, we used a third application named Ensemble, which combines the applications Cruciform and Plate3D.

The application Cruciform allows exploring the behavior of a computational model which simulates the elastoplastic buckling of cruciform columns [12]. The exploration of this model is utilized to compare the theories of total deformation and plasticity incremental [12]. Figure 2a exhibits the geometry of the column utilized in this model. The initial geometry has a length L equal to 50 mm, and a cross section of 10 mm and

1 mm of width and thickness, respectively. The number of mesh elements is equal to 2176. The rotation of the cross section is imposed as initial imperfection along the length of the column. The angle parameter α is selected as the variation parameter to generate the tasks of the application. This parameter is varied as $\alpha_n = \alpha_{n-1} + 0.25$, where $\alpha_0 = 0.5$ and n is an integer value higher than 0. The variation of this parameter is useful to many other applications including seismic protection of structures, where it is essential to know the sensitivity to the imperfection's size.

The application Plate3D allows exploring the behavior of a computational model which simulates the expansion of a plane strain plate with a central circular hole [1]. Figure 2b shows the geometry of this plate, the scheme of spatial discretization, and the boundary conditions that are utilized in this model. The plate has a dimension of 18×10 m, with a radius equals to 5 m. The number of elements of the 3D finite element mesh used is 1152. In the simulation, displacements at 18 m were applied until a final displacement of 2 m was reached, in 400 equal time steps each. A material viscosity parameter η was selected as the variation parameter to generate the tasks of the application. This parameter was varied as $\eta = \{1.10^8 \text{ Mpa}\} \cup \{x \times 10^y \text{ Mpa}$, where x and y are integer numbers higher than $0\}$. The variation of this parameter is useful in areas such as industrial design, to know the flexibility of materials.

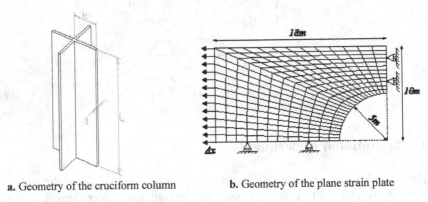

a. Geometry of the cruciform column   **b.** Geometry of the plane strain plate

Fig. 2. PSE applications

The application Ensemble combines the applications Cruciform and Plate3D. This application consists of executing the tasks of both Cruciform and Plate3D at the same time.

For each of the mentioned applications, we defined 3 different sizes based on the number of tasks composing the application. For Plate3D and Cruciform, we defined the sizes 30, 100, and 300 tasks. For Ensemble, we defined the sizes 60, 200 and 600 tasks. Note that the higher the number of tasks that compose an application, the wider the variation parameter exploration developed by the application.

5.2 On-Demand Instances of VMs

We considered five types of VMs available in Amazon EC2, which were utilized in [8], and the characteristics of the on-demand instances of these types. Table 1 shows the main characteristics of the instances of each type. Column 1 presents the name of each type of VM. Column 2 details the number of available virtual CPUs of each instance. Column 3 details the relative computing power of each instance, considering all the virtual CPUs. Column 4 presents the relative performance of one of the CPUs of each instance. Column 5 details the monetary cost of each instance for one hour of computation, in USD.

Table 1. Instances characteristics (Amazon EC2 instances from the US-west region)

VM type	vCPU	ECUtot	ECU	Price [USD]
t2.micro	1	1	1	0.013
m3.medium	1	3	2	0.07
c3.2xlarge	8	28	3.5	0.42
r3.xlarge	4	13	3.25	0.35
m3.2xlarge	8	26	3.25	0.56

5.3 Experimental Settings

We evaluated the autoscaler MIA which is based on the algorithm NSGA-III, on each PSE application and size presented in Sect. 5.1. Then, we evaluated the autoscaler MIA endowed with each of the algorithms described in Sect. 4, namely SMS-EMOA and SMPSO, on each application and size. For simplicity, the autoscaler MIA endowed with the algorithm SMS-EMOA will be referred as autoscaler EMOA-MIA. Besides, the autoscaler MIA endowed with the algorithm SMPSO will be referred as autoscaler PSO-MIA.

To evaluate the autoscalers MIA, EMOA-MIA and PSO-MIA on each application and size, we utilized the CloudSim simulator [13] which is the most popular and used simulator in the Cloud scheduling area, considering the types of VMs presented in Sect. 5.2 and the characteristics of their on-demand instances.

Given that these autoscalers are based on non-deterministic algorithms, we run each autoscaler 30 times on each application and size. For each run, we recorded the results achieved including makespan in seconds and cost in USD.

In order to run the autoscaler EMOA-MIA, we used the parameter settings detailed in Table 2 for the algorithm SMS-EMOA regarding each application and size. These settings were selected based on an exhaustive preliminary sensitivity analysis. In such analysis, to determine the best setting of SMS-EMOA for each application and size, we considered 192 different settings, then we developed 30 runs of SMS-EMOA for each setting, and finally we selected the setting maximizing the average hypervolume (HV) value of the Pareto sets obtained by SMS-EMOA.

To run the autoscaler PSO-MIA, we used the parameter settings detailed in Table 3 for the algorithm SMPSO regarding each application and size. These settings were selected based on an exhaustive preliminary sensitivity analysis similar to that developed for the algorithm SMS-EMOA.

To run the autoscaler MIA, we used the parameter settings recommended in [8] for the algorithm NSGA-III regarding each application/size. However, we defined a new value for the parameter *maxEvaluations* of NSGA-III (i.e., a higher value). We set such parameter with the value used for the parameter *maxEvaluations* of SMS-EMOA and SMPSO. Thus, the three algorithms used the same termination condition for each application/size. This is essential to guarantee a fair evaluation of the autoscalers.

Table 2. Selected parameter settings for SMS-EMOA.

Application	Size	*max Eval.*	*pop. size*	*sbx rate*	*sbx dist.*	*pm rate*	*pm dist.*
Plate3D	30	24932	92	0.94	24.43	0.33	23.96
	100	24932	92	0.99	9.43	0.46	55.60
	300	24932	92	0.94	35.80	0.33	10.60
Cruciform	30	24932	92	0.99	15.00	0.20	20.00
	100	24932	92	0.99	15.00	0.20	20.00
	300	24932	92	0.99	15.00	0.20	20.00
Ensemble	60	24932	92	0.99	15.00	0.20	20.00
	200	24932	92	0.86	5.80	0.04	53.96
	600	24932	92	0.99	15.00	0.20	20.00

Table 3. Selected parameter settings for SMPSO.

Application	Size	*max Eval.*	*pop. size*	*archive size*	*pm rate*	*pm dist.*
Plate3D	30	24932	92	92	0.20	20.00
	100	24932	92	92	0.23	37.68
	300	24932	92	92	0.01	29.21
Cruciform	30	24932	92	92	0.46	20.80
	100	24932	92	92	0.23	36.70
	300	24932	92	92	0.16	53.73
Ensemble	60	24932	92	92	0.07	20.00
	200	24932	92	92	0.14	5.80
	600	24932	92	92	0.02	35.80

5.4 Experimental Results

In Table 4, we present the results obtained from the computational experiments. In this table, columns 4 and 5 detail the average values for makespan in seconds and cost in USD, respectively. Then, columns 6 and 7 detail the average makespan relative percentage difference regarding MIA (average makespan RPD) and the average cost relative percentage difference regarding MIA (average cost RPD), respectively. These two metrics are described in detail below for the sake of clarity. Finally, column 8 details the average value for the L_2-norm metric. This metric analyzes simultaneously the makespan and cost resulting from the experiments. Then, the value of this metric is interpreted as the trade-off between the resulting makespan and cost.

Regarding the metric average makespan RPD, this metric allows to calculate the percentage difference of the average makespan of EMOA-MIA (or PSO-MIA) in respect of the average makespan of MIA, by the formula $((m^t - m)/m^t)$x100. In this formula, m^t represents the average makespan of MIA, and m represents the average makespan of EMOA-MIA (or PSO-MIA). If this difference is positive, this means that EMOA-MIA (or PSO-MIA) has obtained a makespan saving (a decrease in average makespan) respecting MIA. Higher positive values represent better average makespan savings. If this difference is negative, this means that EMOA-MIA (or PSO-MIA) has obtained an increase in average makespan respecting MIA. Higher negative values represent higher average makespan increases. In the same way, the metric average cost RPD allows to calculate the percentage difference of the average cost of EMOA-MIA (or PSO-MIA) in respect of the average cost of MIA, by the formula $((c^t - c)/c^t)$x100. In this formula, c^t represents the average cost of MIA, and c represents the average cost of EMOA-MIA (or PSO-MIA).

As shown in Table 4, regarding the average makespan values, EMOA-MIA and PSO-MIA reached a better performance than MIA in six out of nine applications and sizes (i.e., Plate3D with 30, 100, and 300 tasks, and Ensemble with 60, 200 and 600 tasks), achieving good makespan savings between 17–19% in some cases. The three autoscalers reached the same performance for Cruciform with the different sizes.

Regarding the average cost values, EMOA-MIA and PSO-MIA reached a better performance than MIA in six of the nine applications and sizes (i.e., Plate3D with 30, 100, and 300 tasks, and Ensemble with 60, 200 and 600 tasks), achieving good cost savings (around 27% in some cases, 24% in some other cases, and 15% in some others). In the case of the application Cruciform with 30 tasks, EMOA-MIA reached a better performance than MIA, whereas PSO-MIA was outperformed by MIA. In the case of Cruciform with 100 and 300 tasks, although EMOA-MIA and PSO-MIA were outperformed by MIA, EMOA-MIA obtained better values than PSO-MIA.

Regarding the average values obtained in relation to the metric L_2-norm, EMOA-MIA reached a better (equal) performance than MIA in seven (one) of the nine applications and sizes. This is because EMOA-MIA outperformed (equaled) MIA in seven (one) of such cases considering cost, and outperformed (equaled) MIA in six (two) of such cases considering makespan. On the other hand, PSO-MIA reached a better performance than MIA in six of the nine applications and sizes. This is because PSO-MIA outperformed MIA in such cases considering cost, and outperformed MIA in such cases

Table 4. Results obtained from the experiments. For the metrics average makespan, average cost and average L_2-norm, lower values represent better results. Note that bold values are better than those obtained by the autoscaler MIA, and the symbol * indicates the best value. For the metrics average makespan RPD and average cost RPD, positive values represent good results.

Application	Size	Autoscaler	Makespan	Cost	Makespan RPD (%)	Cost RPD (%)	L_2-norm
Plate3D	30	MIA	1235.522	0.633	–	–	0.51
		EMOA-MIA	**1019.300**	**0.459**	17.50	27.49	**0.42**
		PSO-MIA	**1019.300**	**0.459**	17.50	27.49	**0.42**
	100	MIA	1436.633	1.703	–	–	0.48
		EMOA-MIA	**1351.700**	**1.295***	5.91	23.96	**0.46**
		PSO-MIA	**1308.322***	**1.303**	8.93	23.49	**0.44***
	300	MIA	1592.078	4.243	–	–	0.41
		EMOA-MIA	**1291.056**	**4.046***	18.91	4.64	**0.36***
		PSO-MIA	**1285.756***	**4.076**	19.24	3.95	**0.38**
Cruciform	30	MIA	2766.078	1.670	–	–	0.64
		EMOA-MIA	2766.078	**1.667***	0.00	0.16	**0.56***
		PSO-MIA	2766.078	1.679	0.00	−0.56	0.65
	100	MIA	2777.167	5.370*	–	–	0.39*
		EMOA-MIA	2777.167	5.373	0.00	−0.06	0.40
		PSO-MIA	2777.167	5.388	0.00	−0.33	0.42
	300	MIA	2789.822	15.870*	–	–	0.68*
		EMOA-MIA	2789.822	15.871	0.00	−0.01	0.68*
		PSO-MIA	2789.822	15.893	0.00	−0.15	0.76
Ensemble	60	MIA	3028.511	2.430	–	–	0.43
		EMOA-MIA	**3017.867**	**2.167**	0.35	10.82	**0.40**
		PSO-MIA	**3017.867**	**2.167**	0.35	10.82	**0.40**
	200	MIA	3248.811	7.900	–	–	0.51
		EMOA-MIA	**3136.133***	**6.709**	3.47	15.07	**0.46***
		PSO-MIA	**3148.233**	**6.617***	3.10	16.23	**0.47**
	600	MIA	3220.833	22.355	–	–	0.48
		EMOA-MIA	**3212.670**	**19.101**	0.25	14.56	**0.42**
		PSO-MIA	**3212.670**	**19.101**	0.25	14.56	**0.42**

considering makespan. Moreover, note that EMOA-MIA achieved a better (equal) performance than PSO-MIA in five (three) of the nine applications and sizes, regarding the metric L_2-norm. This indicates that EMOA-MIA reached a better trade-off between the makespan and cost resulting from the experiments, in such five cases.

To determine if the improvements reached by EMOA-MIA and PSO-MIA regarding MIA are significant, we applied a statistical significance test on the results achieved from the experiments for the metric L_2-norm. This metric provides a joint analysis of the makespan and cost resulting from the experiments. Thus, we considered that this metric is suitable and useful to develop the statistical significance test. In relation to the results achieved from the experiments for the metric L_2-norm, each autoscaler was run 30 times on each application and size, and thus obtained 30 results for the metric in relation to each application and size. We applied the Mann-Whitney test [14] on the results obtained by MIA, EMOA-MIA and PSO-MIA in relation to each application and size, using $\alpha = 0.001$. In accordance with this test, EMOA-MIA and PSO-MIA reached significant improvements in terms of the metric L_2-norm. It is necessary to mention that we applied such test since the results obtained by each autoscaler for the metric L_2-norm regarding each application and size do not follow the normal distribution, as determined with the Shapiro-Wilk test which was applied with $\alpha = 0.001$.

6 Related Work

In the literature, there are many approaches addressing the efficient management of scientific applications on the Cloud, in order to deal with resource provisioning and scheduling. However, very few cloud autoscaling approaches have been proposed for PSEs. One of the most recent approaches is the autoscaler MIA [8], a multi-objective autoscaler based on the algorithm NSGA-III for executing PSEs on public Clouds.

First of all, we can mention some approaches that address autoscaling in the Cloud and that, like [8], focus on the use of on-demand instances [5, 15–17, 19]. In [15], a Robust Hybrid Auto-Scaler (RHAS) for executing web applications in Cloud is presented. RHAS was designed for cost saving while met QoS. RHAS is composed of a monitor loop that was designed to implement the scaling system and save the renting cost of on-demand computing resources. The goal of this approach is to estimate the number of required resources in horizontal scaling for the incoming workload. In [16], a novel hybrid autoscaling technique based on a combination of a reactive approach and a proactive approach to scale resources based on user demand was proposed. The approach uses a price model that can lead to an increase in profit of a broker (intermediary enterprise) and a cost reduction of the users at the same. The technique examines the scale-up condition which, in a purely reactive auto-scaling environment, is used to acquire new resources, and builds incrementally updateable predictive models to enable a system to proactively scale up before this condition is met. In [17], a big data-oriented cost-effective autoscaling scheme for smart computing in private Cloud was proposed. The autoscaling algorithm was designed with a type-aware resource provision method to alter the configuration of scaled batch nodes according to the job types and the environment of physical resources. Then, in [19] the authors have presented MLscale, an application-agnostic autoscaler that requires minimal application knowledge and manual tuning. MLscale employs neural networks to online build the application performance model, and then leverages multiple linear regression to predict the post-scaling state of the system. This combination enables accurate autoscaling under MLscale. The approach is able to accurately model the (average and tail) response time while minimizing the resource cost of web applications. Moreover, in [5] the authors have introduced a dynamic cloud resource provisioning algorithm called delay-based dynamic scheduling to minimize the monetary cost while meeting Bag-of-Tasks (BoT) workflow deadlines. BoT, as well as PSEs, are applications composed of many independent tasks that can be processed in parallel. Although these approaches are based on the use of on-demand instances, none of them exploit the use of metaheuristics.

There are other works that, unlike [8], consider spot instances, for example the one presented in [7]. The main tasks characteristics in the workflow structure are learned over time, i.e., the autoscaler dynamically adapts the number of allocated resources in order to meet the deadlines of all tasks without knowing the workflow structure itself and without any information of the execution time. The optimization objective was to minimize makespan. In [18], a scheduling algorithm for a Bag of Tasks (BoT) was proposed, which uses a particle swarm optimization algorithm for achieving load balancing combined with an artificial Neural Network to predict the future values of Spot instances. The algorithm then validates these predicted values with respect to the current values of spot instances to minimize cost of executing BoT applications in Cloud. However, it is worth mentioning that although in this work they consider bag-of-tasks applications, the authors only focused on the task scheduling problem without considering the automatic scaling.

7 Conclusions

In this paper, we proposed to endow the autoscaler MIA with others multi-objective optimization algorithms, with the aim of improving its performance regarding the optimization objectives. To do that, we considered two well-known multi-objective optimization algorithms named SMS-EMOA and SMPSO, which have significant differences with the algorithm NSGA-III currently used by the autoscaler MIA.

We evaluated the autoscaler MIA endowed with SMS-EMOA, and the autoscaler MIA endowed with SMPSO, on three real-world PSE applications, considering three sizes per application. Besides, we considered five different types of VMs available in Amazon EC2, and the characteristics of their on-demand instances. We used such applications and VM types in order to provide diverse realistic experimental settings. After that, we compared the performance of the autoscaler MIA endowed with SMS-EMOA, and the performance of the autoscaler MIA endowed with SMPSO, with that of the autoscaler MIA based on NSGA-III. According to the performance comparison developed, the autoscaler MIA endowed with each one of the algorithms considered outperformed MIA based on NSGA-III, in both optimization objectives, in most of the applications and sizes used. Thus, we conclude that the autoscaler MIA endowed with SMS-EMOA, or SMPSO, may be considered as a better alternative for solving the multi-objective autoscaling problem addressed.

In future works, we will analyze the incorporation of other relevant optimization objectives into the problem addressed. Besides, we will evaluate other multi-objective optimization algorithms in the context of the autoscaler.

References

1. García Garino, C., Ribero Vairo, M.S., Andía Fagés, S., Mirasso, A.E., Ponthot, J.-P.: Numerical simulation of finite strain viscoplastic problems. J. Comput. Appl. Math. **246**, 174–184 (2013)
2. Mauch, V., Kunze, M., Hillenbrand, M.: High performance cloud computing. Futur. Gener. Comput. Syst. **29**(6), 1408–1416 (2013)
3. Monge, D., Garí, Y., Mateos, C., García Garino, C.: Autoscaling scientific workflows on the cloud by combining on-demand and spot instances. Comput. Syst. Sci. Eng. **32**(4), 291–306 (2017)
4. Mao, M., Humphrey, M.: Scaling and scheduling to maximize application performance within budget constraints in cloud workflows. In: 27th International Symposium on Parallel and Distributed Processing, pp. 67–78 (2013)
5. Cai, Z., Li, X., Ruiz, R., Li, Q.: A delay-based dynamic scheduling algorithm for bag-of-task workflows with stochastic task execution times in clouds. Futur. Gener. Comput. Syst. **71**, 57–72 (2017)
6. Li, J., Su, S., Cheng, X., Song, M., Ma, L., Wang, J.: Cost-efficient coordinated scheduling for leasing cloud resources on hybrid workloads. Parallel Comput. **44**, 1–17 (2015)
7. De Coninck, E., Verbelen, T., Vankeirsbilck, B., Bohez, S., Simoens, P., Dhoedt, B.: Dynamic autoscaling and scheduling of deadline constrained service workloads on IaaS clouds. J. Syst. Softw. **118**, 101–114 (2016)
8. Yannibelli, V., Pacini, E., Monge, D., Mateos, C., Rodriguez, G.: An NSGA-III-Based Multi-objective Intelligent Autoscaler for Executing Engineering Applications in Cloud Infrastructures. In: Martínez-Villaseñor, L., Herrera-Alcántara, O., Ponce, H., Castro-Espinoza, F.A. (eds.) MICAI 2020. LNCS (LNAI), vol. 12468, pp. 249–263. Springer, Cham (2020). https://doi.org/10.1007/978-3-030-60884-2_19
9. Beume, N., Naujoks, B., Emmerich, M.: SMS-EMOA: Multiobjective selection based on dominated hypervolume. Eur. J. Oper. Res. **181**(3), 1653–1669 (2007)
10. Nebro, A.J., Durillo, J.J., Garcia-Nieto, J., Coello Coello, C.A., Luna, F., Alba, E.: SMPSO: a new PSO-based metaheuristic for multi-objective optimization. In: 2009 IEEE Symposium on Computational Intelligence in Multi-Criteria Decision-Making (MCDM), pp. 66–73 (2009)
11. Deb, K., Jain, H.: An evolutionary many-objective optimization algorithm using reference-point based non-dominated sorting approach, Part I: Solving problems with box constraints. IEEE Trans. Evol. Comput. **18**(4), 577–601 (2014)
12. Makris, N.: Plastic torsional buckling of cruciform compression members. J. Eng. Mech. **129**(6), 689–696 (2003)
13. Silva Filho, M.C., Oliveira, R.L., Monteiro, C.C., Inácio, P.R.M., Freire, M.M.: CloudSim Plus: a cloud computing simulation framework pursuing software engineering principles for improved modularity, extensibility and correctness. In: 2017 IFIP/IEEE Symposium on Integrated Network and Service Management (IM), pp. 400–406 (2017)
14. Mann, H.B., Whitney, D.R.: On a test of whether one of two random variables is stochastically larger than the other. Ann. Math. Stat. **18**(1), 50–60 (1947)
15. Singh, P., Kaur, A., Gupta, P., Gill, S.S., Jyoti, K.: RHAS: robust hybrid auto-scaling for web applications in cloud computing. Clust. Comput. **24**(2), 717–737 (2020). https://doi.org/10.1007/s10586-020-03148-5

16. Biswas, A., Majumdar, S., Nandy, B., El-Haraki, A.: A hybrid auto-scaling technique for clouds processing applications with service level agreements. J. Cloud Comput. **6**, 29 (2017)
17. Lu, Z., Wang, X., Wu, J.: InSTechAH: Cost-Effectively Autoscaling Smart Computing Hadoop Cluster in Private Cloud. J. Syst. Architect. **80**, 1–16 (2017)
18. Domanal, S.G., Reddy, G.R.M.: An efficient cost optimized scheduling for spot instances in heterogeneous cloud environment. Futur. Gener. Comput. Syst. **84**, 11–21 (2018)
19. Wajahat, M., Karve, A., Kochut, A., Gandhi, A.: MLscale: a machine learning based application-agnostic autoscaler. Sustain. Comput. Inform. Syst. **22**(287), 299 (2017)

Best Paper Award, First Place

Multi-objective Release Plan Rescheduling in Agile Software Development

Abel García-Nájera[1]([⊠])(iD), Saúl Zapotecas-Martínez[1](iD),
Jesús Guillermo Falcón-Cardona[1](iD), and Humberto Cervantes[2](iD)

[1] Universidad Autónoma Metropolitana Unidad Cuajimalpa, Avenida Vasco de
Quiroga 4871, 05348 Col. Santa Fe Cuajimalpa, México
{agarcian,szapotecas,jfalcon}@cua.uam.mx
[2] Universidad Autónoma Metropolitana Unidad Iztapalapa,
Avenida, San Rafael Atlixco 186, 09340 Col. Vicentina, México
hcm@xanum.uam.mx

Abstract. Scrum is an agile software development framework followed
nowadays by many software companies worldwide. Since it is an iterative and incremental methodology, the software is developed in releases.
For each release, the software development team and the customer agree
upon a development plan. However, the context of the software project
may change due to unpredicted circumstances that generally arise, for
example, new software requirements or changes in the development team.
Consequently, these factors force the release plan to be adjusted. When
the release plan is modified, it is necessary to consider at least four criteria to minimize the economic and operational impact of these changes.
Therefore, this activity can be analyzed as a multi-objective problem.
In the last three decades, multi-objective evolutionary algorithms have
become an effective and efficient tool to solve multi-objective problems.
In this paper, we evaluate three multi-objective optimization approaches
when solving the release plan rescheduling problem. Mainly, we focus our
investigation on analyzing the conflict between the considered objectives
and on the performance of the Pareto-based, the indicator-based, and
the decomposition-based multi-objective optimization approaches.

Keywords: Multi-objective optimization · Evolutionary algorithms ·
Release plan rescheduling problem · Scrum

1 Introduction

Scrum is a popular agile project management framework, which has been
adopted in the software development community since two decades ago [12].
Scrum promotes the idea of developing a project in an iterative and incremental way, where iterations are referred to as *sprints*. Scrum sprints have a
fixed duration, generally between two and four weeks. During a sprint, software

© Springer Nature Switzerland AG 2021
I. Batyrshin et al. (Eds.): MICAI 2021, LNAI 13067, pp. 403–414, 2021.
https://doi.org/10.1007/978-3-030-89817-5_30

requirements are transformed into a product increment that may or may not be released. Such requirements are expressed using a technique called *user stories*, which describes the functionality to be achieved along with acceptance criteria. The set of requirements for the product are collected in the *product backlog* and during a meeting at the beginning of the sprint. The team, along with a business representative known as the *Product Owner*, agree on a subset of requirements from the product backlog to be implemented during the sprint. These requirements are collected into the *sprint backlog* and, once it is established, the team defines a detailed plan for the sprint, which includes the tasks needed to complete the user stories, their duration, and their assignment to specific team members.

The sprint backlog establishes the scope for a sprint and it cannot be changed during the development of the sprint. An adequate scope for a specific sprint is important since the goal is to successfully and completely implement the user stories that were selected for that sprint. In order to establish an achievable scope, user stories are associated with an estimated effort for completing them. This effort is measured in a unit called *story points*, and the estimation is made by the team, usually before the start of the execution of sprints. The average number of story points the team is capable of developing in a sprint is called the team's *velocity*. When the team and the Product Owner agree on the scope of the sprint, they consider the total number of story points for the selected user stories, and they make sure that this total does not exceed the team's velocity.

Some projects require some level of planning that goes beyond a single sprint, although planning is not required at the level of detail of older plan-based approaches. This type of planning is useful in less changing contexts and has the advantage that it can provide more precise delivery dates. In order to support this approach, it is possible to create a *release plan* that, as its name implies, concludes with the release of a new version of the system. A release plan is typically composed of a number of sprints, hence, instead of weeks, it may span across a few months. In a release plan, user stories are assigned to particular sprints according to their *priority* and, in general, high priority user stories are allocated to the initial sprints to minimize risks in case of delays.

As it is common in any other types of project plans, the execution of a release plan may not go according to what was scheduled due to different unexpected or *disruptive* events that can occur. We assume that these disruptive events occur before or during the execution of the release plan. In this context, we can identify two categories of disruptive events. In the first category, we can find the events that affect the originally planned number of story points. Some examples of these events are: (i) the addition of a new user story to the release plan and (ii) a user story from the current sprint could not be finished, consequently, it will have to be finished in a subsequent sprint. The second category includes events that change the team's velocity, for example: (i) a developer will leave after the current sprint and (ii) a developer is substituted by another that does not have a similar level of expertise.

If an event that falls into any of the previous two categories occurs, the original release delivery time and cost will probably be compromised. Hence, it is required that the original release plan be adjusted in order to allow the release to be delivered on time as long as possible and within budget. Consequently, the Scrum Master should immediately perform a release plan rescheduling that minimizes the economic and operational impacts, and that meets, as long as possible, the delivery dates that were already set.

In addition to the minimization of the project cost and to maintain the delivery date, there are other important criteria that should be considered when rescheduling a release plan. Firstly, it is desirable that the rescheduled release plan does not differ too much from the original one, since initial planning is considered to be the best option for the project. Secondly, since software development is an expensive task, the team's development capacity should be used at the maximum. Therefore, when carrying out a release plan rescheduling, at least four objectives can be considered for optimization: (1) release delivery time, (2) release cost, (3) release plan stability, and (4) development capacity misspending. Therefore, release plan rescheduling can be tackled as a multi-objective optimization problem. The release plan rescheduling problem can be seen as a kind of generalized assignment problem, which is well known to be NP-hard [11]. In addition to the problem complexity, some of the objectives considered are naturally in conflict. For instance, if the development team is increased by one member, the release delivery time will be shorten, however the release cost will be higher. For this reason, this paper introduces a four-objective optimization model of the release plan rescheduling problem.

In order to solve the multi-objective release plan rescheduling problem, three evolutionary algorithms based on three multi-objective optimization approaches are considered. Particularly, we evaluate the Pareto-based, indicator-based, and decomposition-based approaches. The obtained results give rise to discuss the performance of these approaches. On the other hand, we also analyze the nature of the multi-objective problem formulated in this paper.

The structure of this paper is as follows. Section 2 makes a literature review of a number of related studies, explains the multi-objective release plan rescheduling problem, and introduces its mathematical formulation. Section 3 describes multi-objective optimization problems and the performace metric used in this study. The experimental set-up is presented in Sect. 4. In Sect. 5, the summary of the results is presented and analyzed. Finally, conclusions and future work are described in Sect. 6.

2   The Multi-objective Release Plan Rescheduling Problem

The problem addressed in this paper, referred to as the release plan rescheduling problem (RPRP), is related to the software project scheduling problem (SPSP), which has been studied thoroughly [13]. These two project planning problems seek to produce an appropriate schedule that allows the project to be finished on time and with the lowest cost. However, there are two important differences between them: (1) RPRP assumes that an initial release plan exists, while SPSP is focused on creating an initial project plan. (2) RPRP is based on release plans and, consequently, it is focused on the assignment of user stories to particular sprints. It does not deal with the assignment of individual tasks to team members, as SPSP does.

In the following, we review a number of related studies and formulate the multi-objective release plan rescheduling problem.

2.1   Previous Studies

To the best of our knowledge, the multi-objective RPRP (MORPRP) has not been the subject of systematic investigation, nevertheless, there is some previous related work that deserves a review for the present study.

The work of Jahr [8] work does not consider planning since the author mentions that the planning is done in each sprint, and, for this, previous sprints must be taken into account. In order to assign developers to project tasks, the author presents an integer programming model to minimize the development time of each sprint and uses GAMS software to find solutions to small test instances.

Roque et al. [10] concentrate on the assignment of employees to tasks in agile software projects. Authors argue that allocating a group of similar tasks suited to employee skills may contribute to achieve an efficient and profitable release planning. Consequently, similar tasks performed by the same employee tend to be performed more efficiently. In this work, the authors present a multi-objective optimization problem, which aims at minimizing project duration and total cost, subject to two constraints. The first constraint ensures that the amount of overtime worked by the employee cannot be greater than the allowed and the second prevents a task that requires a specific skill to be allocated to an employee who does not possess that particular skill.

The preliminary work of Nigar [9] formulates the software project scheduling problem as an optimization problem under uncertainties and dynamics for hybrid scRUmP software model (a combination of Scrum and RUP, the Rational Unified Process), and focuses on agility and quality, taking into account the dynamic environment. The problem formulation considers five objective functions to be optimized: project cost, project duration, task fragmentation (the dependency of one task on another), robustness, and stability.

Additionally to the few studies reviewed above, there are some recent studies that do analyze the MORPRP. Zapotecas et al. [15] applied three decomposition-based multi-objective algorithms for analyzing the multi-objective release plan rescheduling problem. They consider three objectives, namely release delivery time, release cost, and release plan rescheduling instability.

More recently, Escandon et al. [5] analyzed the MORPRP that considered five objectives: release delivery time, release cost, release plan rescheduling instability, velocity misspending, and release value. They used three multi-objective evolutionary algorithms based on the three most common approaches, that is, Pareto-based, indicator-based, and decomposition-based.

According to our literature review, the work of Jahr [8] does not consider planning, the work of Roque et al. [10] does not take into account the dynamic nature of software projects, and that of Nigar [9] only mentions what objectives could be considered for optimization. On the other hand, the two most recent studies of Zapotecas et al. [15] and Escandon et al. [5], proposed a mathematical formulation where it is not totally clear what the meaning of the objective functions are. Moreover, it is not completely explicit how they managed the constraint of the problem.

The present study aims at addressing the identified drawbacks from the reviewed studies and focuses on the following goals. Firstly, to propose an adequate multi-objective formulation for the release plan rescheduling problem that considers aspects of agile methodologies, particularly from Scrum, and four objective functions. Secondly, to appropriately undertake the constraint of the problem and properly describe how it is handle. Finally, to provide an analysis on the performance three evolutionary algorithms based on the three most common multi-objective optimization approaches.

2.2 The Multi-objective Release Plan Rescheduling Problem

The release plan rescheduling problem can be formulated as follows. Let us define a release plan L as the set $L = \{S_1, S_2, \ldots, S_s\}$ of s sprints. Let each sprint S_i be the set $S_i = \{h_1^i, h_2^i, \ldots, h_{n_i}^i\}$ of n_i user stories that have to be developed in the i-th sprint. All sprints are disjoint sets of user stories, so that $H = \{S_1 \cup S_2 \cup \cdots \cup S_s\} = \{h_1, h_2, \ldots, h_n\}$ is the set of n user stories h_j that must be developed in the release plan L. Each user story h_i is given a story points estimation sp_i.

There is a team $E = \{e_1, e_2, \ldots, e_m\}$ of m software engineers who are in charge of developing the release plan L. Each engineer e_j contributes with a velocity v_j to the i-th sprint velocity V_i. If overtime is required, each engineer e_j contributes v_j^{over} to the overtime i-th sprint velocity V_i^{over}.

According to our literature review, several objective functions have been taken into account in related problems, however, four of them are the most commonly considered.

Since the RPRP includes concepts that arise in agile methodologies, particularly from Scrum, four criteria can be taken into account when rescheduling a release plan.

A release plan establishes an estimated release delivery date. This date is given by the number of sprints planned for the release and rescheduling the release plan should try to maintain the number of sprints. Thus, the first objective is the release L *delivery time*, defined simply as the number of sprints in which the rescheduled release plan is divided. That is,

$$f_1 = |L| = s. \tag{1}$$

Cost depends on the time and the quantity of resources. The model proposed here considers solutions that can include a certain amount of overtime work or additional sprints or both. However, these options incur an additional cost, which is generally not desirable. Hence, the second objective function is the release *cost*, which is the sum of the estimated overtime translated to velocity, i.e.

$$f_2 = \sum_{i=1}^{s} \theta(SP_i - V_i)(SP_i - V_i). \tag{2}$$

where

$$SP_i = \sum_{h_k \in S_i} sp_k \tag{3}$$

and function $\theta(\cdot)$ is the unit step function.

In the RPRP, it is assumed that the original release plan was created carefully with the best interests of the project in mind. Stability concerns modifying the original plan as little as possible when it is rescheduled. In this way, the third objective function is the release plan *instability*, which counts the number of user stories that were reassigned to a different sprint after the rescheduling. That is

$$f_3 = \sum_{i=1}^{n} \theta\left(|sprint(i,t) - sprint(i,t-1)|\right) \tag{4}$$

where function $sprint(i,t)$ indicates the sprint to which user story h_i is assigned at time t.

Finally, when the total number of story points assigned to a sprint is lower than the team's velocity, the development capacity is misspent. Hence, the release plan rescheduling seeks to minimize the not used development capacity. This is why the fourth objective function is the *velocity misspending*, which seeks to utilize, in every sprint, the total sprint velocity. This function adds, for each sprint, the difference between the sprint velocity and the story points assigned to that sprint, i.e.

$$f_4 = \sum_{i=1}^{s} \theta(V_i - SP_i)\left(V_i - SP_i\right). \tag{5}$$

Additionally to the four objectives functions, the problem considers one constraint, which is that the story points assigned to each sprint must not exceed the team's velocity including overtime. Thus, the multi-objective release plan rescheduling problem can be stated as:

$$\begin{aligned} \text{minimize} \quad & (f_1, f_2, f_3, f_4) & \text{(6a)} \\ \text{subject to} \quad & SP_i \leq V_i^{over}, \ \forall \ S_i \in L. & \text{(6b)} \end{aligned}$$

3   Multi-objective Optimization Problems

Without loss of generality, let us define a multi-objective optimization problem (MOP) as the minimization problem:

$$\begin{aligned} \underset{\vec{x} \in \mathcal{X}}{\text{minimize}} \quad & \vec{f}(\vec{x}) = (f_1(\vec{x}), \dots, f_m(\vec{x})), \\ \text{subject to} \quad & g_i(\vec{x}) \leq 0, \ \forall \ i \in \{1, \dots, p\}, \\ & h_j(\vec{x}) = 0, \ \forall \ j \in \{1, \dots, q\}, \end{aligned} \tag{7}$$

where $\vec{x}$ is a potential solution to the problem, $\mathcal{X}$ is the domain of solutions, and $f_i : \mathcal{X} \to \mathbb{R}, \forall\, i \in \{1, \ldots, m\}$, are m objective functions. The constraint functions $g_i, h_j : \mathcal{X} \to \mathbb{R}$ delimit the feasible search space.

In this context, it is said that a solution $\vec{x} \in \mathcal{X}$ *weakly dominates* (or *covers*) solution $\vec{y} \in \mathcal{X}$, written as $\vec{x} \preceq \vec{y}$, if and only if $f_i(\vec{x}) \leq f_i(\vec{y}), \forall\, i \in \{1, \ldots, m\}$. Solution $\vec{x}$ *dominates* solution $\vec{y}$, written as $\vec{x} \prec \vec{y}$, if and only if $\vec{x} \preceq \vec{y}$, and $\exists\, j : f_j(\vec{x}) < f_j(\vec{y})$. Consequently, solution $\vec{x} \in \mathcal{S} \subseteq \mathcal{X}$ is *non-dominated* with respect to $\mathcal{S}$ if $\nexists\, \vec{y} \in \mathcal{S} : \vec{y} \prec \vec{x}$. A solution $\vec{x} \in \mathcal{X}$ is *Pareto optimal* if it is non-dominated with respect to the entire domain of solutions $\mathcal{X}$. The *Pareto optimal set* $\mathcal{P}_s$ is the set of all Pareto optimal solutions, that is, $\mathcal{P}_s = \{\vec{x} \in \mathcal{X} \mid \vec{x}$ is Pareto optimal$\}$. The *Pareto front* $\mathcal{P}_f$ is the set of the evaluations of the function vector $\vec{f}$ at all solutions in the Pareto optimal set, that is, $\mathcal{P}_f = \{\vec{f}(\vec{x}) \in \mathbb{R}^m \mid \vec{x} \in \mathcal{P}_s\}$.

3.1 The IGD$^+$ Performance Metric

Multi-objective problems have whole sets of solutions to compare with at least two aims: to minimize the distance from the generated solutions, called the *Pareto approximation, approximation set, non-dominated set,* or *non-dominated solutions,* to the true Pareto front, and to maximize the diversity among them, i.e. the coverage of the Pareto front. For this reason, the definition and use of appropriate performance metrics or quality indicators is crucial. In the present study, the inverted generational distance plus was used to compare performance.

Given the set of reference points $\mathcal{Z} = \{\vec{z}_1, \vec{z}_2, \ldots, \vec{z}_{|\mathcal{Z}|}\}$, where $\vec{z}_j = (z_{j1}, z_{j2}, \ldots, z_{jm})$ is a point on the Pareto front in an m-dimensional objective space, the inverted generational distance plus IGD$^+(\mathcal{A}, \mathcal{Z})$ is the average distance from each reference point $\vec{z}_j$ to the nearest solution $\vec{a}_i \in \mathcal{A}$ [6], where $\vec{a}_i = (a_{i1}, a_{i2}, \ldots, a_{im})$ is a solution in the m-dimensional objective space. The IGD$^+$ value of a non-dominated solution set $\mathcal{A} = \{\vec{a}_1, \vec{a}_2, \ldots, \vec{a}_{|\mathcal{A}|}\}$ is calculated as follows [7]:

$$\mathrm{IGD}^+(\mathcal{A}, \mathcal{Z}) = \frac{1}{|\mathcal{Z}|} \sum_{j=1}^{|\mathcal{Z}|} \min_{\vec{a}_i \in \mathcal{A}} d_{\mathrm{IGD}^+}(\vec{a}_i, \vec{z}_j) \tag{8}$$

where $d_{\mathrm{IGD}^+}(\vec{a}, \vec{z})$ is the maximum of the Euclidian distance between $\vec{a}$ and $\vec{z}$, and zero, according to Eq. (9).

$$d_{\mathrm{IGD}^+}(\vec{a}, \vec{z}) = \sqrt{\sum_{k=1}^{m} (\max\{a_k - z_k, 0\})^2} \tag{9}$$

4 Experimental Set-Up

4.1 Solution Encoding, Evolutionary Operators and Repair Operators

Encoding and Representation. Since the multi-objective release plan rescheduling problem requires solutions to assign user stories to sprints, the most natural way

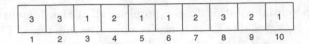

Fig. 1. Integer representation: example solution encoding.

to encode such solutions is the *integer* representation. This encoding considers a vector of size n, the number of user stories and each dimension of the vector is assigned an integer in the interval $[1..s]$, where s is the number of sprints. An example of this representation is shown in Fig. 1. In this example, a release plan with 10 user stories, which are estimated to be developed in three sprints. Then, the example solution is a vector of size 10, the number of user stories and each dimension (user story) is assigned an integer (sprint) in the set $\{1, 2, 3\}$.

Evolutionary Operators. Regarding the evolutionary operators, for the recombination process, the one-point crossover has been adopted: A crossover point on the parent solutions is randomly selected. All information before that point is copied from one of the parents to the child. To complete the solution, all data beyond that point is copied from the second parent. In the case of the mutation operator, the uniform mutation has been selected, which, for each user story, with probability P_m reassigns it to a random sprint.

Repair Operators. It could be the case that infeasible solutions are generated after the crossover and mutation operators, specifically, solutions with empty sprints or with sprints with more story points than the allowed. In the former case, the empty sprint is removed from the solution. In the latter case, user stories are removed from the infeasible sprint and reassigned to a new sprint.

4.2   Test Instances

In this study we have taken nine test instances from the proposed set of Escandon et al. [5]. These nine test instances are classified according to the number of user stories $n \in \{40, 70, 100\}$ and the number of employees $e \in \{4, 5, 6\}$ into three groups: H40-E4-k, H70-E5-k, and H100-E6-k, where $k \in \{1, 2, 3\}$ identifies three different instances for each instance size.

4.3   Experimental Design and Parameter Settings

We analyze the performance of three MOEAs based on three different multi-objective optimization approaches, namely NSGA-II [4], SMS-EMOA [1], and MOEA/D [16]. We use the same evolutionary operators in the mating process of the MOEAs. The population size was defined by $N = 300$ individuals. For each MOEA, the search was restricted to perform $N \times 1,000$ fitness function evaluations. A fair comparison is carried out by defining the same the crossover and mutation probabilities, which are $P_c = 1$ (crossover probability) and $P_m = 1/n$ (mutation probability), where n is the number of decision variables.

For each test problem, 30 independent runs with each MOEA were performed. The algorithms were evaluated using the IGD$^+$ [3] performance indicator. The reference set to compute the IGD$^+$ indicator was obtained by capturing all the non-dominated solutions produced by each MOEA in all the experiments for the test instance under consideration. Statistical analysis was performed over all the runs for each test instance considering the adopted performance indicator. To identify significant differences among the results obtained by the EMOAs, we adopt the Mann-Whitney-Wilcoxon [14] non-parametric statistical test with a p-value of 0.05 and Bonferroni correction [2].

5   Analysis of Results

The results obtained by the three MOEAs were analyzed to determine the best choice to solve the release plan rescheduling problem. Table 1 show the results achieved by the algorithms in the test instance, for the IGD$^+$ performance indicator. In each cell, the number on the left is the average indicator value, and the number on the right (in small font size) is the standard deviation. The best values for each performance indicator and test problem are reported in **bold-face**. An algorithm statistically better than all others can be considered as the best algorithm in the concerned test problem regarding the performance indicator under consideration, in such case, this value is underlined. If an algorithm obtained statistically the worst results, its values are wavy underlined.

In Table 1, it can be seen that the results obtained by NSGA-II outperformed those achieved by the other two MOEAs in seven out of the nine instances, and, in fact, it was statistically better in two of them. Table 1 provides a quantitative assessment of the performance of NSGA-II in terms of the IGD$^+$ performance indicator. This means that the solutions obtained by NSGA-II achieved a better approximation and spread along the reference set than those produced by the other two MOEAs.

To complement our study, we also analyzed the anytime behavior among the algorithms considered in our comparative study. For this task, we extracted the convergence plots of the IGD$^+$ performance indicator for each test instance along the generations of each MOEA. Figure 2 shows the convergence plots of the IGD$^+$ performance indicator for each test problem. Each plot shows the convergence of the averaged IGD$^+$ value with a confidence interval of 0.95 which is captured through the generations of each MOEA. From this figure, it is possible to see that, in most of the cases, the NSGA-II reached a lower IGD$^+$ value in comparison to the other MOEAs. This means that the NSGA-II approximated, in a better way, the reference set of each test problem faster than the other MOEAs. On the other hand, from these plots, it can be observed that SMS-EMOA obtained the worst performance in all the test problems. This observation suggests that the reference point is not adequate to compute the hypervolume in this type of test problems. In other words, the reference point is not suitable for the proper performance of SMS-EMOA.

Table 1. Table of results obtained by the decomposition-based MOEAs using the IGD$^+$ performance indicator

Instance	NSGA-II	SMS-EMOA	MOEA/D
H40-E4-01	**2.6428**±0.941	4.4515±4.066	2.7796±0.755
H40-E4-02	**0.9436**±0.504	2.2171±2.596	1.1670±0.604
H40-E4-03	2.7008±4.164	3.3224±3.780	**1.7612**±2.029
H70-E5-01	**5.6511**±5.698	10.8105±6.500	7.2163±1.910
H70-E5-02	**6.5726**±2.643	12.4626±6.994	7.1226±2.195
H70-E5-03	**12.5605**±7.299	18.4076±5.309	14.0794±3.609
H100-E6-01	**12.7954**±11.419	24.6548±7.767	14.1216±4.074
H100-E6-02	**16.5530**±8.178	26.3130±5.943	23.9340±6.584
H100-E6-03	21.8756±14.631	24.9204±10.650	**18.5398**±1.804

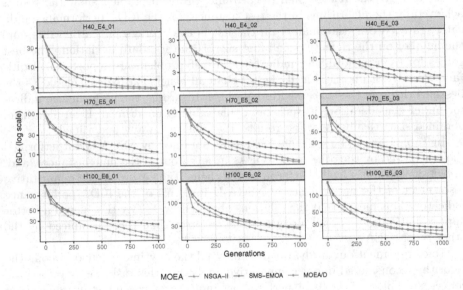

Fig. 2. Convergence plots of the IGD$^+$ performance indicator for the adopted test instances.

On the other hand, in order to investigate the nature of the problem, we observe the parallel coordinates of the reference set, which was obtained by capturing all non-dominated solutions found by the algorithm at each test instance. Thus, in Fig. 3 it is possible to see the conflict between pairs of objectives. Besides, the discretization of objective values for f_1 (the release delivery time), which has an integer value in the objective space, see the formulated objective in Sect. 2.2. Although we have observed different characteristics of the adopted test problems, we noticed that there are more details that deserve to be investigated. Nonetheless, the analysis of the multi-objective release plan rescheduling problem continues to be a path for future research.

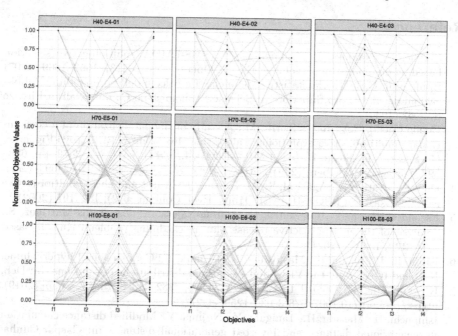

Fig. 3. Parallel plots for the reference sets of the adopted test problems.

6   Conclusion and Future Work

In this paper, we have introduced an improved mathematical formulation for the release plan rescheduling problem in the multi-objective context. In order to approximate the solution to the test instance presented in this study, we evaluate three representative MOEAs: Pareto-based, indicator-based, and decomposition-based. We showed the potential of the adopted MOEAs when approximating nine test instances introduced in this study. These results are promising since no exhaustive effort regarding parameter tuning was carried out (for instance, the choice of a scalarizing function, the neighborhood size, crossover and mutation probabilities, etc.). Nonetheless, using standard parameters, we found that NSGA-II was able to outperform the indicator-based and the decomposition-based MOEAs. This suggests that NSGA-II appropriately balance the exploration and exploitation for the test problems adopted in our experimental study.

With regard to future work, as a consequence of the analysis related to the release plan rescheduling problem, we envision a clear possible topic for research. This has to do with considering more objective functions in the model introduced here, which could include the release value and other concepts in the context of agile software development. This, however, is part of our ongoing research.

References

1. Beume, N., Naujoks, B., Emmerich, M.: SMS-EMOA: multiobjective selection based on dominated hypervolume. Eur. J. Operat. Res. **181**(3), 1653–1669 (2007)
2. Bonferroni, C.E.: Teoria statistica delle classi e calcolo delle probabilita. Pubblicazioni del R Istituto Superiore di Scienze Economiche e Commerciali di Firenze **8**, 3–62 (1936)
3. Coello Coello, C.A., Reyes Sierra, M.: A study of the parallelization of a coevolutionary multi-objective evolutionary algorithm. In: Monroy, R., Arroyo-Figueroa, G., Sucar, L.E., Sossa, H. (eds.) MICAI 2004. LNCS (LNAI), vol. 2972, pp. 688–697. Springer, Heidelberg (2004). https://doi.org/10.1007/978-3-540-24694-7_71
4. Deb, K., Pratap, A., Agarwal, S., Meyarivan, T.: A fast and elitist multiobjective genetic algorithm: NSGA-II. IEEE Trans. Evol. Comput. **6**(2), 182–197 (2002)
5. Escandon-Bailon, V., Cervantes, H., García-Nájera, A., Zapotecas-Martínez, S.: Analysis of the multi-objective release plan rescheduling problem. Knowl.-Based Syst. **220**, 106922 (2021)
6. Ishibuchi, H., Imada, R., Masuyama, N., Nojima, Y.: Comparison of hypervolume, IGD and IGD$^+$ from the Viewpoint of optimal distributions of solutions. In: Deb, K., et al. (eds.) EMO 2019. LNCS, vol. 11411, pp. 332–345. Springer, Cham (2019). https://doi.org/10.1007/978-3-030-12598-1_27
7. Ishibuchi, H., Masuda, H., Tanigaki, Y., Nojima, Y.: Modified distance calculation in generational distance and inverted generational distance. In: Gaspar-Cunha, A., Henggeler Antunes, C., Coello, C.C. (eds.) EMO 2015. LNCS, vol. 9019, pp. 110–125. Springer, Cham (2015). https://doi.org/10.1007/978-3-319-15892-1_8
8. Jahr, M.: A hybrid approach to quantitative software project scheduling within agile frameworks. Project Manag. J. **45**(3), 35–45 (2014)
9. Nigar, N.: Model-based dynamic software project scheduling. In: Proceedings of the 2017 11th Joint Meeting on Foundations of Software Engineering, pp. 1042–1045. ACM (2017)
10. Roque, L., Araújo, A.A., Dantas, A., Saraiva, R., Souza, J.: Human resource allocation in agile software projects based on task similarities. In: Sarro, F., Deb, K. (eds.) SSBSE 2016. LNCS, vol. 9962, pp. 291–297. Springer, Cham (2016). https://doi.org/10.1007/978-3-319-47106-8_25
11. Savelsbergh, M.: A branch-and-price algorithm for the generalized assignment problem. Operat. Res. **45**(6), 831–841 (1997)
12. Schwaber, K., Sutherland, J.: The scrum guide. the definitive guide to scrum: The rules of the game. (2020)
13. Vega-Velázquez, M.Á., García-Nájera, A., Cervantes, H.: A survey on the software project scheduling problem. Int. J. Prod. Econ. **202**, 145–161 (2018)
14. Wilcoxon, F.: Individual comparisons by ranking methods. Biomet. Bull. **1**(6), 80–83 (1945)
15. Zapotecas-Martínez, S., García-Nájera, A., Cervantes, H.: Multi-objective optimization in the agile software project scheduling using decomposition. In: 2020 Genetic and Evolutionary Computation Conference Companion, pp. 1495–1502 (2020)
16. Zhang, Q., Li, H.: MOEA/D: a multiobjective evolutionary algorithm based on decomposition. IEEE Trans. Evol. Comput. **11**(6), 712–731 (2007)

Author Index

Printed in the United States
by Baker & Taylor Publisher Services